Inside the Victories of Manfred von Richthofen

Comprehensive Victory Summaries and Combat Statistics

Volume 1

James F. Miller

Inside the Victories of Manfred von Richthofen

Comprehensive Victory Summaries and Combat Statistics

Volume 1

James F. Miller

Facing Page: "What a man he was. Certainly, the others fought too. But they had wives or children, a mother or a profession. This they could forget only on rare occasions. But he constantly lived beyond those boundaries that we only cross during great moments. His personal life was blotted out when he fought at the front. And he always fought when he was at the front. Food, drink, and sleep was all he was willing to concede to life. Only that which was necessary to keep that machine of flesh and blood going. He was the least complicated man I ever knew." – Ernst Udet

ISBN: 978-1-935881-42-1

Design and layout: Jack Herris
Cover design: Aaron Weaver
Digital photo editing: Aaron Weaver & Jack Herris

www.aeronautbooks.com

Table of Contents

Volume 1

Volume 2

Acknowledgements

The author wishes to thank Michael Backus, Jill Bush, Eileen Barrow, Gene Buckmaster, Chris & Charyn Cordry, Mary Jo Darrah, Dave Douglass, Tom & Karen Dillon, Pia Götz at Ullstein-Buchverlage, Jon Guttman, Jack Herris, Bert Hughlett, Reinhard Kastner, Herb & Sarah Kilmer, Dr. Craig Miller, Jim & Judy Miller, Dr. Gary Ordog, Colin Owers, David Peace, Alex Revell, Bruno Schmaeling, Dr. Mark Senft, Russell Smith, Pierre Vandervelden, Aaron Weaver, JR Williams, Reinhard Zankl and so many others who have offered their time and thoughts during continual discussions and debates throughout the last decade. Special thanks are extended to Peter Kilduff, whose generous help and encouragement ultimately led to this book; Lance Bronnenkant and Greg VanWyngarden, for their boundless photograph generosity and enormous patience fielding my rambling and way-too-long emails; Jack & Nicôl Herris, for creating Aeronaut Books and for their personal patience, understanding, and support far beyond the call of duty; Marton Szigeti for his friendship, hospitality (Claudia, too), photographic and information generosity, and the Mach-2 *Red Rocket* rides around France, Belgium, and Germany; and Pekka Jääskeläinen and Detlev Mahlo, both for their constructive criticism, advice, and feedback on almost a daily level, and Detlev for his tireless patience and friendly cooperation with scores of pestering requests for German-to-English translations. And very special and deep heartfelt thanks go to LaFonda, Jim and Major Miller. Your patience, encouragement and love are my fuel for living. A million Shakespeares writing for a million years could neither describe my luck nor how much love and gratitude I feel for you every day.

Goodbye, Stripey-girl. We love you. Tell Charlotte we love her, too. M, D, J & M.

Fly as high as the sun! A E F#

Left: A *Jasta* 2 Albatros D.II warms up at Lagnicourt airfield, late autumn 1916. The white stripe on the nose is tentatively associated with Richthofen as an early personal identification marking.

Foreword

Inside the Victories of Manfred von Richthofen began years ago as a small, personal compendium designed to separate the wheat from the chaff regarding the details of Richthofen's life and achievements. It was intended for organization, not publication. However, since its birth it has grown so vast and now contains so much detail that publication was sought for those who may find access to this information interesting and beneficial. The catalyst for the original compendium was frustration with sundry Richthofen books containing contradictory information that left one wondering which was the accurate representation, book a or book b? Or c? Or m? Because of this I delved into as many Richthofen books as possible, extracted the presented information, and then meticulously cross-checked it all against itself, Richthofen's combat records and autobiography, RFC/RNAS/RAF combat reports, RFC/RNAS/RAF squadron record books, RFC Communiques, Kofl Armee reports, RFC/RAF combat casualties, British and German pilot anecdotes, medical information, photographs, etc. From there, the statistics were compiled. Throughout there are still gray areas that need to be conquered, and hopefully they *can* be conquered across the echoes of a century, but on the whole the process has been a distillation that removed impurities to produce as unclouded, unbiased, accurate, and factually consistent presentation of Richthofen's life, health, victories, strategies, and achievements as possible. Of course, this distillation never stops; research is always underway. And although errata and addenda can be compiled and made available, one of the benefits of a print-on-demand book is new or updated information can be entered directly into the text, *now*. Because of this process, *Inside the Victories of Manfred von Richthofen* is its own filter against becoming the kind of book it was designed to overcome.

Although source endnotes are used throughout, they are not with the 1916–1918 victory details chapters. In a nutshell, doing so was too much. The information within is so vast and varied that to explain and source it all would have cluttered the text to an unproductive degree (as are the original compendium notes) and bloated this work to a size exceeding the limits of print-on-demand publication – i.e., the book could not be published in this format. One solution was a companion volume of endnotes, but the realities of limited customer demand and interest rendered such publication impractical. Therefore, be advised that the details and information in these three chapters were distilled entirely from the sources listed in both the previous paragraph and the bibliography, although many of the combat victory entries do include notes that explain how particular points or details were derived. Overall, the information is systematic and consistent, and the format has become a template for future volumes focused on other pilots.

Frequently this work employs abbreviations. For clarity, the most common are presented here:

Ltn – *Leutnant*	Oblt – *Oberleutnant*
Hptm – *Hauptman*	Rittm – *Rittmeister*
Vzfw – *Vizefeldwebel*	Flt Sgt – Flight Sergeant
2nd Lt – Second Lieutenant	Lt – Lieutenant
Capt – Captain	Maj – Major
u/c – Uncertain	u/k – Unknown
n/a – Not applicable	PoW – Prisoner of War
MiA – Missing in Action	KiA – Killed in Action
DoW – Died of Wounds.	

Because print-on-demand exigencies dictate a 400-page print limitation, necessarily this large work was separated into two volumes, with Richthofen's 6 July 1917 head wound serving as the dividing point. This first volume chronicles his ascension to the pinnacle of success and includes two detail-laden chapters regarding his famous air battle with Lanoe Hawker and the circumstances of his previously mentioned wounding. Volume 2 covers the second half of his career and includes detailed statistical and mythical outcomes of not only his post-wound career, but of his entire career. Thus, both volumes dovetail into a complete historical and statistical overview of all of Richthofen's victories, the details inside these victories, and the outcomes of both.

I realize some of the information within *Inside the Victories of Manfred von Richthofen* roughly cuts across the grain of the accepted understanding of Richthofen's character, actions, and health; particularly the latter. One should not blame the mirror for what it reflects but occasional unwillingness to accept this new tack, or outright disagreement with it, regardless of how deeply sourced, is inevitable. Although I stand by this work and its conclusions, I am always receptive to and welcome constructive discussions about any of the subjects and details within.

Lastly, I, too, certainly am not immune to error. Any within are solely my responsibility.

Jim Miller
11 December 2015
Naples, FL
JamesF_Miller@yahoo.com

Biography to First Credited Victory

Above: Manfred, purportedly at age 7. His facial features are well recognizable.

Left: Manfred von Richthofen, age 2. His mother recalled, "as a child, he had such wonderful curls that shimmered like spun sunlight."

Known today the world over as "The Red Baron," Manfred Albrecht *Freiherr* von Richthofen was born Monday, 2 May 1892 in Breslau, Silesia, Germany (today Wroclaw, Poland). Richthofen and his two younger brothers and older sister grew up in nearby Schweidnitz (today Swidnica, Poland), where as a young child Manfred spent much of his time hunting. At his father's behest he entered military academy in 1903, attending the Cadet Academy at Walhlstatt (today Legnickie Pole, Poland) until 1909 and the Cadet Academy at Gross-Lichterfelde until 1911. In these schools he excelled in sports and equestrian activities but less so scholastically, later confessing he "did just enough work to pass." Upon graduation he was accepted as an officer candidate with the cavalry unit *Ulanen-Reg "Kaiser Alexander III. von Russland (westpreussisches)" Nr* 1, and after advanced officer training at the Berlin's *Kriegschule* (War School) he received his officer's patent on 19 November 1912.

Riding with *Ulanen-Reg Nr 1*when the First World War broke in August 1914, Richthofen engaged various Russian and French forces throughout Poland and Belgium, using the swift mobile warfare for which the years of military schools and horseback riding had prepared him. However, *Ulanen-Reg Nr 1's* transfer to Verdun France that autumn mired the unit in the network of trenches, dugouts and fortifications created when the 1914 German advance had bogged down and the opposing forces were forced to dig in. The resultant stalemate throttled the usefulness of swift cavalry and eroded Richthofen's long-standing desire for a dashing war-on-horseback. His morale waned. "We are now in and out of the trenches, like the infantry," he wrote his mother 2 November 1914. "It is no fun to lie quietly for the duration of 24 boring hours. Some shells come back and forth in singular exchange; that is all I have experienced in the last four weeks. For weeks the position before Verdun has not shifted fifty meters. We are camped in a burned-out village.... I live in a house in which you must hold your nose. We seldom if ever ride, as Antithesis [his horse] is sick and my chestnut bay is dead. In other words: There is no movement at all. Eating does little good: Nothing agrees with me—even though I am fat as a barrel. If I should ride again, I will need to exercise until I get back to my normal weight

Above: The Richthofen boys in the backyard of their family home. (Left to Right) Bolko, Manfred, Lothar.

again. I would like very much to have gotten the Iron Cross, First Class, but there is no opportunity here."[1]

A leave in May provided blessed respite from the half-year of monotony, but upon his return Richthofen was assigned administrative chores rather than participate in a "minor offensive on our front[2]." This was the last straw for the frustrated *Uhlan* and prompted his written request for transfer into the *Fliegertruppe* (Air Service). Although he "hadn't the slightest idea what our fliers did" when the war began and had "considered every flier an enormous fraud[3]," the months of running messages and answering telephones in the trenches eventually softened his position enough to consider joining the *Fliegertruppe*—not as a pilot, but as an observer. Time was the motivating factor in this choice: fearful the war would end before the completion of a three-month pilot training course—popular belief is that it would be over by Christmas—Richthofen reasoned the two-week observer's course would enable him see action almost immediately. Moreover, he felt his cavalry reconnoitering skills would serve him well as an aerial observer.

His decision was fortuitous. Many young officers were being drawn from cavalry units to supply observers for the growing Air Service and—to his great excitement and relief—his request was granted. By month's end he and 29 others reported to *Flieger-Ersatz-Abteilung* 7 (Aviation Replacement Unit 7, or *FEA* 7) in Cologne for training, where he survived undocumented "difficult and doubtful circumstances" to be among those selected and retained for continued training[4] with *FEA* 6 in Grossenhain, Germany, beginning 10 June 1915. The specific aspects of his training are unknown but likely he received instruction regarding engine and airplane familiarization, the handling and firing of machine guns, the use of a compass and telescope, map reading, bomb-dropping, photography, some meteorology, and practiced drawing while flying (e.g., notations of enemy positions)[5]. By 21 June he had completed the course and was assigned to *Feldfleiger-Abteilung* 69 (Field Aviation Unit 69, or *FFA* 69) on the Eastern front. This posting delighted him because Richthofen felt Russia was "the only place where there was still a chance to get in the active war[6]." Soon he and his pilot *Oberleutnant* Georg Zeumer flew morning and afternoon reconnaissance sorties nearly every day, gathering valuable intelligence information in support of the German advance on Brest-Litovsk.

Above: AEG G.II 6/15, purportedly in which Richthofen flew as an observer.

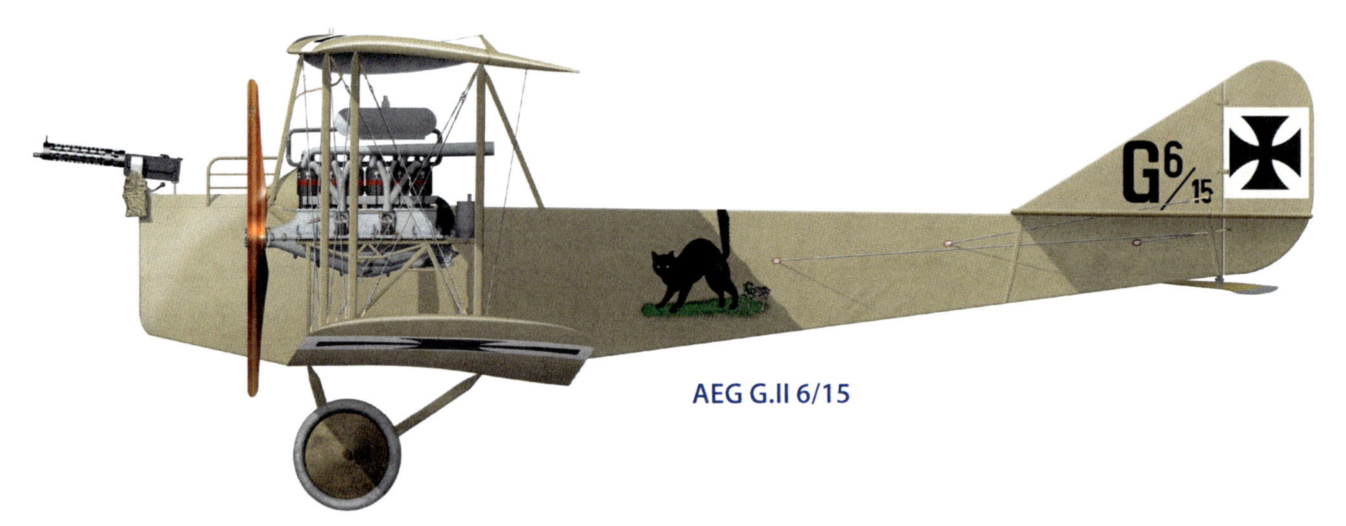

AEG G.II 6/15

Initially feeling "foolish" as a neophyte observer, he quickly settled into his new role after discovering: "Life in the Air Service was very much like life in the cavalry. Reconnaissance had been my business in the cavalry, so I now served in my specialty and had great fun in the far-ranging reconnaissance flights we undertook...[7]"

His association with Zeumer ended with the latter's transfer to *Breiftauben-Abteilung-Ostende* (Carrier Pigeon Section at Ostende, or *BAO*, a cover name to conceal the unit's real identity as Germany's first mobile combat squadron and from whence *FFA 69* had been created), after which Richthofen was paired with *Rittmeister* Erich *Graf* von Holck. Manfred was impressed with his new pilot's airmanship and determination, and in Zeumer's absence he missed not a beat prosecuting his new warfare to the fullest. However, shortly thereafter on 21 August, Richthofen was transferred to *BAO* on the Western front and reunited with his first pilot Georg Zeumer, who was now flying a twin-engine "*Grosskampfflugzeug*" [large battle plane], the AEG (*Allgemeine Elektrizitäts Gesellschaft*) G.II. With their *Grosskampfflugzeug* the pair picked up where they had left off, flying patrols five or six hours a day during which they searched for enemy airplanes and "entertained" the English with bombs. Normally they encountered no aerial opposition but during a morning sortie 1 September they chanced upon a British-flown Farman—an ungainly French twin-engine machine of lattice construction—and flew toward it. Ungainly

Above: This closeup of an AEG G.II illustrates the close proximity between the observer and spinning propellers. Accidents such as the one Richthofen suffered when a propeller nicked his little finger resulted in the installation of various guards to keep hands and arms clear.

Above: Richthofen (at left) with a bandaged finger after his propeller accident. After Richthofen (left to right): *Freiherr* von Könitz, *Ltn.* Hans Haller von Hallerstein, unknown.

though it was, ostensibly the Farman was crewed by experienced airmen who detected the stalking G.II and turned to face their attackers. As the two airplanes approached head-on Richthofen managed to fire just four shots before the combatants flew past each other and began circling to attain a firing position. Unfortunately for the Germans the Farman got the better of them, gaining their six o'clock and "shooting the whole works at us," scoring many hits until it suddenly disengaged and then flew away. Chagrined and irritated post-landing, Richthofen realized that despite his marksmanship and Zeumer's stick-and-rudder ability it was no easy matter to shoot down an airplane, especially from a G.II. His enthusiasm for the *Grosskampfflugzeug* suffered a further blow some days later when he stuck his outstretched hand into one of the propeller arcs while gesticulating to Zeumer to bank or slip the G.II to observe their fall of bombs, nicking his right little fingertip. Although minor, the injury was enough to ground him for a week.

In September Zeumer and Richthofen transferred to the *3. Armee* Champagne front and assisted flying cover for other *3. Armee* airplanes. Richthofen recalled they took their G.II with them and although he made no further mention of sorties involving aerial combat with this machine there likely was at least one other, because he later remarked he and Zeumer "soon realized that our old packing crate was a grand airplane, but it would never make a fighter[8]." Thus, Zeumer often availed himself of one of the two available Fokker *Eindecker* fighters and "sailed over the earth alone," at times leaving Richthofen to fly with Oblt Paul Henning von Osterroht. During one such flight in a machine "somewhat smaller than our [Richthofen and Zeumer's] old barge[9]," reportedly an Albatros or Aviatik C-type, he and Osterroht encountered another Farman five kilometers behind the lines. Osterroht overtook the enemy plane unawares and skillfully flew so close aboard that Richthofen easily served the enemy airplane with machine gun fire, and not until his gun jammed did the Farman fire back. But it was too little too late; Richthofen cleared the jam and exhausted the rest of his 100-round[10] magazine. The Farman slanted away and entered a steep spiral with either its controls destroyed or its crew wounded or killed, and after its terrain impact Richthofen could see the fallen machine's tail sticking up out of a bomb crater. Due to its fall across the lines he and Osterroht did not receive credit for their victory, but this did not sour Manfred: "...I was very proud of my success, for the main thing is that a fellow is brought down not that one is credited for it[11]."

By the start of the following month Richthofen and Zeumer had been redeployed to Rethel, some 30km northeast. There in the *BAO* support train dining car Richthofen happened to meet *Ltn* .Oswald Boelcke, whose then-incredible four victories flying a Fokker *Eindecker* were well known to Richthofen. He was impressed by this man's accomplishments and after honestly assessing his own inability to shoot down enemy airplanes, save for his uncredited Farman, he asked Boelcke: "Tell me honestly,

Above: Albatros B.I.

how do you do it[12]?" Boelcke laughed and then provided an answer that became the cornerstone of Richthofen's future success: "Yes, good God, it is quite simple. I fly right up to him and take good aim, then he falls down[13]." Such simplicity puzzled Richthofen; he had done the same thing but mostly without reward. He strove to become better acquainted with Boelcke and throughout the next few days questioned him further during walks and card games, learning as much as he could about aerial combat straight from the source. Finally, he realized that the biggest difference between their successes was that Boelcke piloted a single-engine fighter while he flew in a twin-engine *Grosskampfflugzeug.* Richthofen resolved that if he learned to fly a Fokker for himself then perhaps he could emulate Boelcke's success.

Toward that end Richthofen requested and attained the tutelage of Zeumer, who agreed to give him informal flight instruction between and after their assigned combat sorties. Likely using a B-type two-seater, Richthofen's flight training progressed at the break-neck pace of 25 hours in less than ten days—an impressive workload for he and Zeumer, even without their concurrent combat sorties—and on 10 October Zeumer informed Richthofen he was ready for his first solo. Despite his initial fear his single left-hand airfield circuit went well until the landing approach, when at some point he lost control of the machine and crash landed. Although the airplane was damaged to an unknown extent—his only reference to the damage was "slight"[14]—Richthofen was uninjured, save for an ego bruised by the laughter of onlookers.

Undeterred by this minor setback, two days later his second solo flight went "wonderfully well," and after continued training he flew his first of three required examinations. Proud and confident after flying the "prescribed figure eight and the ordered number of landings[15]," he was surprised when informed of his failure. It is not known when he retested but obviously he did so successfully; 15 November he was ordered to *FEA* 2 Döberitz to receive further training, and on Christmas Day he passed his third and final examination to become a licensed pilot. His training had taken some three months and although it had begun with a shaky start, Richthofen had survived the 50% pilot washout rate[16] and finished strongly. Not much is known of his flying activities immediately thereafter, but presumably for much of the winter he was assigned to an *Armee Flugpark* to gain experience and practice flying the latest machines[17]. Finally, on 16 March 1916 he reported to *Kasta* 8/ *Kagohl* II to begin combat sorties as a two-seater pilot based at Mont airfield, located 35 km northeast of Verdun, on the border between the tiny villages of Murville and Mont-Bonvillers, France.

Richthofen's months with *KG2* were pivotal toward his growth as a pilot. He saw enough action to generate combat experience but not enough to be shot down while still a neophyte. Primarily

Above: *Kasta* 8 LVG C.IIs (left) and LFG Roland C.IIs (right) on display at the unit's airfield in Mont, France.

Above: Closeup of *Kasta* 8's LFG Roland C.II *Walfish*. In years past photo examination of the men atop the planes fostered a belief Richthofen flew the machine adorned with circles, but recent examination of large, high-quality scans reveal neither man on that machine was Richthofen (Inset A). Instead, the men on the *Walfisch* at far left (Inset B) bear the closest resemblance to Richthofen, especially the man at right. However, due to distance and the grainy resolution of this area, faultless positive identification cannot be gained.

Inset A

Inset B

Above: LVG C.II.

he flew escort and aerial blockade missions, first in an LVG C.II and then later that spring in an LFG Roland C.II[18], the fat fuselage of which filled the gap between the wings and engendered the sobriquet "*Walfisch* [whale]." Despite its girth, the streamlined *Walfisch* was an exceptionally fast two-seater—20 kmh faster than the LVG C.II, it was faster than the single-seat AMC DH.2, Nieuport 10, and as fast as a Nieuport 11—this speed resulting from a successful effort to reduce drag rather than increase engine power. It came with a price, however: the fuselage reduced control effectiveness by disturbing airflow across the empennage, and the narrow wing gap created aerodynamic interference that increased stall speed and precluded climbing turns. Limited downward visibility and a high sink rate increased pilot workload on landing, and it was presumed every *Walfisch* pilot would crash-land on his first landing. Despite these difficulties there are no reports of Richthofen having any trouble handling or landing the tricky machine and, indeed, its speed bode well for the offensive minded pilot.

Above: Richthofen and *Kasta* 8 *Staffelführer Hptm.* Victor Carganico.

Inspired by Boelcke and Immelmann's success flying the Fokker *Eindecker* (with which each had shot down over ten enemy airplanes by this time), but without a Fokker of his own, in late April Richthofen had a forward-firing machine gun rigged to his *Walfisch* rollover cage in such a way that its line-of-fire cleared the propeller arc. With his airplane so equipped he began flying pursuit flights over the lines in search of hostile airplanes to attack, and on 26 April he and his observer chanced upon a Nieuport flying near Ft. Douaumont. For whatever reason—be it engine trouble, gun jam, a disinclination to fight or that he simply did not see the approaching Germans—the Nieuport turned away instead of engaging; a poor tactical decision but perhaps understandable if the Frenchman reasoned

Above: LFG Roland C.II *Walfisch*.

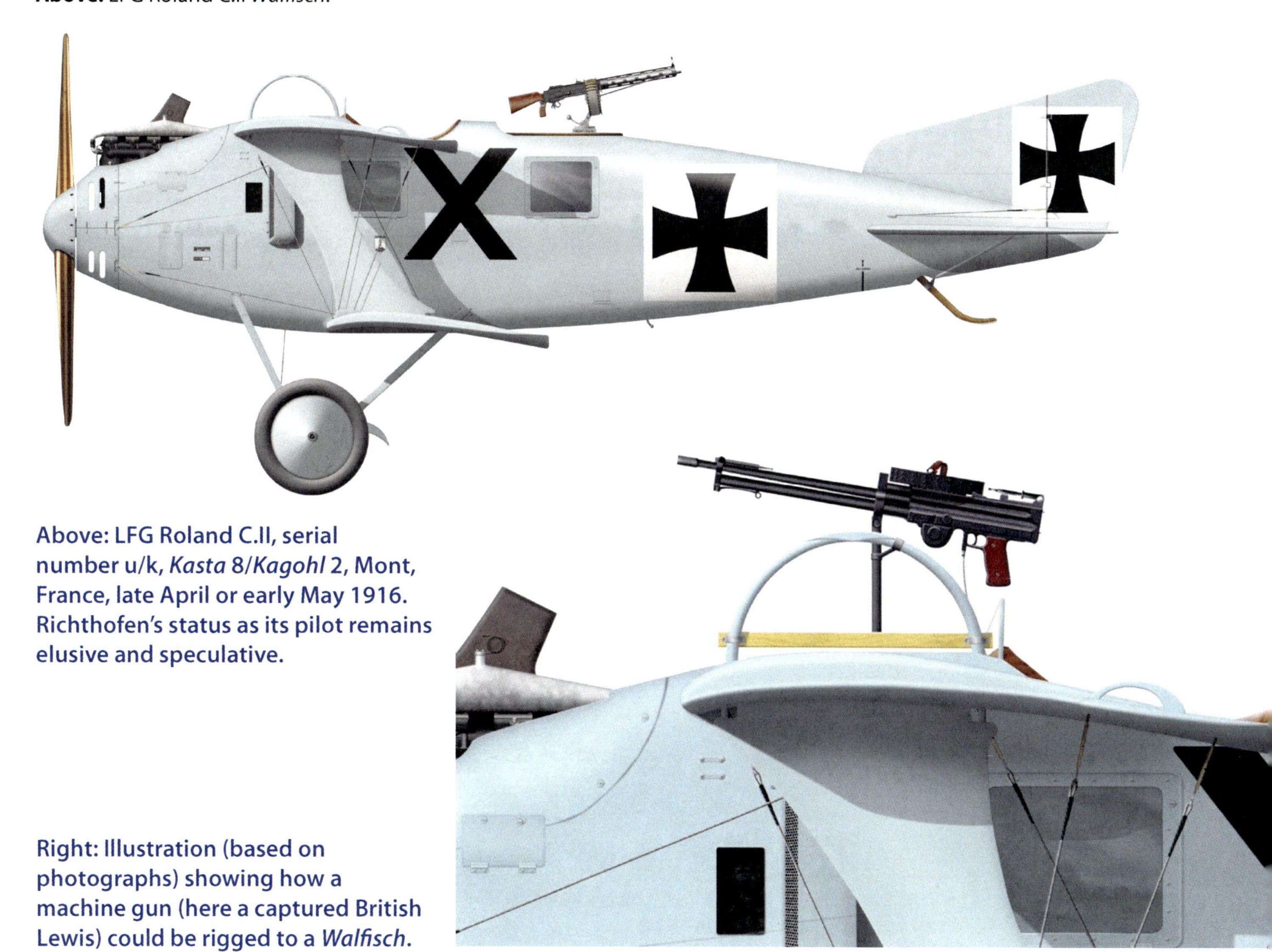

Above: LFG Roland C.II, serial number u/k, *Kasta* 8/*Kagohl* 2, Mont, France, late April or early May 1916. Richthofen's status as its pilot remains elusive and speculative.

Right: Illustration (based on photographs) showing how a machine gun (here a captured British Lewis) could be rigged to a *Walfisch*.

Above: The remains of Richthofen's Fokker *Eindecker* after a takeoff engine failure and subsequent crash landing at Mont, June 1916.

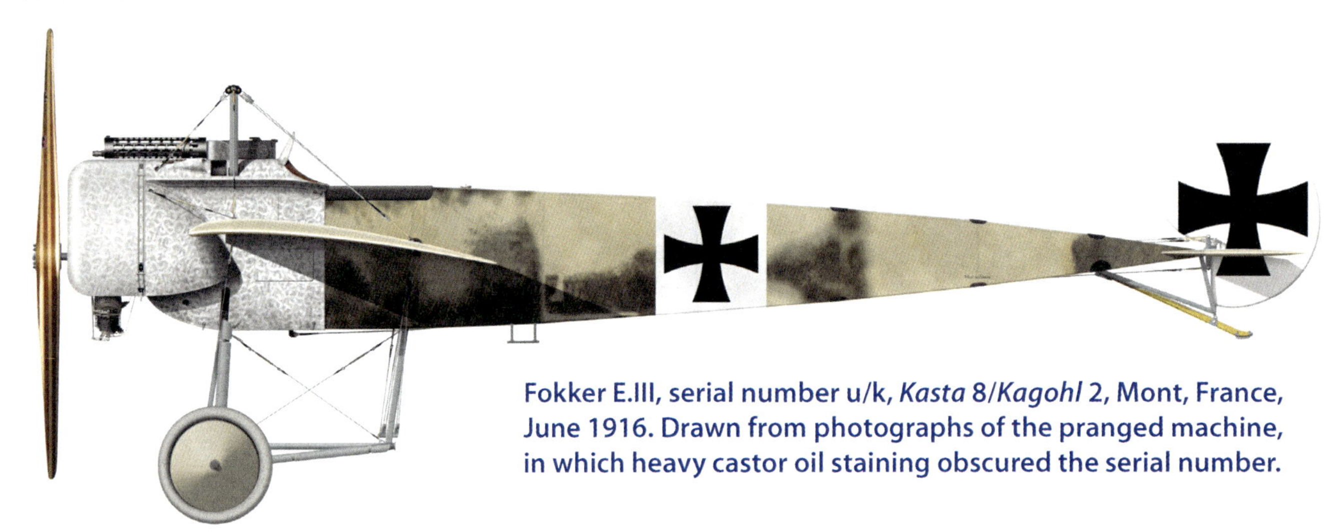

Fokker E.III, serial number u/k, *Kasta* 8/*Kagohl* 2, Mont, France, June 1916. Drawn from photographs of the pranged machine, in which heavy castor oil staining obscured the serial number.

he could easily outpace a German two-seater. Being new to the front[19] the *Walfisch's* speed was still unrealized—speed that Richthofen used to pursue the Nieuport and steadily close from astern. Based on Richthofen's account the Frenchman apparently affected no evasive maneuvers and presented Richthofen with an easy straight-and-level target. He could hardly miss; a few well-aimed bursts from his rigged machine gun and the Nieuport reared up (a nearly ubiquitous sign of pilot injury) and then fell away. At first Richthofen thought this was an escape maneuver, but to his joy the airplane did not recover and crashed into a forest near Fleury. Ultimately he was not credited with this victory, despite its official

Above: *Kasta* 8 at Köwel, Russia, summer 1916. Richthofen is at center, saluting. The *Walfisch* has been replaced by the Albatros C.III.

Right: Closeup of Richthofen at Köwel.

Below: Albatros C.III.

Above: The *Orden Pour le Mérite*. Popularly known as the "Blue Max," the Pour le Mérite was a coveted and highly respected Prussian order of merit awarded as recognition of extraordinary personal achievement. Richthofen was the 14th pilot to earn this decoration in WW1. (Lance Bronnenkant)

recognition, but his success further piqued his interest to fly single-seat fighters[20].

This interest compelled Richthofen to continually "annoy" *Kasta* 8 *Staffelführer Hauptman* Victor Carganico to send him to the Air Park in Montmédy for single-seat fighter instruction. Carganico finally relented and afterward, in very early May, Richthofen began flying Fokker E.III air-defense flights[21] concurrently with his normal *Walfisch* sorties. These flights were few. On the second day of single-seater sorties the pilot with whom Richthofen shared the Fokker[22] was brought down with engine damage and forced to burn the machine, and with its replacement in late June Richthofen experienced a takeoff engine failure on his third sortie and crash-landed heavily, destroying the machine. He emerged from the wreck unscathed but was once again without a Fokker[23]. These single-seater forays must have been somewhat disappointing: he had engaged no enemy airplanes, and if he managed three flights on both mornings he flew the machine in May then he logged a total of only eight *Eindecker* flights (not including his takeoff crash-landing) with some six weeks of down time between the two periods of its availability[24].

On 28 June *KG2* departed Mont for the Eastern

Above: Anthony Fokker in front of the unarmed M18, a prototype for the Fokker D.I.

Front, traveling four days by train to arrive 1 July at the Ukrainian city Köwel (today Kovel), an important railhead threatened by the recent Russian offensive. Having left their *Walfisch* and LVG C.IIs in France, *KG2* now flew bombing and armed reconnaissance sorties equipped with the Albatros C.III, a single-engine two-seater multi-purpose biplane. Armed with a fixed Maxim lMG 08/15 machine gun synchronized to fire through the propeller arc—eliminating the need for Richthofen to rig a machine gun above the wings—and a flexible Parabellum lMG 14 for rear defense, it carried 60 kg of bombs in vertical containers between the cockpits and could accommodate an additional 100 kg of bombs externally—which Richthofen likely did, since he wrote "many times I hauled one-hundred-fifty-kilogram bombs with a normal C-type airplane[25]." With its excellent flying qualities, the C.III was well regarded by the combat and training units with which it served throughout the war.

Richthofen often flew a C.III on two bombing sorties per day, attacking trains and train stations before seeking targets of opportunity to "harass" with machine guns. During one sortie he roared down to the "lowest possible altitude[26]" and attacked Russian troops crossing the Stokhod river, catching them cold as they crossed a narrow footbridge. Via repeated passes he and his observer bombed and strafed the cavalry using tactics similar to those employed by the future *Schlachtstaffeln*, causing wild disorder that halted traffic and scattered the surviving men and horses in all directions. Richthofen wrote about his role as bomber pilot. "It is beautiful to fly straight ahead, with a definite target and firm orders. After a bombing flight one has the feeling he has accomplished something...." Still, his interest in single-seat fighters persisted, as he considered there was "nothing finer for a young cavalry officer than flying off on a hunt[27]."

On 12 August Oswald Boelcke arrived at Köwel to visit his brother Wilhelm, *Staffelführer* of *Kasta* 10. Boelcke had been on an "inspection tour" in the Balkans (designed to keep him out of harm's way after the death of Immelmann 18 June) and while

Above: Boelcke's Fokker D.III 352/16.

in Köwel received a telegram ordering his return to develop and command one of the *Fliegertruppe's* new *Jadgstaffeln* implemented to support *Armees* of the western front. Boelcke freely availed himself to the pilots of *KG2*, regaling them with anecdotes of his recent trip and describing his need of pilots for his new command, and although the prospect of being selected for Boelcke's new *Jagdstaffel* secretly excited Richthofen he dared not ask to be chosen—so he was beside himself one morning when Boelcke showed up at his door and sought his accompaniment to the Somme to fly with *Jasta* 2. He accepted immediately and three days later departed Köwel as a friend bade farewell: "Don't come back without the Pour le Merite!" It had taken 11 months since his first instruction flights with Zeumer, but at last Richthofen's "ardent desire was fulfilled and now for me began the most beautiful time of my life[28]."

However, upon arriving in Bertincourt France 1 September Richthofen discovered the *Staffel* was woefully short of fighters and would be for some time—by mid-month the entire *Staffel* still comprised just four machines—one Fokker D.I, one Fokker D.III, one "refurbished" Halberstadt D-type, and one Albatros D.I[29]. By 10 September *Jasta 2* boasted a complement of 12 pilots but had received no additional airplanes, and since Boelcke flew the Fokker D.III on frequent "lone wolf" sorties the *Staffel* often had just three fighters available for the other eleven pilots[30]. This likely necessitated a rotation schedule of some sort and left scant time for thorough transition and proficiency flights. Regardless, Boelcke imparted his vast knowledge and experience during ground lectures and used his daily sorties to illustrate his points.

In mid-September Boelcke began taking two pilots with him during his combat sorties. Richthofen flew missions with him on 14 and 15 September and although he fired upon Sopwith 1 ½ Strutters, he shot down none. In a letter to his mother he commented that "I had to fly a reserve plane with which I could not do much, being beaten in most encounters," suggesting he flew the Halberstadt or more likely the Fokker D.I during these early sorties, the overall performance of each being less than that of the Albatros D.I he eventually flew—although as a neophyte fighter pilot the blame for being "beaten"

Above: Halberstadt D.II.

cannot rest solely with the airplane. Finally, on 16 September—the day before Richthofen's first aerial victory—five Albatros D.Is and one Albatros D.II became available in Cambrai, at which time several *Jasta* 2 pilots drove over and then flew them back to Bertincourt.

During a sortie the next day Richthofen attacked and traded shots with an RFC No.11 Squadron FE.2b, which ultimately he shot down for his first credited victory. The details of that encounter, as well as the events and details of the 79 credited victories that came afterwards during the next twenty months, are chronicled within the following pages.

Endnotes

(1) von Richthofen, *The Red Baron*, (1969), p.23.
(2) ibid., p.24.
(3) ibid., p.11.
(4) According to the Air Ministry's *Handbook of German Military and Naval Aviation 1918*, p.18, there was a 30% washout rate among those who entered the Air Service for training in any of its branches. 15% through accidents, 10% through ill-health, and 5% through lack of moral qualifications. ('Serious' venereal disease may fall within the latter two categories, the contraction of which resulted in immediate disqualification and banishment from further course participation, with no consideration of rank or time spent.)
(5) *Air Ministry, Handbook of German Military and Naval Aviation*, (1995), p.25.
(6) *The Red Baron*, op. cit., p.26.
(7) ibid., p. 27.
(8) von Richthofen, *The Red Baron*, (1969), p.36.
(9) von Richthofen, *The Red Baron*, (1969), p.36.
(10) Albatros and Aviatik observers were normally armed with a single Parabellum lMG Modell 14 machine gun, a belt-fed weapon that utilized several magazine types, one of which held 100 rounds.
(11) von Richthofen, *The Red Baron*, (1969), p.36.
(12) von Richthofen, *The Red Baron*, (1969), p.37.
(13) Kilduff, *Richthofen: Beyond the Legend of the Red Baron*, (1993), p.40.
(14) von Richthofen, *The Red Baron*, (1969), p.38.
(15) von Richthofen, *The Red Baron*, (1969), p.38.
(16) Air Ministry, *Handbook of German Military and Naval Aviation (War) 1914–1918*, (1995 facsimile reprint, original 1918), p.25.
(17) Air Ministry, *Handbook of German Military and Naval Aviation (War) 1914–1918*, (1995 facsimile reprint, original 1918), p.25. A 1916 order was likely applicable, whereby no pilot was sent to the front with less than six months training. This dovetails with the practice of receiving 'one to three months' of subsequent training at an *Armee Flugpark*.
(18) Ferko, *Richthofen*, (1995), pp.8, 9.

Above: A *Jasta* 2 Albatros D.I. As was common with *Jasta* 2 machines, note that the upper wing cross fields have been overpainted in such a way that mostly camouflaged them while leaving enough of the original white to create borders around the crosses. Although regularly flown by *Ltn.* Diether Collin (hence the "Co"), ownership passed to *Prinz* Friedrich Karl (suiting up near the empennage at left), acting commander of *Fl. Abt. (A)* 258 and second cousin to *Kaiser* Wilhelm. Eventually the Prince overpainted Collin's "Co" with the "Death's Head" emblem of his former Hussar unit, but Collin's overall green overpainting and white rudder remained. On 21 March 1917, after tangling with No.32 Squadron DH.2s, the Prince was forced to land this machine between the trenches and make a run for German lines, but he was shot in the spine by troops and taken prisoner (the Albatros was also captured). Although he received good hospital care, the Prince died of his wounds on 6 April.

(19) Grosz, *LFG Roland C.II*, Windsock Datafile 49, (1995), p.35. On 30 April the frontline inventory of total C-types was 1,029. Of these, only 17 were *Walfisch*.

(20) Richthofen never specifically identified his airplane as being a *Walfisch*. However, an LVG C.II could not catch a Nieuport in the manner Richthofen described—i.e., overtaken from astern. It is unknown when Richthofen began flying the *Walfisch* but Peter Grosz noted they arrived at the front in March and 17 were inventoried 30 April. AE Ferko contends *Kasta* 8 received them 'about May.' Although not an exact match, the last week of April can certainly be considered 'about May,' and it falls within the period of the *Walfisch's* frontline availability.

(21) Kilduff, *Richthofen: Beyond the Legend of the Red Baron*, (1993), p.46.

(22) *Ltn.d.Res.* Hans Reimann, a future *Jasta* 2 comrade.

(23) Carganico stated that after Richthofen's crash landing he was allowed to use Carganico's *Eindecker*. However, Richthofen makes no mention of flying a third Fokker. If any sorties were flown with this machine their numbers and frequencies are unknown.

(24) Kilduff, *Richthofen: Beyond the Legend of the Red Baron*, (1993), pp.46, 47. The early May/late June period of *Eindecker* availability was derived from Richthofen's letters home. On 3 May he wrote 'I am flying a Fokker, which is the airplane with which Boelcke and Immelmann have had their tremendous success.' 'Am flying'; i.e., present tense. Since it was flown on only two different days before being lost in combat, this period of availability must have been the first few days in May, on or about the third. On 6 July he wrote: 'A few days ago I crashed my Fokker right on its nose.' Although the letter states his location as 'Before Verdun,' *Kasta* 8 had departed for Russia 28 June. Regardless, photographs confirm the Fokker crash-landed at Mont. Since this occurred on Richthofen's third flight and 'a few days' before 6 July, this second period of Fokker availability must have been about the last week of June.

(25) von Richthofen, *The Red Baron*, (1969), p.49.

(26) Ibid., p.51.

(27) Ibid., p.52.

(28) von Richthofen, *Der Rote Kampfflieger*, (1917), p.89.

(29) VanWyngarden, *Jasta Boelcke*, (2007), p.12.

(30) Franks, *Jasta Boelcke*, (2004), p.16, 17, 23. The eleven pilots, excluding Boelcke: *Ltn.* Wolfgang Günther, *Ltn.* Otto Walter Höhne; *Ltn.* Ernst Diener, *Ltn.* Winand Grafe, *Ltn.* Manfred von Richthofen, *Offizierstellvertreter* Leopold Rudolf Reimann, *Offizierstellvertreter* Max Müller, *Ltn.d.R.* Hans Reimman, *Ltn.* Erwin Böhme, Obltn Günther Viehweger; *Ltn.* Herwarth Philipps.

Victory Details 1916

Victory No.1

Above: FE.2b 7018 in German hands, after Morris's encounter with Richthofen and subsequent forced landing. (H. Kilmer)

Above: Another photo of FE.2b 7018 in German hands.

17 September
FE.2b 7018
No.11 Squadron RFC

Day/Date: Sunday, 17 September
Time: 11:00AM
Weather: Bright morning with clouds in the afternoon.
Attack Location: Near Villers Plouich
Crash Location: Force landed near Flesquières
Side of Lines: Friendly

RFC Communique No.54: "A bombing raid carried out by machines of the 3rd Brigade was heavily engaged by about 20 hostile machines on its return from Marcoing Station. No details of the fighting are procurable. Four machines of No.11 Squadron and 2 of No.12 Squadron, which took part in the raid, did not return."

KOFL 1. Armee Weekly Activity Report: "An der Spitze seiner Jagdstaffel griff Hauptmann Boelke im Verein mit anderen deutschen Fliegern ein Geschwader von 13 Flugzeugen an und brachte nach hartem Kampfe den feindlichen Führer zum Absturz; fünf weitere Flugzeuge wurden von den anderen deutschen Fliegern niedergekämpft."
"At the head of his *Jagdstaffel Hauptmann* Boelcke in conjunction with other German flyers attacked a formation of 13 aircraft and brought the enemy leader to crash after hard fighting; five more aircraft were fought down by other German flyers."

MvR Combat Report: "Vickers Nr. 7018. Motor No. 701. Machine guns Nos. 17314, 10372.

When patrol flying I detected shrapnel clouds in the direction [of] Cambrai. I hurried forth and met a squad which I attacked shortly after 11 a.m. I singled out the last machine and fired several times at closest range (10 meters). Suddenly the enemy propeller stood stock still. The machine went down gliding and I followed until I had killed the observer who had not stopped shooting until the last moment.

Now my opponent went downwards in sharp curves. In approx. 1200 meters a second German machine came along and attacked my victim right down to the ground and then landed next to the English plane.

Frhr. v. Richthofen
Leutnant

Witnesses: Capt. Boelcke from above and Capt. Gaede, Lieut. Pelser and other officers from below.
Pilot: N.C.O. Rees [*sic*], wounded, hospital at Cambrai.
Observer: Killed, buried by *Jagdstaffel* 4."

RFC Combat Casualties Report: "Left aerodrome 9.10AM. Two F.Es seen to go down under control west of MARCOING."

RFC A/C:
Make/model/serial number: Royal Aircraft Factory FE.2b 7018
Manufacturer: Subcontracted to and built by Boulton & Paul, Ltd., Norwich
Unit: No.11 Squadron RFC
Aerodrome: Izel-le-Hameau
Sortie: Bombing Escort
Colors/markings: Presentation a/c Punjab No.32, Montgomery. PC10 nacelle with white

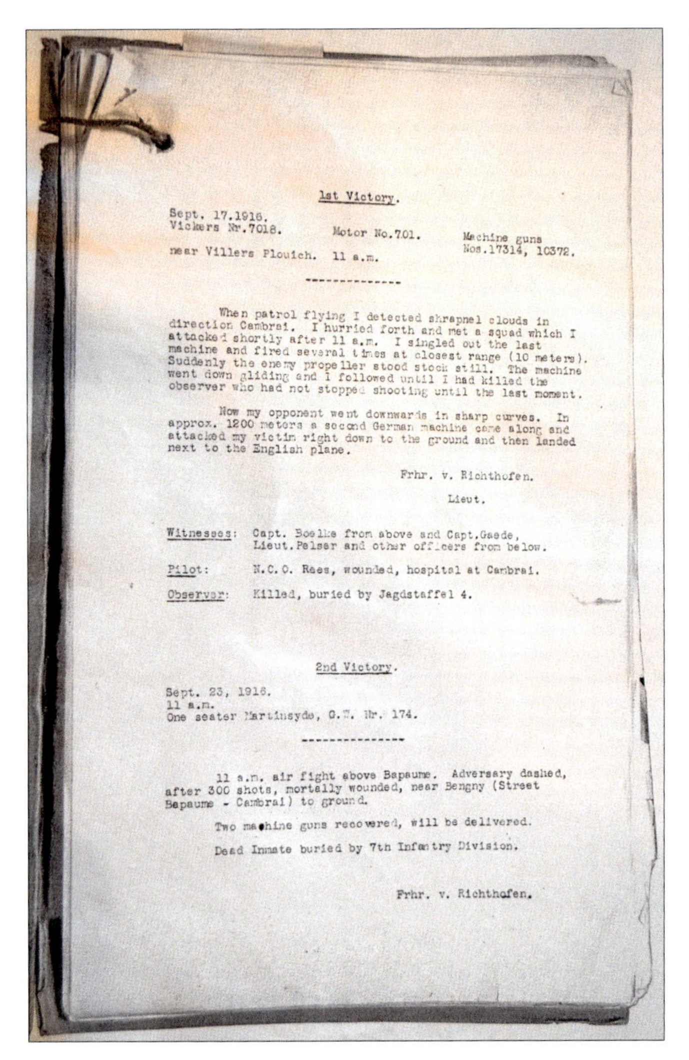

1st Victory.

Sept. 17.1916.
Vickers Nr.7018. Motor No.701. Machine guns
near Villers Plouich. 11 a.m. Nos.17314, 10372.

When patrol flying I detected shrapnel clouds in direction Cambrai. I hurried forth and met a squad which I attacked shortly after 11 a.m. I singled out the last machine and fired several times at closest range (10 meters). Suddenly the enemy propeller stood stock still. The machine went down gliding and I followed until I had killed the observer who had not stopped shooting until the last moment.

Now my opponent went downwards in sharp curves. In approx. 1200 meters a second German machine came along and attacked my victim right down to the ground and then landed next to the English plane.

Frhr. v. Richthofen.

Lieut.

Witnesses: Capt. Boelke from above and Capt.Gaede, Lieut.Pelser and other officers from below.

Pilot: N.C.O. Rees, wounded, hospital at Cambrai.

Observer: Killed, buried by Jagdstaffel 4.

2nd Victory.

Sept. 23, 1916.
11 a.m.
One seater Martinsyde, G.W. Nr. 174.

----

11 a.m. air fight above Bapaume. Adversary dashed, after 300 shots, mortally wounded, near Bengny (Street Bapaume - Cambrai) to ground.

Two machine guns recovered, will be delivered.

Dead Inmate buried by 7th Infantry Division.

Frhr. v. Richthofen.

Above: Translated versions of Richthofen's first two combat reports, as procured from the National Archives, London.

"PUNJAB 32 MONTGOMERY" on sides; thick white horizontal line running low across front of the nacelle, intersected by thick vertical white lines at its midpoint and near both ends; wing uppersurfaces PC10 with CDL undersurfaces; upper wing roundels had white surrounds. Description derived from photographs of downed machine.
Engine: 160 hp Beardmore
Engine Number: 701 WD 7061 (RFC Loss Report), 701 (MvR Combat Report)
Guns: Two Lewis. 16382; 12795 (RFC Loss Report), 17314; 10372 (MvR Combat Report)
Manner of Victory: Gunfire caused engine failure, precipitating spiral glide and controlled dead-stick landing.
Items Souvenired: Boulton & Paul/Howes & Sons manufacture placard. Photographed on display at Roucourt and the post-war Richthofen Museum in Schweidnitz.

Above: 7018's manufacture placard, the first of many such placards seized as souvenirs and mounted together on a wooden plaque.

Right: Lionel Morris.

RFC Crew:
Pilot: 2nd Lt. Lionel Bertram Frank Morris, 19, PoW/DoW
Cause of Death: Shot by machine gun
Burial: Porte-de-Paris Cemetery, Cambrai, Grave I A 16
Obs: Capt. Tom Rees, 21, KiA
Cause of Death: Shot by machine gun
RFC Combat Casualties Report: "Information received from 2/Lt. Pinkerton [ostensibly in the same letter that advised of his 16.9.16 downing and PoW status] that both Morris and Rees were killed."
Burial: Villers Plouich Communal Cemetery, Grave C 2

MvR A/C:
Make/model/serial number: Albatros D.I, serial no. u/k
Unit: *Jagdstaffel* 2
Commander: *Hptm.* Oswald Boelcke
Airfield: Vélu/Bertincourt
Colors/markings: u/k. Likely factory finish shellacked and varnished 'warm straw yellow' wooden fuselage with pale greenish-gray spinner/cowl/fittings/vents/hatches/struts. Upper surfaces of wings/ailerons and horizontal stabilizers/elevator were factory three-color camouflage of pale green, olive green and Venetian red. Wheel covers/undersurface of wings/ailerons

and horizontal stabilizers/elevators light blue. Rudder was either clear doped fabric or one of the camouflage colors. All crosses were black on square white cross fields.
Damage: u/k

Initial Attack Altitude: Reportedly ca. 3,000 meters
Gunnery Range: 10 meters
Rounds Fired: u/k
Known Staffel Participants: Hptm. Oswald Boelcke (#27), *Ltn.d.R.* Hans Reimman (#2)

Notes:

1. After Richthofen damaged 7018's engine he continually attacked the gliding machine at least until its return gunfire ceased, which presumably was the moment when Rees was shot and killed. This event is the first example of Richthofen's fixated no-quarter-attack *modus operandi* against obviously incapacitated machines, a methodology he employed throughout his combat career.
2. A discrepancy exists between Richthofen's combat report and the account in his autobiography *Der rote Kampfflieger*. The combat report reveals that after Rees was KiA the FE.2 "went downwards in sharp curves" and that at "approx. 1200 meters a second German machine came along and attacked my victim right down to the ground and then landed next to the English plane." The autobiography indicates Richthofen "was so excited that I could not resist coming down, and I landed with such eagerness…that I almost went over on my nose," and there is no mention of a second machine. Thus it is unclear if Richthofen flew away at 1200 meters (ca. 4,000 feet) and left the claim up for grabs later or if the threat of the second machine stealing his victory prompted an eager landing to contest/secure the claim on the spot—although why not include this event in his combat report? In the end, Richthofen received credit for the claim and did retain a manufacture placard from 7018 as a souvenir, although his personal removal of this placard remains conjectural. Regardless, possession of this placard is not evidence of any landing by Richthofen.
3. A discrepancy exists regarding where Morris died. Richthofen's *Der rote Kampfflieger* indicates "the pilot died while being transported to the nearest field hospital," while the RFC Combat Casualty report states "information received from a private source that 2/Lt. Morris died at CAMBRAI Hospital on Sept.17th."
4. A discrepancy exists regarding when Morris died. A letter written by No.11 Sqn Capt. D.S. Gray, who during the same sortie was shot down by

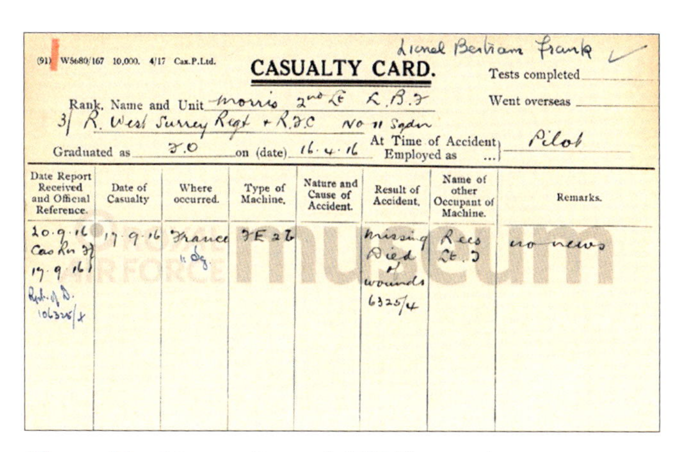
(91) W5680/167 10,000. 4/17 Cax.P.Ltd.

Lionel Bertram Frank ✓

CASUALTY CARD.

Tests completed

Went overseas

Rank, Name and Unit Morris 2nd Lt L.B.F
3/ R. West Surrey Regt + R.F.C No 11 Sqdn
Graduated as F.O on (date) 16.4.16 At Time of Accident Employed as Pilot

Date Report Received and Official Reference.	Date of Casualty	Where occurred.	Type of Machine.	Nature and Cause of Accident.	Result of Accident.	Name of other Occupant of Machine.	Remarks.
20.9.16 Cas Rn 37 17.9.16 Rept. of D. 106325/x	17.9.16	France 11 Sq.	FE 2b		Missing Died of wounds 6325/4	Rees Lt. J	no news

Above: Morris's casualty card. (RAF Museum)

Oswald Boelcke, revealed: "About a minute after landing I saw another one of our machines come down and disappear behind some trees and houses not far about 500 yards from us… It was crashed beside the road which was here on an embankment and a crowd of people round. The car drew up and I gathered from the German officer that the observer was killed and the pilot injured and had already been removed in an ambulance... We arrived that night at midnight at Cambrai and were told that there was a pilot named Morris in hospital brought down that morning, his observer being killed. The man who told us spoke English and walked up from the station with us to the prison and was I believe connected with the hospital. We saw no-one again until I think the next afternoon when we were told Morris had died that morning [i.e., 18 September]." However, Morris's casualty card, headstone, combat casualty report, and grave registrations report form all list the date of death as 17 September.
5. RFC Communiques reveal some of Morris and Rees's earlier actions in 1916:
a. From Communique No. 50, 22 August: "On the front of the 3rd Brigade an offensive patrol of No. 11 Squadron encountered a formation of about 15 German machines, chiefly Rolands and L.V.G's. These were engaged by our F.E's, assisted by one Nieuport Scout. The engagement became of a general nature, all of our machines being engaged. 2/Lt. Morris, Pilot and Lt. Rees, Observer, singled out one machine which was seen to side-slip and plunge to earth out of control, and was seen on the ground in a wrecked condition."
b. From Communique No. 53, 14 September: "There was a considerable amount of fighting on the fronts of the 3rd, 4th and 5th Brigades. Capt. Price and Lieut. Libby, No. 11 Squadron, drove down a hostile machine apparently out of

Albatros D.I 391/16. It is unknown if Richthofen flew this machine during this sortie but its appearance reflects that of a typical factory-fresh Albatros D.I flown by *Jasta* 2.

control. 2nd Lieut. Morris and Lieut. Rees, No. 11 Squadron, also drove down a hostile machine which appeared to be out of control."

Victory No.1 Statistics:

Pushers⋆
1st of 2 September total pusher victories
1st of 2 September two-seater pusher victories
1st of 5 1916 two-seater pusher victories
1st of 12 total two-seater pusher victories
1st of 18 total pusher victories

Two-Seaters
1st of 2 September two-seater victories
1st of 7 1916 two-seater victories
1st of 39 pre-wound two-seater victories
1st of 45 total two-seater victories

FE.2 Δ
1st of 2 September FE.2 victories
1st of 5 1916 FE.2 victories
1st of 12 total FE.2 victories

No.11 Squadron ±
1st of 4 No.11 Squadron FE.2 victories
1st of 2 September No.11 Squadron victories
1st of 2 1916 No.11 Squadron victories
1st of 4 total No.11 Squadron victories

Deaths†
No.11 Squadron:
1st and 2nd of 4 September No.11 Squadron crewmen KiA
1st and 2nd of 4 1916 No.11 Squadron crewmen KiA
1st and 2nd of 4 total No.11 Squadron crewmen KiA

Pushers:
1st and 2nd of 4 September two-seater pusher crewmen KiA
1st and 2nd of 10 1916 two-seater pusher crewmen KiA
1st and 2nd of 17 total two-seater pusher crewmen KiA

Two-Seaters
1st and 2nd of 4 September two-seater crewmen KiA
1st and 2nd of 11 1916 two-seater crewmen KiA
1st and 2nd of 54 pre-wound two-seater crewmen KiA
1st and 2nd of 63 total two-seater crewmen KiA

All:
1st and 2nd of 5 total September crewmen KiA
1st and 2nd of 17 1916 crewmen KiA
1st and 2nd of 66 total pre-wound crewmen KiA
1st and 2nd of 84 total crewmen KiA

Etc.
1st of 3 total September victories
1st of 15 total 1916 victories
1st of 57 total pre-wound victories
1st victory in Albatros D.I or D.II
1st victory forced down with dead engine
1st and only two-seater victory flying from Vélu/Bertincourt
1st of 2 total victories flying from Vélu/Bertincourt
1st of 2 two-seater victories flying under Boelcke
1st of 6 total victories flying under Boelcke

Statistics Notes:
⋆ *All pusher victories are pre-wound*
Δ *All FE.2 victories are pre-wound*
± *All No.11 Squadron victories are pre-wound*
† *All No.11 Squadron deaths are pre-wound; all pusher deaths are pre-wound; KiA includes DoW*

Victory No.2

Above: Martinsyde G.100 7488. Although not the plane flown by Bellerby (7488 was with No.14 Squadron in the Middle East), this view provides a general impression of the make/model Richthofen shot down for his second victory.

23 September
Martinsyde G.100 7481
No.27 Squadron RFC

Day/Date: Saturday, 23 September
Time: 11AM (MvR Report), 9.50 (*Jasta* 2 Report)
Weather: Bright and clear all day; ground mist in early morning.
Attack Location: Above Bapaume
Crash Location: Beugny
Side of Lines: Friendly

RFC Communique No. 55: "...a sharp engagement... resulted in the loss of three Martinsydes of No. 27 Squadron."

MvR Combat Report: "One seater Martinsyde, G.W. Nr. 174. Two machine guns recovered, will be delivered. Dead Inmate buried by 7th Infantry Division.

11AM air fight above Bapaume. Adversary dashed, after 300 shots, mortally wounded, near Bengny [*sic*] (Street Bapaume – Cambrai) to ground.

Frhr. v. Richthofen."

RFC Combat Casualties Report: "Left aerodrome 8.40AM. Patrol leader's report states that one Martinsyde was observed to collide with a hostile machine and both fell to the ground out of control. Two other Martinsydes were seen to go down, one over MARCOING and one over LE TRANSLOY with smoke emitting from fuselage."

RFC A/C:
Make/model/serial number: Martinsyde G.100 7481
Manufacturer: Martinsyde, Ltd., Brooklands, Byfleet, Weybridge, Surrey.
Unit: No.27 Squadron RFC
Aerodrome: Fienvillers
Sortie: Offensive Patrol, BAPAUME-CAMBRAI
Colors/markings: BB gray cowls; PC10 fuselage; CDL wings, hor/vert stabilizers; b/w/r rudder; black 7481 on CDL vertical stabilizer. Confirmed by photographed of downed machine.
Engine: 120 hp Beardmore
Engine Number: 417 WD 2389
Guns: Two Lewis. Nos. 4060, 17074
Manner of Victory: "Dashed to ground."
Items Souvenired:
a. Machine gun, which "has a bullet of mine in the lock and is useless." ("*Es hat eine Kugel von mir im Schloss und ist unbrauchbar.*") Unknown if this refers to the forward- or rearward-firing gun.
b. Martinsyde manufacture placard. Photographed on display at Roucourt and the post-war Richthofen Museum.
c. No. 27 Squadron elephant plaque. Photographed on display at Roucourt, Richthofen's home bedroom in Schweidnitz, and in the post-war Richthofen Museum.

Albatros D.I 391/16. It is unknown if Richthofen flew this machine during this sortie but its appearance reflects that of a typical factory-fresh Albatros D.I flown by *Jasta* 2.

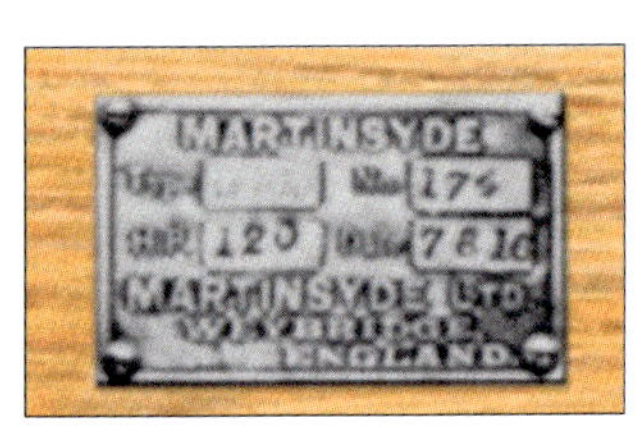

Above: Manufacturing placard souvenired from 7418.

Left: No.27 Squadron Elephant fuselage plaque. Seen here on display in Richthofen's quarters in the Roucourt chateau.

RFC Crew:

Pilot: Sgt. Herbert Bellerby, 28, KiA (possibly DoW, see notes)
Cause of Death: u/k; likely gunshot wound(s)
Burial: In-field; grave lost during turmoil of war.

MvR A/C:

Make/model/serial number: Albatros D.I, serial no. u/k
Unit: *Jagdstaffel* 2
Commander: *Hptm.* Oswald Boelcke
Airfield: Vélu/Bertincourt
Colors/markings: See Vic. #1
Damage: u/k

Initial Attack Altitude: 3,500 meters
Gunnery Range: u/k
Rounds Fired: 300
Known Staffel Participants: Ltn.d.R. Erwin Böhme (u/c), *Ltn.d.R.* Hans Reimann, (#3, then KiA at 0955).

Notes:

1. Photograph reveals 7481's damage was consistent with a hard-landing/overturning, suggesting Bellerby crash-landed machine and then DoW. If so, being buried by an infantry division indicates it would have been shortly after landing.
2. Richthofen's identification of "Nr. 174" originated from the souvenired manufacture placard.

Victory No.2 Statistics:

Single-Seaters
1st and only September single-seater victory
1st of 8 1916 single-seater victories
1st of 18 pre-wound single-seater victories
1st of 35 total single-seater victories

G.100
1st and only G.100 victory

No. 27 Squadron
1st and only No. 27 Squadron victory

Deaths
No. 27 Squadron:
1st and only No.27 Squadron crewman KiA

Single-Seaters
1st and only September single-seater crewman KiA
1st of 6 1916 single-seater crewmen KiA
1st of 12 pre-wound single-seater crewmen KiA
1st of 21 total single-seater crewmen KiA

All:
3rd of 5 total September crewmen KiA

Above: Bellerby's 7418 after his encounter with Richthofen. Its fairly intact condition reveals the airplane did not impact the terrain in a vertical nosedive. Instead, the collapsed landing gear, wings, and damaged nose and rudder suggest Bellerby managed a hard landing that led to a nose-over inversion. The airplane is shown here after righting.

3rd of 17 1916 crewmen KiA
3rd of 66 pre-wound crewmen KiA
3rd of 84 total crewmen KiA

Etc.
2nd of 3 total September victories
2nd of 15 total 1916 victories
2nd of 57 total pre-wound victories
2nd victory in Albatros D.I or D.II
1st and only single-seater victory flying from Vélu/Bertincourt
2nd of 2 victories victory flying from Vélu/Bertincourt
1st of 4 single-seater victories flying under Boelcke
2nd of 6 total victories flying under Boelcke

Below: Albatros D.I 391/16's cockpit, seen here post-capture. Note the C-type cabane struts, strapped to which is plumbing associated with a British-installed pitot tube, a device which measures airspeed. The instrument between the struts appears to be a British-installed whiskey compass; Albatros compasses were located deep within the cockpit.

Victory No.3

Above: Lt. Ernest Conway Lansdale's headstone in 2011.

Above: Lansdale's stained glass memorial in St. John the Evangelist church, Goole. (H. Kilmer)

30 September
FE.2b 6973
No.11 Squadron RFC

Day/Date: Saturday, 30 September
Time: 11.50AM
Weather: Bright and fine all day, with occasional clouds in the afternoon.
Attack Location: Above Lagnicourt
Crash Location: Near Fremicourt
Side of Lines: Friendly

RFC Communique No. 56: "The 3rd Brigade attacked Lagnicourt aerodrome, causing damage to the aerodrome and buildings, and many bombs falling near hangars."

MvR Combat Report: "About 11.50AM I attacked, accompanied by 4 planes of our Staffel, above our airdrome Lagnicourt and in 3000 meters alt. a Vickers squad. I singled out a machine and after some 200 shots enemy plane started gliding down direction Cambrai. Finally it began to draw circles. The shooting had stopped and I saw that the machine was flying uncontrolled. As we were already rather far away from our front lines, I left the broken-down plane and selected a new adversary. Later on I could observe, how the above-mentioned machine, pursued by German Albatross machines, dashed burning to the ground near Fremicourt. The machine burnt to ashes.

Frhr. v. Richthofen.
Leutnant"

RFC Combat Casualties Report: "Left aerodrome 9.10AM. F.E.2b seen to go down in flames at 10.45AM S.E. of BAPAUME."

Right: Closeup of Lansdale's memorial. (H. Kilmer)

RFC A/C:
Make/model/serial number: Royal Aircraft Factory FE.2b 6973
Manufacturer: Subcontracted to and built by Boulton & Paul, Ltd., Norwich.
Unit: No.11 Squadron RFC
Aerodrome: Izel-le-Hameau
Sortie: Bombing Escort
Colors/markings: Presentation a/c *Malaya No.22, The Sime Darby*. Generally, PC10 nacelle; wing uppersurfaces PC10 with CDL undersurfaces; speculative white equilateral triangle on nose with base line extending longitudinally toward the fuselage sides.
Engine: 160 hp Beardmore
Engine Number: 634 WD 6994
Guns: Two Lewis. Nos. 2144, 11003
Manner of Victory: Uncontrolled spiral descent in flames until terrain impact.
Parts Souvenired: *Ich habe mir ein kleines Schild zum Andenken aufgehoben.* ("I have kept me a little sign as a keepsake.") Likely an engine or airplane manufacture placard.

RFC Crew:
Pilot: Lt. Ernest Conway Lansdale, 21, KiA
Cause of Death: Either gunshot wounds, immolation, or trauma associated with airplane crash.
Burial: Bancourt Communal Cemetery, Grave 2
Obs: 3049 Sgt. Albert Clarkson, 22, KiA_
Obs Victories: Reportedly 3 (details u/c, see Notes)
Cause of Death: Either gunshot wounds, immolation, or trauma associated with airplane crash.
Burial: Bancourt Communal Cemetery, Grave 3

MvR A/C:
Make/model/serial number: Albatros D.I, serial no. u/k, possibly 391/16
Unit: *Jasta* 2
Commander: *Hptm*. Oswald Boelcke
Aerodrome: Lagnicourt
Colors/markings: Shellacked and varnished "warm straw yellow" wooden fuselage overpainted green or brown. Extent of this overpainting included the engine cowl panels/radiators and extended forward to the edge of the nose cowl panels, which remained pale greenish-gray. This overpainting was applied thinly over the white fuselage/empennage crossfields and black crosses, which were located well forward of the production Albatros crosses. Pale greenish-gray cabane/interplane/landing gear struts. Upper surfaces of wings/ailerons and horizontal stabilizers/elevator likely remained factory three-color camouflage Venetian red, pale green and olive green, with the square white cross fields likely overpainted green, brown or some combination, perhaps leaving a thin border around the black cross, although photographs show at least some Jasta 2 Albatros D.Is had their wings/horizontal stabilizers/elevator uppersurfaces entirely overpainted. Wheel covers and undersurfaces of wings/ailerons and horizontal stabilizers/elevator were light blue. Rudder was either clear doped fabric or pale green, overpainted green or brown. Spinner was either white or very light gray, and a white or very light grey longitudinal stripe wrapped around the nose, just aft of the forward engine cowl panels.

Above: Clarkson's headstone, 2011.

This description is based on photographs and Lothar von Richthofen's comments regarding the appearance of his brother's machines, which he said after "Manfred began to gain his first successes" employed a "variety of colors to make himself invisible." Lothar described these colors as being initially "earth," which ostensibly refers to either green or brown. It is unknown when these earth colors were first employed, other than after Manfred's "first successes with *Jagdstaffel* Boelcke" (which is what *Jagdstaffel* 2 would eventually be renamed). Although Lothar attributed this camouflage to Manfred, many *Jasta* 2 Albatros D.Is and D.IIs had similar appearances.

Damage: u/k

Initial Attack Altitude: 3000 meters

Gunnery Range: u/k

Rounds Fired: 200

Known Staffel Participants: Richthofen was accompanied by "4 planes of our *Staffel*;" pilot identities u/k.

Notes:

1. Lansdale's record contains two undated Casualty Cards. Each contains identical information under the column *Result of Accident*, handwritten in black ink: "Missing. Died as P of W 9(asc) 3949." On one card, under this information and written in blue ink, is "8.12.16." This suggests Lansdale lingered in ill-health for 69 days and then died of wounds—or was it just the date the RFC received confirmation of his death? His gravestone records the date of death as 30 September 1916. Considering he was shot down in flames and is buried in the grave next to Clarkson, 30 September appears to be the accurate date of death.

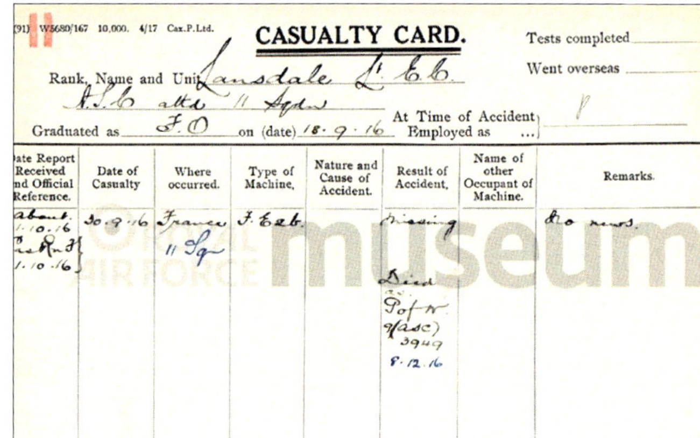

(91) W5680/167 10,000. 4/17 Cax.P.Ltd.

CASUALTY CARD.

Rank, Name and Unit Landsale Lt E.C.
R.F.C. attd 11 Sqdn

Graduated as F.O on (date) 18.9.16

At Time of Accident Employed as ... P

Tests completed

Went overseas

Date Report Received and Official Reference.	Date of Casualty	Where occurred.	Type of Machine.	Nature and Cause of Accident.	Result of Accident.	Name of other Occupant of Machine.	Remarks.
about 1.10.16 Ex R.F.C. 1.10.16	30.9.16	France 11 Sq	F.E.2b.		Missing Died as P of W 9(asc) 3949 8.12.16		No news.

Above: Landsale's casualty card. (RAF Museum)

2. RFC Communiques reveal some of Clarkson's earlier actions in 1916:
   a. 1 August: "On the 3rd Army front a few single machines were seen patrolling behind their own lines. Five were encountered, one of which was seen to dive very steeply after a fight with an F.E. of No. 11 Squadron—pilot, Sgt. Thomson, observer Sgt. Clarkson. It was not however, definitely seen to be out of control."
   b. 17 September: "A Roland was engaged and driven down out of control by Sergt. Thompson and Sergt. Clarkson, of No. 11 Squadron."
3. The colors and markings description of Richthofen's Albatros D.I is the result of "reverse engineering" several photographs of Richthofen standing near an overpainted Albatros D.II with a white stripe around its engine cowl (a machine that will be discussed at length in Victory Nos. 6 and 7). Reasonable circumstantial photographic provenance suggests (but does not prove) this D.II was his personal machine, regardless of its serial number (which cannot be seen). Researcher Lance Bronnenkant generously provided new photographs first published in his *Blue Max Airmen Series Vol. 5* that shows a *Jasta* 2 Albatros D.I overpainted in the same manner, also with a white stripe around its engine cowl. The logical

Albatros D.I 391/16. It is unknown if Richthofen flew this machine on this sortie but its appearance is based on photographs from the same time period. The dark fuselage suggests brown overpainting rather than green. The exact hue of brown is unknown; that shown in the profile is conjectural.

speculative conclusion based on the similar markings is this D.I likely was also Richthofen's.

4. In photographs of this machine, the "3" and the "1" in the serial number can be determined but the middle digit cannot, aside from its round shape suggesting 0, 3, 6, 8, or 9. That the serial number begins with a 3 identifies the machine as from the 12 Albatros D-type pre-production machines, numbered 380/16 to 391/16, which eliminates 0, 3, and 6 as the middle digit and limits the remaining choices to either 381/16 or 391/16. Careful scrutiny via Photoshop reveals the right side of the middle digit is a continuous arc, more closely resembling a 9 than an 8, which would have a midpoint indentation. The digit thickness is consistent with a 9 as well, whereas an 8 would be very thin on its upper right side, and that is not present in the photograph. 391/16 is well-known machine that was studied and photographed extensively after it was brought down by a bullet to the radiator and subsequently captured intact (16 November 1916, pilot *Ltn*. Karl Büttner, PoW). It was repaired and test-flown by the British until it crashed via unknown circumstances and presumably destroyed. If 391/16 were formerly Richthofen's there is no trace of any overpainted white nose stripe in any post-capture photograph, although ostensibly Büttner would have removed Richthofen's personal white stripe when adding his own markings, a large "Bü" on both sides of the fuselage adjacent the cockpit. Additionally, at some point a thin white outline had been added to at least the port fuselage cross.

Victory No.3 Statistics:

Pushers*

2nd of 2 September total pusher victories
2nd of 2 September two-seater pusher victories
2nd of 5 1916 two-seater pusher victories
2nd of 12 total two-seater pusher victories
2nd of 18 total pusher victories

Two-Seaters

2nd of 2 September two-seater victories
2nd of 7 1916 two-seater victories
2nd of 39 pre-wound two-seater victories
2nd of 45 total two-seater victories

FE.2 Δ

2nd of 2 September FE.2 victories
2nd of 5 1916 FE.2 victories
2nd of 12 total FE.2 victories

No.11 Squadron ±

2nd of 4 No.11 Squadron FE.2b victories
2nd of 2 September No.11 Squadron victories
2nd of 2 1916 No.11 Squadron victories
2nd of 4 total No.11 Squadron victories

Deaths†

No.11 Squadron:

3rd and 4th of 4 September No.11 Squadron crewmen KiA
3rd and 4th of 4 1916 No.11 Squadron crewmen KiA
3rd and 4th of 4 total No.11 Squadron crewmen KiA

Pushers:

3rd and 4th of 4 September two-seater pusher crewmen KiA

Above: Lineup of *Jasta* 2 Albatros D.Is at Lagnicourt. Those with the forward fuselage cross and light (presumably light blue) undersurfaces are pre-production machines, the uppersurfaces of which have been overpainted in undetermined darker color(s). By contrast, the second machine at left is 431/16 (the "S" on the fuselage identifies it as that of *Ltn.* Jürgen Sandel) and remains in its factory "warm straw yellow" appearance. Both machines at right have been almost entirely overpainted, including the radiator sides and most of the engine cowls. Second from right, the Albatros with the white spinner and nose stripe is suspected to have been Richthofen's. (Lance Bronnenkant)

Above: Closeup of Richthofen's presumed Albatros D.I, what appears to be 391/16. (Lance Bronnenkant)

3rd and 4th of 10 1916 two-seater pusher crewmen KiA
3rd and 4th of 17 total two-seater pusher crewmen KiA

Two-Seaters:
3rd and 4th of 4 September two-seater crewmen KiA
3rd and 4th of 11 1916 two-seater crewmen KiA
3rd and 4th of 54 pre-wound two-seater crewmen KiA
3rd and 4th of 63 total two-seater crewmen KiA

All:
4th and 5th of 5 total September crewmen KiA
4th and 5th of 17 1916 crewmen KiA
4th and 5th of 66 total pre-wound crewmen KiA
4th and 5th of 84 total crewmen KiA

Etc.
3rd of 3 total September victories
3rd of 15 total 1916 victories
3rd of 57 total pre-wound victories
3rd victory in Albatros D.I or D.II
1st victory shot down in flames
1st of 5 two-seater victories flying from Lagnicourt

Above: 391/16 shortly after being brought down and captured 16 November 1916. If indeed once Richthofen's machine, all traces of the nose stripe had been overpainted, presumably upon possession by *Ltn.* Karl Büttner.

Above: 391/16 meets its apparent end.

1st of 9 total victories flying from Lagnicourt
2nd of 2 two-seater victories flying under Boelcke
3rd of 6 total victories flying under Boelcke

Statistics Notes:
* All pusher victories are pre-wound
Δ All FE.2 victories are pre-wound
± All No.11 Squadron victories are pre-wound
† All No.11 Squadron deaths are pre-wound; all pusher deaths are pre-wound; KiA includes DoW

Victory No.4

Above: Royal Aircraft Factory BE.12 6536 provides a general glimpse of the make/model Richthofen shot down for his fourth, fifth, and sixth victories. The BE.12's BE.2 heritage is quite apparent.

7 October
BE.12 6618
No.21 Squadron RFC

Day/Date: Tuesday, 7 October
Time: 9.10AM
Weather: Low clouds and strong winds—stormy all day. Unfavorable for flying.
Attack Location: Near Rancourt
Location: Near Equancourt
Side of Lines: Friendly

*KOFL 1. Armee Weekly Activity Report: "***Erfolge im Luftkampf:** *Durch Leutnant von Richthofen wurde am 7.10. vorm 9.10 bei Equancourt ein engl. Rumpfdoppeldecker Einsitzer abgeschossen. Insassen Leutnant Fenwich tot. Flugzeug war ein B.E. neuer Art."*

"**Successes in Aerial Combat:** On the 7th of October AM [at] 09:10 h an English fuselage biplane single-seater was shot down by *Leutnant* von Richthofen near Equancourt. Occupants: *Leutnant* Fenwich dead. Aircraft was a B.E. of a new type."

"**Vom Feinde:** Der am 7.10 von Lt.Frh.v.Richthofen bei Equancourt abgeschossene Rumpfdoppeldecker-Einsitzer (Lt.Fenwick) gehörte zur 21.Squ. und zwar den Papieren nach zuneinem B.E. Flight. Das Flugzeug ist anseh scheinend eine neuere Konstrucktion."

"**From the enemy:** On the 7th of October the fuselage biplane single-seater (Lt. Fenwick) [which was shot down] near Equancourt by *Lt. Frh.* von Richthofen belonged to 21 Sqdn, in fact, according to the papers, to a B.E. flight. The aircraft seems to be a newer construction [version]."

MvR Combat Report: "Machine Type: New and not seen up till now. Plane No.6618. A two deck plane (byplane)[*sic*] with 12 cyl. Daimler Motor No.25 226."

About 9 a.m. I attacked in 3000 meters alt. and accompanied by 2 other machines an English plane near Rancourt. After 400 shots enemy plane dashed downwards, the pilot having been mortally wounded. Inmate: Lieut. Fenwick, killed by shot in the head.

Frhr. v. Richthofen
Leutnant"

Right: The remains of BE.12 6618.

RFC Combat Casualties Report: "Left aerodrome 7.30AM." [No other information.]

RFC A/C:
Make/model/serial number: Royal Aircraft Factory BE.12 6618
Manufacturer: Subcontracted to and built by Daimler Co., Ltd., Coventry
Unit: No.21 Squadron RFC
Aerodrome: Bertangles
Sortie: Offensive Patrol
Colors/markings: PC10 uppersurfaces; CDL lower surfaces; b/w/r rudder; 6618 (black w/white border) on vertical stabilizer. Description based on photo of crashed machine.
Engine: 150 hp RAF4a
Engine Number: 25226 WD 5730
Guns: One Vickers, one Lewis. Nos. 16442, L6920
Manner of Victory: Unchecked descent ("dashed downwards") until terrain impact.
Parts Souvenired: a."6618" from either port (likely) or starboard vertical stabilizer. Black numbers/ white borders on PC10 background. Photographed on display at Roucourt, his home bedroom in Schweidnitz, and in the post-war Richthofen Museum.
b. Daimler Co. Ltd engine placard. Photographed on display at Roucourt and the post-war Richthofen Museum.

RFC Crew:
Pilot: 2nd Lt. William Cecil Fenwick, 19 or 20 (DoB 1897), KiA
Cause of Death: Shot in head by machine gun.
Burial: Either in field or laid where fallen; location lost during turmoil of war.

MvR A/C:
Make/model/serial number: Albatros D.I, possibly 391/16
Unit: *Jasta* 2
Commander: *Hptm.* Oswald Boelcke
Aerodrome: Lagnicourt
Colors/markings: See Vic. #3.
Damage: u/k
Initial Attack Altitude: 3000 meters
Gunnery Range: u/k
Rounds Fired: 400
Known Staffel Participants:

Victory No.4 Statistics:
Single-Seaters
1st of 3 October single-seater victories
2nd of 8 1916 single-seater victories
2nd of 18 pre-wound single-seater victories
2nd of 35 total single-seater victories

BE.12
1st of 3 October BE.12 victories
1st of 3 1916 BE.12 victories
1st of 3 total BE.12 victories

No.21 Squadron *
1st of 2 No.21 Squadron BE.12 victories
1st of 2 October No.21 Squadron victories
1st of 2 1916 No.21 Squadron victories
1st of 2 total No.21 Squadron victories

Deaths †
No.21 Squadron:
1st of 2 October No.21 Squadron crewmen KiA
1st of 2 total No.21 Squadron crewmen KiA

Above: Another view of BE.12's demise. The shattered and tangled wreckage reveals the violence of its terrain impact.

Above: 6618's souvenired fabric on display in the post-war Richthofen Museum. Note that while once tacked directly to the wall at Roucourt, now the fabric has been mounted onto a rigid backing of unknown material and affixed to the wall via two eyelets, suspended (presumably) by nails.

Left: Souvenired fabric from 6618's vertical stabilizer on display at Roucourt.

Left: 6618's souvenired engine placard.

Albatros D.I 391/16. It is unknown if Richthofen flew this machine on this sortie but its appearance is based on photographs from the same time period. The dark fuselage suggests brown overpainting rather than green. The exact hue of brown is unknown; that shown in the profile is conjectural.

Single-Seaters:
1st of 3 October single-seater crewmen KiA
2nd of 6 1916 single-seater crewmen KiA
2nd of 12 pre-wound single-seater crewmen KiA
2nd of 21 total single-seater crewmen KiA

All:
1st of 3 total October crewmen KiA
6th of 17 1916 crewmen KiA
6th of 66 pre-wound crewmen KiA
6th of 84 total crewmen KiA

Etc.
1st of 3 total October victories
4th of 15 total 1916 victories
4th of 57 total pre-wound victories
4th victory in Albatros D.I or D.II
1st of 4 single-seater victories flying from Lagnicourt
2nd of 9 total victories flying from Lagnicourt
2nd of 4 single-seater victories flying under Boelcke
4th of 6 total victories flying under Boelcke

Statistics Notes:
**All No. 21 Squadron victories are pre-wound*
†All No. 21 Squadron deaths are pre-wound; KiA includes DoW

Victory No.5

Above: 6580's souvenired fabric on the wall at Roucourt.

Above: 6580's souvenired fabric at the Richthofen Museum.

16 October
BE.12 6580
No.19 Squadron RFC

Day/Date: Monday, 16 October
Time: 5AM (MvR report) 17.00 (*Jasta* 2 report)
Weather: Fine with occasional clouds.
Attack Location: Above Bertincourt
Crash Location: Near Ytres
Side of Lines: Friendly

MvR Combat Report: "B.E. One seater Nr. 6580. Daimler motor, Nr. 25188. Inmate: Lieut. Capper.

Together with 4 planes I singled out above Bertincourt enemy squad in 2800 m. alt. After 350 shots I brought down an enemy plane. Plane crashed to the ground, smashed. Motor can probably be secured.

Frhr. v. Richthofen
Leutnant"

RFC A/C:
Make/model/serial number: Royal Aircraft Factory BE.12 6580
Manufacturer: Subcontracted to and built by Daimler Co. Ltd., Coventry
Unit: No.19 Squadron RFC
Aerodrome: Fienvillers
Sortie: Bomb Raid
Colors/markings: u/k. Typically for the BE.12, PC10 uppersurfaces; CDL lower surfaces; b/w/r rudder; black serial number with white border on vertical stabilizer.
Engine: 150 hp RAF4a
Engine Number: 25188 WD 5693
Guns: One Vickers, one Lewis. Nos. 14543, L6632
Manner of Victory: "Crashed to the ground, smashed."

Right: 6580's souvenired engine placard.

Parts Souvenired:
a. "6580" from both the port and starboard sides of vertical stabilizer. Black numbers/ white borders on PC10 background. Photographed on display at Roucourt, his home bedroom in Schweidnitz, and in the post-war Richthofen Museum.
b. Daimler Co. Ltd engine placard. Photographed on display at Roucourt and the post-war Richthofen Museum.

RFC Crew:
Pilot: 2nd Lt. John Thompson, 23, KiA
RFC Combat Casualties Report: "Death accepted by the Army Council as having occurred on 16/10/16 (3rd Echelon part II Orders, AFO 1810, dated 4/8/17) on the evidence of a report by the O.C. a Graves Registration unit."
Cause of Death: u/k. Either gunshot wounds or trauma associated with airplane crash.
Burial: Lebucquière Communal Cemetery Extension, Grave III. B. 27.

MvR A/C:
Make/model/serial number: Albatros D.I, possibly 391/16
Unit: *Jasta* 2
Commander: *Hptm.* Oswald Boelcke
Airfield: Lagnicourt
Colors/markings: See Vic. #3.
Damage: u/k

Initial Attack Altitude: 2800 meters
Gunnery Range: u/k
Rounds Fired: 350
Known Staffel *Participants:* u/k

Notes:
1. The *Jasta* 2 and No. 19 Squadron reports agree their engagement took place in the afternoon. Thus, 5"AM" in Richthofen's combat report likely is a typo for 5PM (if so, it is unknown if said typo originated on the German report or occurred during translation.)
2. Richthofen misidentified the pilot he shot down as No. 19 Sqn's Ltn. E. W. Capper. Capper was actually shot down and killed on 14 April 1917 by *Jasta* 11 pilot *Ltn.* Kurt Wolff, his 15th victory. It has been theorized that Thompson possessed

Albatros D.I 391/16. It is unknown if Richthofen flew this machine on this sortie but its appearance is based on photographs from the same time period. The dark fuselage suggests brown overpainting rather than green. The exact hue of brown is unknown; that shown in the profile is conjectural.

some article of clothing or flight gear that was marked with Capper's name.

Victory No.5 Statistics:

Single-Seaters
2nd of 3 October single-seater victories
3rd of 8 1916 single-seater victories
3rd of 18 pre-wound single-seater victories
3rd of 35 total single-seater victories

BE.12
2nd of 3 October BE.12 victories
2nd of 3 1916 BE.12 victories
2nd of 3 total BE.12 victories

No.19 Squadron
1st and only No.19 Squadron BE.12 victory
1st and only October No.19 Squadron victory
1st and only 1916 No.19 Squadron victory
1st of 3 pre-wound No.19 Squadron victories
1st of 4 total No.19 Squadron victories

Deaths*
No.19 Squadron:
1st and only October No.19 Squadron crewman KiA
1st and only 1916 No.19 Squadron crewman KiA
1st of 2 pre-wound No.19 Squadron crewmen KiA
1st of 3 total No.19 Squadron crewmen KiA

Single-Seaters:
2nd of 3 October single-seater crewmen KiA
3rd of 6 1916 single-seater crewmen KiA
3rd of 12 pre-wound single-seater crewmen KiA
3rd of 21 total single-seater crewmen KiA

All:
2nd of 3 total October crewmen KiA
7th of 17 1916 crewmen KiA
7th of 66 pre-wound crewmen KiA
7th of 84 total crewmen KiA

Etc.
2nd of 3 total October victories
5th of 15 total 1916 victories
5th of 57 total pre-wound victories
5th victory in Albatros D.I or D.II
2nd of 4 single-seater victories flying from Lagnicourt
3rd of 9 total victories flying from Lagnicourt
3rd of 4 single-seater victories flying under Boelcke
5th of 6 total victories flying under Boelcke

Statistics Notes:
*KiA includes DoW

Above: Jürgen Sandel, Richthofen, Bodo von Lyncker, and Hans Immelman (L-R) on the undercarriage of Sandel's Albatros D.I 431/16. (Greg VanWyngarden)

Victory No.6

Left: German cross that marked 2nd Lt. Arthur James Fischer's initial grave. (RAF Museum)

Right: Fischer's headstone, 2011.

25 October
BE.12 6629
No.21 Squadron RFC

Day/Date: Wednesday, 25 October
Time: 9.35AM (MvR report) 9.50 (*Jasta 2* report)
Weather: Clear intervals in morning, then mostly rain. (*KOFL 1.* Report)
Attack Location: Above trenches near Lesboefs
Crash Location: South end of Bapaume
Side of Lines: Friendly

KOFL 1. Armee Weekly Activity Report: "Successes in Aerial Combat: At 9.50 by Frhr. von Richthofen, Jasta 2, B.E. biplane, Bapaume-Riencourt."

MvR Combat Report: "B.E. Two seater. Nr. 6629. Motor dashed into earth, therefore number not legible. Inmate, a Lieut. Seriously wounded by shot in bowels. Plane itself cannot be brought back, as under heavy fire."

About 9AM I attacked enemy plane above trenches near Lesboefs. Unbroken cover of clouds in 2000 m.alt. Plane came from the German side and was just approaching own lines. I attacked and after some 200 shots he went down in large right hand curves and was forced back by the strong wind to the south end of Bapaume. Finally the machine crashed to the ground.

As I first saw the enemy plane there was no other German machine in the neighborhood, and also during the fight no machine approached the scene of action. As the enemy plane started to go down, I saw a German Rumpler machine and several Halberstadter planes. One of these machines came down to the ground. It was Sergt.Major Mueller of *Jagdstaffel* 5. He claims to have discharged first at 300 meters and then at 1000 meters distance some 500 shots at enemy plane. Afterwards his gun jammed and the sight of his gun flew away. Quite apart from these curious circumstances, a child knows that one cannot hit a plane from such a ridiculous distance. Then a second plane, a Rumpler, came down, also claiming his share of the loot. But all other planes were perfectly sure that he had not taken part in the fight.

Frhr. v. Richthofen
Leutnant"

RFC Combat Casualties Report: "Left aerodrome 7.45AM. A B.E.12 was reported by 11th A.A.Battery to be seen diving down 15,000 yds N.E. of MARICOURT, apparently under control pursued by a German biplane."

RFC A/C:
Make/model/serial number: Royal Aircraft Factory BE.12 6629

Albatros D.I 391/16. It is unknown if Richthofen flew this machine on this sortie but its appearance is based on photographs from the same time period. The dark fuselage suggests brown overpainting rather than green. The exact hue of brown is unknown; that shown in the profile is conjectural.

Manufacturer: Subcontracted to and built by Daimler Co. Ltd., Coventry
Unit: No.21 Squadron RFC
Aerodrome: Bertangles
Sortie: Patrol, Beaulencourt
Colors/markings: u/k. Typically for the B.E.12, PC10 uppersurfaces; CDL lower surfaces; b/w/r rudder; black serial number with white border on vertical stabilizer.
Engine: 150 hp RAF4a
Engine Number: 303 WD 3583
Guns: One Vickers, one Lewis. Nos. 17334, L6327
Manner of Victory: Descending right spiral until terrain impact.
Parts Souvenired: None known, due to the wreckage being targeted by artillery fire.

RFC Crew:

Pilot: 2nd Lt. Arthur James Fischer, 21, PoW/DoW
RFC Combat Casualties Report: "Letter from 5th Aus.Divn.No.A.67/121 [dated 27.3.17] states that the grave of 2/Lt. A.J.Fisher has been located in cemetery 1.31.a.6.6… The grave is marked with a regulation German wooden cross giving particulars."
Cause of Death: u/k. Either gunshot wound(s) and/or trauma associated with airplane crash.
Burial: Bancourt Communal Cemetery, Grave 9

MvR A/C:

Make/model/serial number: Albatros D.I, possibly 391/16; or Albatros D.II, possibly 481/16.
Unit: *Jasta* 2
Commander: *Hptm.* Oswald Boelcke
Airfield: Lagnicourt
Colors/markings: Albatros D.I 391/16 – See Vic. #3.
Damage: u/k

Initial Attack Altitude: 2000 meters
Gunnery Range: u/k
Rounds Fired: 200
Known Staffel Participants: u/c

Notes:

1. It is possible Richthofen was flying a new Albatros D.II, a model ordered into production August 1916. Regarding Albatros manufacture norms, generally there was an approximate window of two months between production order and arrival of the first machines, which in this case identifies October as when the D.IIs appeared with *Jasta* 2. However, the exact date of the August production order is unknown and if inked at the end of August, the first D.II arrivals likely would have been closer to November. An album of *Jasta* 2 photographs includes shots of the D.II presumed to be Richthofen's (darkly overpainted with a white stripe on the nose, similar to the appearance of Albatros D.I 391/16) amongst photographs of Oswald Boelcke, which speculatively dates these D.II photos as being taken before or around the time of Boelcke's 28 October death. Beingthat Richthofen's Albatros D.II is thought to have been 481/16 (a subject to be addressed in Victory No. 7), this means that D.II was a very early production machine (the tenth out of a batch of fifty) and it would have been one of the earliest to arrive. Still, while it is possible *Jasta* 2 began receiving Albatros D.IIs in mid-October, suggesting Richthofen could have flown one during his fifth victory on the 15th of that month, their arrival is unknown precisely. Additionally, it is possible that for a brief period Richthofen flew both the Albatros D.I *and* D.II.

Albatros D.II 481/16. It is unknown if *Jasta* 2 had received their first production Albatros D.II by this date and, if so, if it had been overpainted. Upon arrival, Albatros D.II 481/16 would have appeared similar to a new Albatros D.I, with factory finish shellacked and varnished "warm straw yellow" wooden fuselage and pale greenish-gray spinner/cowl hatches and struts. Upper surfaces of wings/ailerons and horizontal stabilizers/elevator were factory three-color camouflage of pale green, olive green and Venetian red. Wheel covers/undersurface of wings/ailerons and horizontal stabilizers/elevators light blue. Rudder was either clear doped fabric or one of the camouflage colors. All crosses were black on square white cross fields, which on the fuselage were located further aft than on the pre-production Albatros D.Is.

The arrival of the D.II did not mean the extent D.Is just "went away." Many *Jasta* 2 photographs show both models were used simultaneously, and Richthofen's later combat reports reveal he often switched between machines during consecutive sorties.

2. The combat report reference "Sergt.Major Mueller" refers to Vfw Heinrich Müller, *Jasta* 5.

Victory No.6 Statistics:

Single-Seaters
3rd of 3 October single-seater victories
4th of 8 1916 single-seater victories
4th of 18 pre-wound single-seater victories
4th of 35 total single-seater victories

BE.12
3rd of 3 October BE.12 victories
3rd of 3 1916 BE.12 victories
3rd of 3 total B.E.12 victories

No.21 Squadron *
2nd of 2 No.21 Squadron BE.12 victories
2nd of 2 October No.21 Squadron victories
2nd of 2 1916 No.21 Squadron victories
2nd of 2 total No.21 Squadron victories

Deaths †
No.21 Squadron:
2nd of 2 October No.21 Squadron crewmen KiA
2nd of 2 total No.21 Squadron crewmen KiA

Single-Seaters:
3rd of 3 October single-seater crewmen KiA
4th of 6 1916 single-seater crewmen KiA
4th of 12 pre-wound single-seater crewmen KiA
4th of 21 total single-seater crewmen KiA

All:
3rd of 3 total October crewmen KiA
8th of 17 1916 crewmen KiA
8th of 66 pre-wound crewmen KiA
8th of 84 total crewmen KiA

Etc.
3rd of 3 total October victories
6th of 15 total 1916 victories
6th of 57 total pre-wound victories
6th victory in Albatros D.I or D.II
3rd of 4 single-seater victories flying from Lagnicourt
4th of 9 total victories flying from Lagnicourt
4th of 4 single-seater victories flying under Boelcke
6th of 6 total victories flying under Boelcke

Statistics Notes:
*All No.21 Squadron victories are pre-wound
†All No.21 Squadron deaths are pre-wound; KiA includes DoW

Victory No.7

Above: Sporting binoculars and rather casual attire, Richthofen enjoys a jovial moment with fellow *Jasta* 2 pilots (L-R) Stefan Kirmaier, Hans Imelmann, and Hans Wortmann at Lagnicourt. The appearance of Kirmaier (KiA 22 November) and Wortmann (arrived at *Jasta* 2 early November; certainly by the 9th, when he shot down his first plane) in the same photograph dates the occasion as having been early-to-mid-month. The darkly-overpainted and white-striped Albatros D.II behind them is believed to have been Richthofen's, but absolute proof of this is lacking.

3 November
FE.2b 7010
No.18 Squadron RFC

Day/Date: Friday, 3 November
Time: 2.10PM
Weather: Misty, afternoon clear intervals (*KOFL 1.* report)
Attack Location: Over Gommecourt
Crash Location: Northeast of Grevillers Wood
Side of Lines: Friendly

RFC Communique No. 60: "There was considerable hostile aerial activity on the fronts of the Third, Fourth and Fifth Armies."

KOFL 1. Armee Weekly Activity Report: "*Weiter angemeldete Abschüsse:* über welche eine Entscheidung noch aussteht. *3.11.16. Leutnant frhr.v. Richthofen meldet 2.10 Nachm. nordöstl. Gréviller Wald diesseits der Linie den Abschuss eines Vickers Zweisitzers. Insassen 2 Engländer tot.*

Machines Claimed but Awaiting Confirmation: 3 November [19]16. *Leutnant Frhr.* v. Richthofen reports the kill of a Vickers two-seater at 2:10PM northeast of Grévillers Wood this side of the line. Crew 2 Englishmen dead."

"*Erfolge im Luftkampf: Der Bericht des Leutnants Frhr.v.Richthofen, Jagdstaffel 2, uber einen am 3.11.16 nachm. 2.10 beim Gré-viller-Wald abgeschossenen Vickers Zweisitzer wird dem Kogen. Luft. zur Anerkennung vorgeleg.*

Successes in aerial combat: The report of the

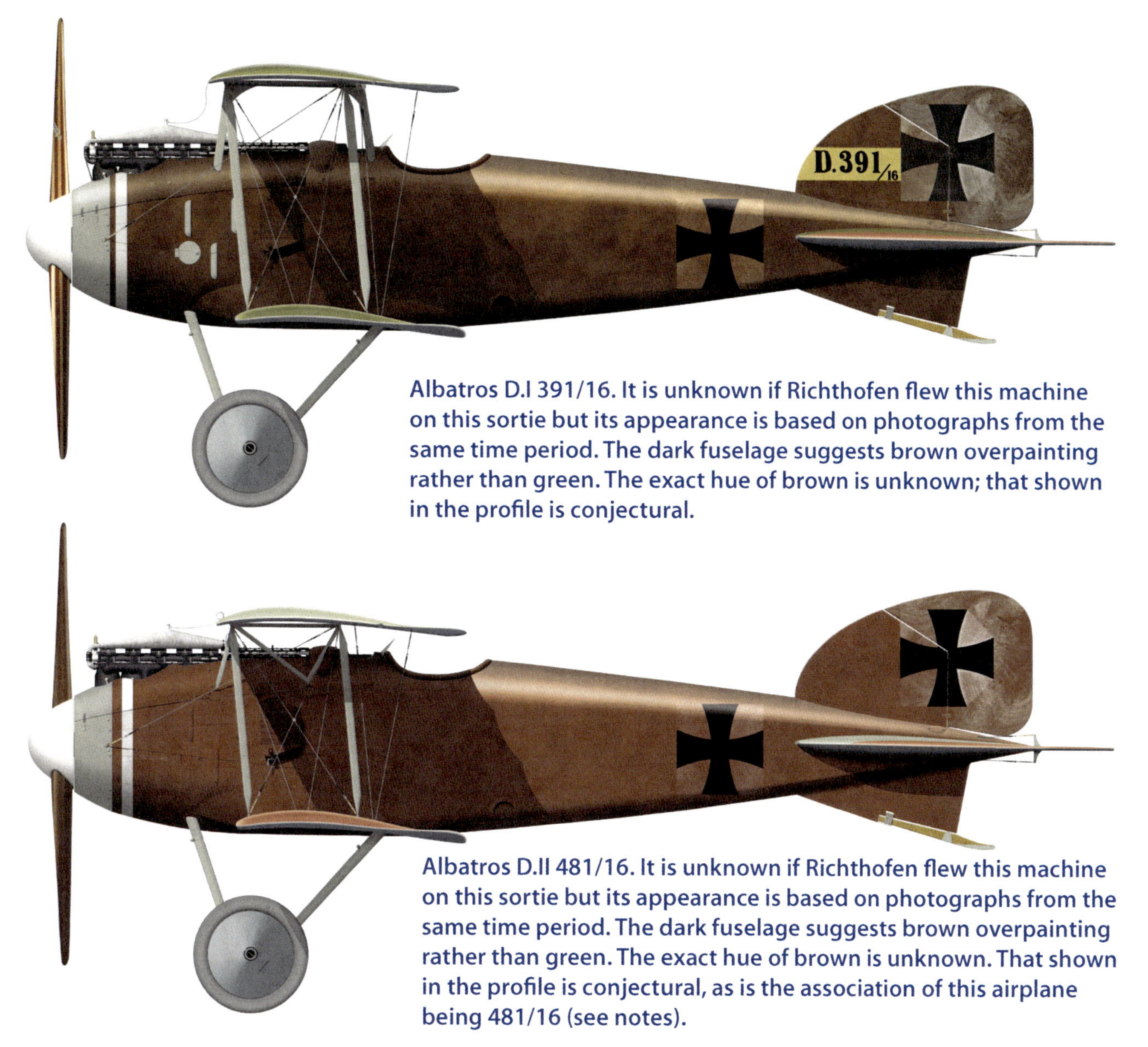

Albatros D.I 391/16. It is unknown if Richthofen flew this machine on this sortie but its appearance is based on photographs from the same time period. The dark fuselage suggests brown overpainting rather than green. The exact hue of brown is unknown; that shown in the profile is conjectural.

Albatros D.II 481/16. It is unknown if Richthofen flew this machine on this sortie but its appearance is based on photographs from the same time period. The dark fuselage suggests brown overpainting rather than green. The exact hue of brown is unknown. That shown in the profile is conjectural, as is the association of this airplane being 481/16 (see notes).

Leutnant Frhr. v. Richthofen, *Jagdstaffel* 2, about one Vickers two-seater, shot down [by him] on 3 November 16, afternoon, at 2.10 near Grevillez [*sic*]Wood, will be submitted to *Kogenluft* for confirmation."

MvR Combat Report: "Vickers Two Seater: No.7010.

Accompanied by two machines of the *Staffel* I attacked a low flying enemy plane in 1800 meters alt. After 400 shots adversary dashed to the ground. The plane was smashed to pieces, inmates killed.

As the place where the plane fell is under heavy fire, no details could be ascertained yet.

Frhr. v. Richthofen
Leutnant"

RFC Combat Casualties Report: "Left aerodrome 11.35 a.m. 5th Brigade report that A.A.Battery saw an F.E.2.b. with two hostile machines above, and under heavy shell fire, finally come down, apparently out of control 7,000 yds E. of ENGLEBELMER and beyond BEAUMONT HAMEL."

RFC A/C:
- Make/model/serial number: Royal Aircraft Factory FE.2b 7010
- Manufacturer: Subcontracted to and built by Boulton & Paul Co. Ltd., Norwich
- Unit: No.18 Squadron RFC
- Aerodrome: Laviéville
- Sortie: Hostile Aircraft Patrol
- Colors/markings: u/k. Typically for the make/

model, PC10 nacelle/wings/ailerons/horizontal stabilizers/elevator uppersurfaces; CDL undersurfaces; r/w/b rudder with black serial numbers, usually bordered in white when atop the red and blue.
Engine: 160 hp Beardmore
Engine Number: 399 WD 2371
Guns: Two Lewis. Nos. 1405, 4438
Manner of Victory: Unchecked descent until terrain impact
Parts Souvenired: u/k

RFC Crew:

Pilot: Sgt. Cuthbert Godfrey Baldwin, 28, KiA
Cause of Death: u/k Either gunshot wound(s) or trauma associated with airplane crash.
Burial: Either in-field or laid where fallen; location lost in the turmoil of war.
Obs: 2nd Lt. George Andrew Bentham, 21, KiA
Cause of Death: u/k Either gunshot wound(s) or trauma associated with airplane crash.
Burial: Either in-field or laid where fallen; location lost in the turmoil of war.
RFC Combat Casualties Report: "Death of 2/Lt. G.A. Bentham accepted by the Army Council as having occurred on or since 3/11/16 on lapse of time."

MvR A/C:

Make/model/serial number: Albatros D.I, possibly 391/16; or Albatros D.II, possibly 481/16.
Unit: *Jasta* 2
Commander: *Oblt.* Stefan Kirmaier
Airfield: Lagnicourt
Colors/markings: Albatros D.I 391/16 – See Vic. #3.
Albatros D.II 481/16: Shellacked and varnished "warm straw yellow" wooden fuselage completely overpainted in green or more likely brown, including the engine cowl panels/vents/access hatches and extending forward to the edge of the nose cowl panels, which remained pale greenish-gray. This brown overpainting was either thinly applied or faded atop the white empennage crossfield and was liberally applied atop the white fuselage crossfield. Pale greenish-gray cabane/interplane/landing gear struts, with a white or very light gray spinner and white stripe on nose, adjacent second engine cylinder. Upper surfaces of wings/ailerons and horizontal stabilizers/elevator were factory three-color camouflage Venetian red, pale green and olive green, with the square white crossfields most likely overpainted green, brown or some combination, possibly leaving a thin border around the black cross, although the entire uppersurface may have been overpainted in the same manner as other *Jasta* 2 Albatros D.Is. Wheel covers and undersurfaces of wings/ailerons and horizontal stabilizers/elevator were light blue. Rudder was either clear doped fabric or pale green, overpainted green or more likely brown. All crosses on overpainted square white cross fields. Propeller appears to be an Axial with an early "dagger" manufacture decal. The absence of visible stamped data and the unusually dark and matte finish of the blades suggests they had been either greased or overpainted, although this might have resulted from orthochromatic film which darkened yellows (see notes).
This description is based on photographs of Richthofen posing with this particular machine (see notes), although confirmation of ownership has not been ascertained absolutely. Photos clearly illustrate this Albatros D.II had been overpainted, perhaps a result of the "variety of colors [Manfred used] to make himself invisible" prior to his employment of red to promote conspicuousness. Lothar von Richthofen described these colors as being initially "earth," which ostensibly refers to either green or brown. Brown is believed to be the most likely color, based on the dark nature of the Albatros's appearance, eyewitness descriptions by RFC airmen who fought *Jasta* 2 machines in autumn 1916.
Damage: u/k

Initial Attack Altitude: 1800 metres
Gunnery Range: u/k
Rounds Fired: 400
Known Staffel Participants:

Notes:

1. Although at this time it is possible Richthofen flew both the Albatros D.I and D.II, based on the ca. two month span between production order and in-field machine arrival it is presumed that Richthofen primarily flew a D.II by November. Supporting this are several photographs that show Richthofen standing near an overpainted Albatros D.II with a white stripe around its engine cowl; reasonable circumstantial photographic provenance that suggests (but does not prove) this was his personal machine, regardless of its serial number (which cannot be seen). These photographs are also the keystone for associating Richthofen with the similarly marked Albatros D.I 391/16.

One photograph features Richthofen, *Ltn.d.R.* Hans Wortmann, *Oblt.* Stefan Kirmaier, and *Ltn.d.R.* Hans Imelmann standing next to the overpainted/white-striped Albatros D.II. Since Wortmann arrived at *Jasta* 2 in early November and Kirmaier was KiA 22 November, these events date the photograph as having been taken sometime during the first three weeks of November—certainly between 9 November (Wortmann's first victory) and Kirmaier's death 22 November. *When* during this period is unknown but if dovetailed with the previous paragraph it is reasonably certain that by November 1916 Richthofen had beenissued and was flying this Albatros D.II. Although photographed next to this airplane, other photographs associate Kirmaierwith another D.II, and a D.I marked with a "W" on the fuselage is associated with Wortmann.

2. Regardless, it is unknown which D.II Richthofen flew during which sortie. For decades there has been a source-based discrepancy regarding the serial number of his D.II, either 481/16 or 491/16. As far as the author can determine, the first use of 491/16 was in Nowarra and Brown's 1958 *Von Richthofen and the Flying Circus.* Their source for this is uncredited. At the University of Texas Dallas, in Ed Ferko's handwritten manuscript for his 1995 book *Richthofen*, the author found "491/16" listed as Richthofen's D.II. However, in the published book Ferko lists the serial number as "481/16." The source for both is uncredited, although many believe "481/16" merely was a typo for "491/16." But, if so, could not "491" be a type for "492"?

The only source the author can find (which he overlooked for his 2009 book *Manfred von Richthofen: The Aircraft, Myths and Accomplishments of 'The Red Baron'*) is in Richthofen's 15th combat report. In it, for the first time (at least in the translated reports), Richthofen identifies the airplane he flew during a sortie, in this case "Albatros D.481." Be aware that this does not mean he always flew 481/16 and/or never flew 491/16. Furthermore, while Richthofen's association with the overpainted/ white-striped Albatros D.II seems likely, it is unproven conjecture *that* airplane's serial number was 481/16—or 491/16. Corroborating photographs and documents are still being sought.

3. Regarding the appearance of this D.II's propeller. Beginning on page 91 of Bob Gardner's *WW1 Aircraft Propellers Volume 3* (2008) is the *Propellermerkbuch Der Luftschrauben-Abteilung der Prüfanstalt und Werft der Fliegentruppen*; a guide issued at unit level on the management and handling of propellers. From Section 2, *Behandlung der Luftschrauben* (Treatment of Propellers):
"8. The propeller is to be kept clean always. Wood and metal parts are to be greased strongly, particularly after flights in the fog or rain, or in damp weather.
9. Damage caused in flight by raindrops or hail and erosion caused by sand harms the propeller and detracts slightly from the flight performance. The prop should be sanded off and repainted on the aircraft, an easy task."

Victory No.7 Statistics:

Pushers*
1st of 3 November total pusher victories
1st of 2 November two-seater pusher victories
3rd of 5 1916 two-seater pusher victories
3rd of 12 total two-seater pusher victories
3rd of 18 total pusher victories

Two-Seaters
1st of 4 November two-seater victories
3rd of 7 1916 two-seater victories
3rd of 39 pre-wound two-seater victories
3rd of 45 total two-seater victories

FE.2 Δ
1st of 2 November FE.2 victories
3rd of 5 1916 FE.2 victories
3rd of 12 total FE.2 victories

No.18 Squadron ±
1st of 4 No.18 Squadron FE.2 victories
1st of 2 November No.18 Squadron victories
1st of 3 1916 No.18 Squadron victories
1st of 4 total No.18 Squadron victories

Deaths†
No.18 Squadron:
1st and 2nd of 4 November No.18 Squadron crewmen KiA
1st and 2nd of 6 1916 No.18 Squadron crewmen KiA
1st and 2nd of 8 total No.18 Squadron crewmen KiA

Pushers:
1st and 2nd of 4 November two-seater pusher crewmen KiA
5th and 6th of 10 1916 two-seater pusher crewmen KiA
5th and 6th of 17 total two-seater pusher crewmen KiA

Two-Seaters:
1st and 2nd of 5 November two-seater crewmen KiA
5th and 6th of 11 1916 two-seater crewmen KiA

5th and 6th of 54 pre-wound two-seater crewmen KiA
5th and 6th of 63 total two-seater crewmen KiA

All:
1st and 2nd of 6 total November crewmen KiA
9th and 10th of 17 1916 crewmen KiA
9th and 10th of 66 total pre-wound crewmen KiA
9th and 10th of 84 total crewmen KiA

Etc.
1st of 5 total November victories
7th of 15 total 1916 victories
7h of 57 total pre-wound victories
7th victory in Albatros D.I or D.II
2nd of 5 two-seater victories flying from Lagnicourt
5th of 9 total victories flying from Lagnicourt
1st of 4 two-seater victories flying under Kirmaier
1st of 4 total victories flying under Kirmaier

Statistics Notes

* *All pusher victories are pre-wound*
Δ *All FE.2 victories are pre-wound*
± *All No.18 Squadron victories are pre-wound*
† *All No.18 Squadron deaths are pre-wound; KiA includes DoW*

Above: Richthofen stands between fellow *Jasta* 2 pilots Max Müller (L) and Wolfgang Günther (R), November 1916.

Left: 2nd Lt. John (Ian) Gilmour Cameron's headstone, 2011. Cameron was Manfred von Richthofen's eighth victory, see facing page.

Victory No.8

Above: A Royal Aircraft Factory BE.2c. Although slow, stable, and no match for an Albatros scout in aerial combat, nevertheless each BE.2 was a danger to tens of thousands via the artillery for which they spotted and the behind-the-lines reconnaissance photographs they took and brought back to base. Richthofen shot down 17 BE.2s combating these threats, a tally greater than that of any other make/model he shot down.

9 November
BE.2c 2506
No.12 Squadron RFC

Day/Date: Thursday, 9 November
Time: 10.30AM
Weather: Bright and clear nearly all day.
Attack Location: Above Mory
Crash Location: Near Beugny
Side of Lines: Friendly

RFC Communique No. 61: "A bombing raid of the 3rd Brigade consisting of 16 bombing machines and an escort of 14, was attacked on its way to Vraucourt by at least 30 Germans, chiefly fast scouts. The enemy attacked from the front, and our scouts dived and got to close quarters with them. As this fight progressed the escort got gradually below the bombing machines. Meanwhile the enemy was reinforced and the bombers were attacked from both sides Numerous individual fights ensued."

KOFL 1. Armee Weekly Activity Report: "Erfloge im Luftkampf: B.E. Zweisitzer abgeschossen 9.11.16 vorm. 10.30 bei Beugny von Leutnant Frhr.von Richthofen, Jagdstaffel 2.

Successes in aerial combat: B.E. two-seater shot down 9 November 16, 10:30AM near Beugny by *Leutnant* Frhr. von Richthofen, *Jagdstaffel* 2."

MvR Combat Report: "B.E. two-seater, Nr. 2506. Motor: Daimler, Nr. 22082. Inmates: Seriously wounded, pilot very seriously, observer shoulder.

About 10.30AM I attacked with several other planes enemy bomb squad above Nory [sic] in 2500 meters alt. After preceding curve fight my victim crashed to the ground near Bengny [*sic*].

Frhr. v. Richthofen
Leutnant"

MvR Autobiography: "We were at the Front a short time when we saw an enemy bomber squadron flying along impudently. They came in tremendous numbers, just as during the Somme battle. I believe

Above: 2506's souvenired serial number on the wall at Roucourt.

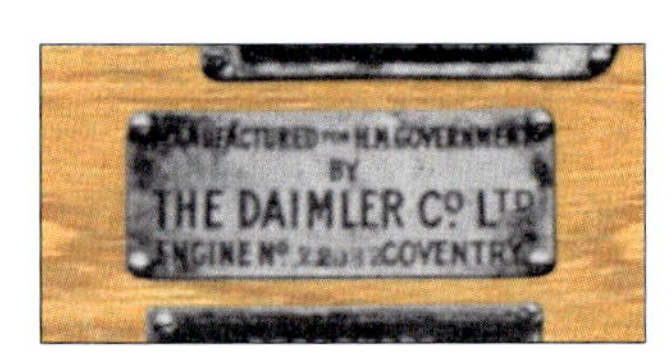

Above: 2506's souvenired engine placard.

Above: 2506's souvenired serial number on display in the Richthofen Museum. Whatever agent was used to adhere the serial number to the new museum backing seems to have marred and darkened the fabric.

there were about forty to fifty in the squadron, although I cannot give the exact number. They had picked a target not far from our airfield. Shortly before they reached the target, I reached the bomber in the rear position of the enemy formation. My first shots put his gunner out of action and probably tickled the pilot as well; in any case, he decided to land with his bombs. I burned him a bit around the edges, and the haste with which he sought to reach the ground was most noticeable; that is to say, he plunged down and fell near our airfield at Lagnicourt."

RFC Combat Casualties Report: "Left aerodrome at 8.50AM. Two B.E's seen to go down near MORY by other pilots in the formation. One was seen to descend out of control near SAPIGNIES."

RFC A/C:
Make/model/serial number: Royal Aircraft Factory BE.2c 2506
Manufacturer: Subcontracted to and built by Wolseley Motors Ltd., Adderley Park, Birmingham
Unit: No.12 Squadron RFC
Aerodrome: Avesnes-le-Comte
Sortie: Bombing Vraucourt Sugar Factory
Colors/markings: Based on the souvenired serial number fabric, likely overall CDL wings/ailerons/vertical and horizontal stabilizers/elevators; battleship gray engine cowl; b/w/r rudder; black serial number on vertical stabilizer.
Engine: RAF 1a
Engine Number: 941 WD 22082
Guns: Two Lewis. Nos. 11375, 15947
Manner of Victory: Unchecked decent until terrain impact.
Parts Souvenired: a. "2506" from both the port and starboard sides of vertical stabilizer. Borderless black numbers on CDL background. Photographed on display at Roucourt, his home bedroom in Schweidnitz, and in the post-war Richthofen Museum.
b. Daimler Co. Ltd engine placard.

RFC Crew:
Pilot: 2nd Lt. John (Ian) Gilmour Cameron, 19, PoW/DoW.
RFC Combat Casualties Report: "Death accepted by the Army Council as having occurred on 9/11/16 on report received through the Geneva Red Cross Society that he died of wounds in enemy hands and on the forwarding of his disc by the German authorities."
Cause of Death: u/k. Either gunshot wound(s) or trauma associated with airplane crash.
Burial: Achiet-le-Grand Communal Cemetery, Grave II. M. 19.
Obs: n/a. Observer omitted in lieu of larger bomb load (see notes).

MvR A/C:
Make/model/serial number: Albatros D.II, possibly 481/16
Unit: *Jasta* 2
Commander: *Oblt*. Stefan Kirmaier
Airfield: Lagnicourt
Colors/markings: See Vic. #7.
Damage: u/k

Initial Attack Altitude: 2500 meters
Gunnery Range: u/k
Rounds Fired: u/k
Known Staffel Participants: *Oblt*. Stefan Kirmaier (#9), *Ltn*. Hans Imelmann (#5)

Notes:
1. Richthofen's reference to an observer being wounded begat the thought that the "observer" was Lt. Gerald Featherstone Knight. Knight was also thought to be the pilot of 2506 and Cameron his observer. However, Knight was pilot of BE.2c 2502 shot down by *Oblt*. Kirmaier, for his 9th victory.
2. In his autobiography *Der Rote Kampfflieger*, Richthofen recalled that after this victory he and Imelmann drove and then hiked through mud to reach the crash site. Upon gaining Cameron's crashed BE.2c, which still carried its bombs, Richthofen was introduced to the Sovereign Duke of Saxe-Coburg and Gotha. Later that evening Richthofen was summoned to appear before the Duke and informed of the belief he had

Albatros D.II 481/16. It is unknown if Richthofen flew this machine on this sortie but its appearance is based on photographs from the same time period. The dark fuselage suggests brown overpainting rather than green. The exact hue of brown is unknown. That shown in the profile is conjectural, as is the association of this airplane being 481/16.

prevented the British from bombing the Duke's headquarters (although No.12's stated objective had been bombing the Vraucourt sugar factory). For this he received the *Saxe- Coburg and Gotha Duke Carl Eduard Medal, 2nd Class with Swords* on 30 December, 1916.

Victory No.8 Statistics:

Two-Seaters

2nd of 4 November two-seater victories
4th of 7 1916 two-seater victories
4th of 39 pre-wound two-seater victories
4th of 45 total two-seater victories

BE.2

1st of 2 November BE.2 victories
1st of 2 1916 BE.2 victories
1st of 17 total BE.2 victories

No.12 Squadron ±

1st of 2 No.12 Squadron BE.2 victories
1st and only November No.12 Squadron victory
1st and only 1916 No.12 Squadron victory
1st of 2 total No.12 Squadron victories

Deaths†

No.12 Squadron:
1st and only November No.12 Squadron crewman KiA
1st and only 1916 No.12 Squadron crewman KiA
1st of 3 total No.12 Squadron crewmen KiA

Two-Seaters:

3rd of 5 November two-seater crewmen KiA
7th of 11 1916 two-seater crewmen KiA
7th of 54 pre-wound two-seater crewmen KiA
7th of 63 total two-seater crewmen KiA

All:

3rd of 6 total November crewmen KiA
11th of 17 1916 crewmen KiA
11th of 66 total pre-wound crewmen KiA
11th of 84 total crewmen KiA

Etc.

2nd of 5 total November victories
8th of 15 total 1916 victories
8th of 57 total pre-wound victories
8th victory in Albatros D.I or D.II
3rd of 5 two-seater victories flying from Lagnicourt
6th of 9 total victories flying from Lagnicourt
2nd of 4 two-seater victories flying under Kirmaier
2nd of 4 total victories flying under Kirmaier

Statistics Notes

Δ All BE.2 victories are pre-wound
± All No.12 Squadron victories are pre-wound
† All No.12 Squadron deaths are pre-wound; KiA includes DoW

Victory No.9

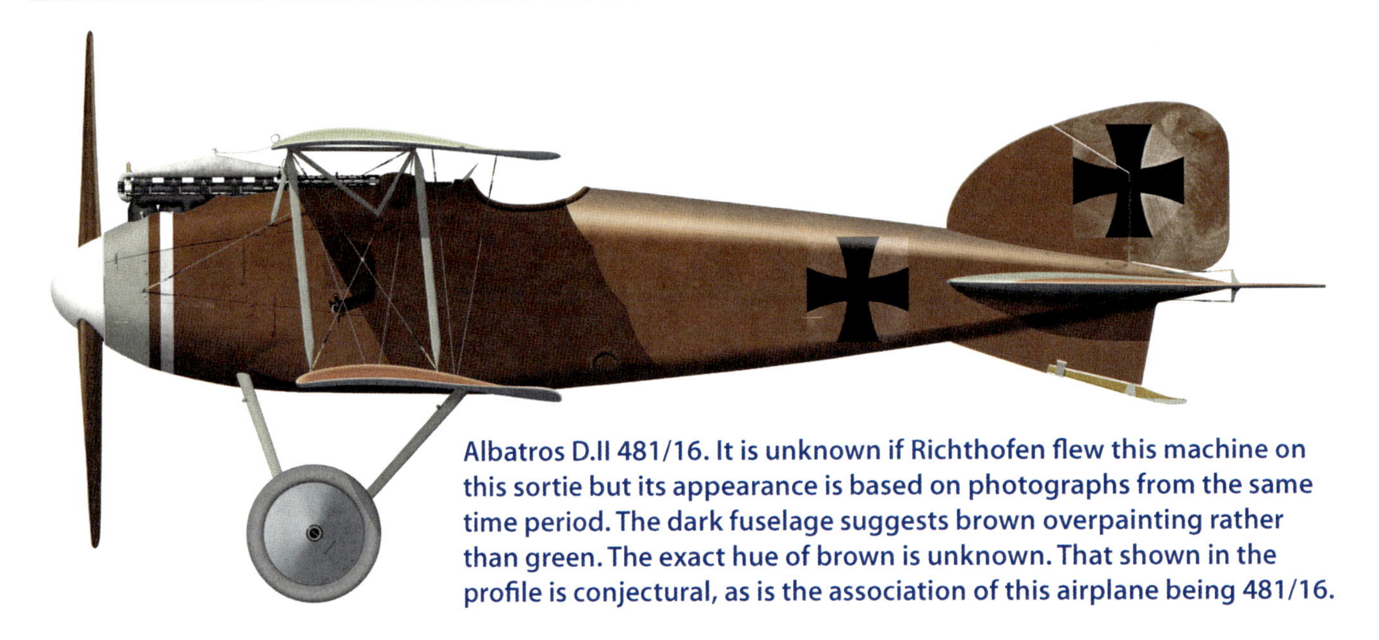

Albatros D.II 481/16. It is unknown if Richthofen flew this machine on this sortie but its appearance is based on photographs from the same time period. The dark fuselage suggests brown overpainting rather than green. The exact hue of brown is unknown. That shown in the profile is conjectural, as is the association of this airplane being 481/16.

20 November
BE.2c 2767
No.15 Squadron RFC

Day/Date: Monday, 20 November
Time: 9.40AM
Weather: Low clouds, strong winds and showers.
Attack Location: Above Grandcourt
Crash Location: Near Miraumont
Side of Lines: Enemy

RFC Communique No. 63: "On the Fifth Army front 1 gun pit was destroyed, 11 others damaged, and 8 ammunition pits blew up. 58 active hostile batteries were reported by zone call, several were successfully engaged, and a large explosion of ammunition was caused."

KOFL 1. Armee Weekly Activity Report: "Erfolge im Luftkampf: am 20.11.16 vorm. 9.40 südl. Grandcourt durch Lt.Frhr.v.Richthofen, Jagdstaffel 2.

Successes in Aerial Combat: On 20 November 16, 9:40AM south [of] Grandcourt by *Lt.Frhr.* v.Richthofen, *Jagdstaffel* 2."

MvR Combat Report: "Vickers Two-seater.

Together with several machines of our *Staffel* we attacked on the enemy side above Grandcourt in 1800 meters alt. Several low flying artillery planes. After having harassed a B.E. two-seater for a time, the plane disappeared in the clouds and then crashed to the ground, between the trenches south of Grandcourt. The machine was taken immediately under artillery fire and destroyed.

Frhr. v. Richthofen
Leutnant"

RFC Combat Casualties Report: "Left aerodrome 6.50AM. Driven down by hostile aircraft, apparently hit, but under control. Was seen gliding down over MIRAUMONT followed by enemy machine. At 9.50AM was seen on its back at R.4a.37 apparently undamaged."

RFC A/C:
Make/model/serial number: Royal Aircraft Factory BE.2c 2767
Manufacturer: Subcontracted to and built by Ruston, Proctor & Co., Ltd., Lincoln
Unit: No.15 Squadron RFC
Aerodrome: Léalvillers
Sortie: Artillery Observation
Colors/markings: u/c. Generally, either overall CDL wings/ailerons/vertical and horizontal stabilizers/elevators or the same in PC10; battleship gray engine cowl; b/w/r rudder; black serial number on vertical stabilizer.
Engine: RAF 1a
Engine Number: 23011 WD 879
Guns: Two Lewis. Nos. 7330, 7258
Manner of Victory: Engine damaged by machine

gun bullets and then failed at 245 meters, precipitating dead-stick landing.
Parts Souvenired: n/k

RFC Crew:
Pilot: 2nd Lt. James Cunningham Lees, WiA/PoW
RFC Combat Casualties Report: "2/Lt. Lees was hit, but his present location is unknown."
Manner of Injury: Gunshot wound in the right leg
Imprisonment: u/c
Repatriation: New Year's Day, 1919
Obs: Lt. Thomas Henry Clarke, PoW
RFC Combat Casualties Report: "Mrs. Clarke received a communication from Lt. Clarke stating that he is a prisoner at OSNABRUCK, and is uninjured."
Imprisonment: Osnabruck
Repatriation: 17 December 1918

MvR A/C:
Make/model/serial number: Albatros D.II, possibly 481/16
Unit: *Jasta* 2
Commander: *Oblt.* Stefan Kirmaier
Airfield: Lagnicourt
Colors/markings: See. Vic. #7.
Damage: u/k
Initial Attack Altitude: 1800 meters
Gunnery Range: u/k
Rounds Fired: u/k
Known Staffel Participants: Stefan Kirmaier (#11)

Notes:
1. Evidence suggests *Jasta* 2 *Staffelführer* Stefan Kirmaier may have been the actual victor over 2767. He and Richthofen were each credited with a victory on this day but RFC records indicate only one machine was lost over the lines in the manner of 2767. Richthofen and Kirmaier's victories were claimed within ten minutes and two miles of each other—i.e., nearly simultaneously and collocated—and nine years after the war Clarke stated that he and Cunningham had been attacked by five German airplanes. It is unknown if Richthofen lost 2767 in the clouds and then presumed its crash, or if Kirmaier attacked 2767 after Richthofen and each had not seen or discounted the other's attack in the fog of war. In any event, it seems both men received credit for downing the same airplane.
2. This attack was initiated across enemy lines.
3. This was the first of Richthofen's defeated RFC crews to survive the event.
4. It is presumed Richthofen was flying an Albatros D.II primarily at this point, but the precise make/model of airplane flown during any sortie, as well as their serial numbers, during autumn 1916, is conjectural.

Victory No.9 Statistics:
Two-Seaters
3rd of 4 November two-seater victories
5th of 7 1916 two-seater victories
5th of 39 pre-wound two-seater victories
5th of 45 total two-seater victories

BE.2 Δ
2nd of 2 November BE.2 victories
2nd of 2 1916 BE.2 victories
2nd of 17 total BE.2 victories

No.15 Squadron
1st and only No.15 Squadron BE.2c victory
1st and only November No.15 Squadron victory
1st and only 1916 No.15 Squadron victory
1st and only pre-wound No.15 Squadron victory
1st of 2 total No.15 Squadron victories

Wounded†
No.15 Squadron:
1st and only No.15 Squadron crewman WiA

Two-Seaters:
1st and only November two-seater crewman WiA
1st and only 1916 two-seater crewman WiA
1st of 16 pre-wound two-seater crewmen WiA
1st of 18 total two-seater crewmen WiA

All:
1st and only November crewman WiA
1st of 2 1916 crewmen WiA
1st of 18 total pre-wound crewmen WiA
1st of 26 total crewmen WiA

Etc.
1st of 14 victories behind enemy lines
3rd of 5 total November victories
9th of 15 total 1916 victories
9th of 57 total pre-wound victories
9th victory in Albatros D.I or D.II
4th of 5 two-seater victories flying from Lagnicourt
7th of 9 total victories flying from Lagnicourt
3rd of 4 two-seater victories flying under Kirmaier
3rd of 4 total victories flying under Kirmaier

Statistics Notes
Δ *All BE.2 victories are pre-wound*
† *All No.15 Squadron WiA are pre-wound*

Victory No.10

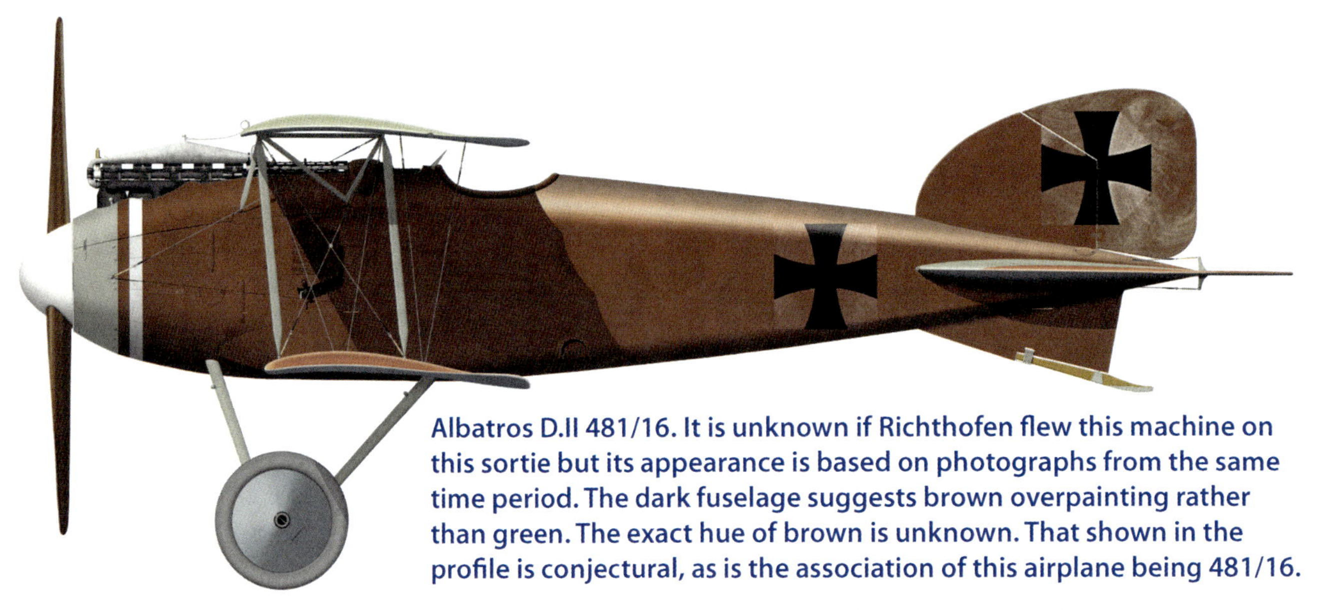

Albatros D.II 481/16. It is unknown if Richthofen flew this machine on this sortie but its appearance is based on photographs from the same time period. The dark fuselage suggests brown overpainting rather than green. The exact hue of brown is unknown. That shown in the profile is conjectural, as is the association of this airplane being 481/16.

20 November
FE.2b 4848
No.18 Squadron RFC

Day/Date: Monday, 20 November
Time: 4.15PM
Weather: Low clouds, strong winds and showers.
Attack Location: Above Grandcourt
Crash Location: Near Grandcourt
Side of Lines: Friendly

KOFL 1. Armee Weekly Activity Report: "*Erfolge im Luftkampf: am 20.11.16 nachm.4.15 bei Gueudecourt* [sic] *durch Lt. Frhr.v.Richthofen Jagdstaffel 2.*

Successes in Aerial Combat: on 20 November 16, 4:15 PM near Gueudecourt [sic] by *Lt.Frhr.* v.Richthofen, *Jagdstaffel* 2."

MvR Combat Report: "Vickers two seater, Nr. 4000. Motor Nr. 36574. Plane cannot be secured as under fire. Inmates: One killed: Lieut. George Doughty. Lieut. Gilbert Stall, seriously wounded, prisoner.

Together with 4 planes I attacked a Vickers two-seater type above the clouds in 2500 meters alt. After 300 shots adversary broke through clouds, pursued by me. Near Grandcourt I shot him down.

Frhr. v. Richthofen
Leutnant"

RFC Combat Casualties Report: "Left aerodrome 1.15PM." [No other information.]

RFC A/C:
Make/model/serial number: Royal Aircraft Factory FE.2b 4848
Manufacturer: G. & J. Weir Ltd., Cathcart, Glasgow
Unit: No.18 Squadron RFC
Aerodrome: Laviéville
Sortie: Defensive Patrol (RFC Combat Casualties Report lists Artillery Observation)
Colors/markings: u/k. Typically for the make/model, PC10 nacelle/wings/ailerons/horizontal stabilizers/elevator uppersurfaces; CDL undersurfaces; r/w/b rudder with black serial numbers, usually bordered in white when atop the red and blue.
Engine: 120 hp Beardmore
Engine Number: 295 WD 1357
Guns: Two Lewis. Nos. 3085, 1365
Manner of Victory: u/k. Details unspecified.
Parts Souvenired: u/k. Likely none, as "plane cannot be secured as under fire."

RFC Crew:
Pilot: 2nd Lt. Gilbert Sudbury Hall, 25, PoW/DoW 30.11.16
Cause of Death: u/k. Either gunshot wound(s) or trauma associated with airplane crash.
Burial: Porte-de-Paris Cemetery, Grave II. A. 1.
Obs: 2nd Lt. George Doughty, 21, KiA
RFC Combat Casualties Report: "A message dropped by the Germans in the French Lines states that 2nd Lt. Doughty was killed, and 2nd Lt.

Hall a wounded prisoner."
Cause of Death: u/k. Either gunshot wound(s) or trauma associated with airplane crash.
Burial: Doughty: Achiet-le-Grand Communal Cemetery, Grave II. M. 6.

MvR A/C:
Make/model/serial number: Albatros D.II, possibly 481/16
Unit: *Jasta* 2
Commander: *Oblt.* Stefan Kirmaier
Airfield: Lagnicourt
Colors/markings: See Vic. #7
Damage: u/k
Initial Attack Altitude: 2500 meters
Gunnery Range: u/k
Rounds Fired: 300+
Known Staffel Participants: u/k

Notes:
1. This victory is Richthofen's first double victory in one day.
2. Richthofen noted an incorrect serial number in his combat report. There was no FE.2b with the serial number "4000." Furthermore, his listed engine number does not match RFC records. It must be pondered how Richthofen would glean any engine number if the "plane cannot be secured as under fire."

Victory No.10 Statistics:
Pushers*
2nd of 3 November total pusher victories
2nd of 2 November two-seater pusher victories
4th of 5 1916 two-seater pusher victories
4th of 12 total two-seater pusher victories
4th of 18 total pusher victories

Two-Seaters
4th of 4 November two-seater victories
6th of 7 1916 two-seater victories
6th of 39 pre-wound two-seater victories
6th of 45 total two-seater victories

FE.2 Δ
2nd of 2 November FE.2 victories
4th of 5 1916 FE.2 victories
4th of 12 total FE.2 victories

No.18 Squadron ±
2nd of 4 No.18 Squadron FE.2 victories
2nd of 2 November No.18 Squadron victories
2nd of 3 1916 No.18 Squadron victories
2nd of 4 total No.18 Squadron victories

Deaths†
No.18 Squadron:
3rd and 4th of 4 November No.18 Squadron crewmen KiA
3rd and 4th of 6 1916 No. 18 Squadron crewmen KiA
3rd and 4th of 8 total No.18 Squadron crewmen KiA

Pushers: ††
3rd and 4th of 4 November two-seater pusher crewmen KiA
7th and 8th of 10 1916 two-seater pusher crewmen KiA
7th and 8th of 17 total two-seater pusher crewmen KiA

Two-Seaters:
4th and 5th of 5 November two-seater crewmen KiA
8th and 9th of 11 1916 two-seater crewmen KiA
8th and 9th of 54 pre-wound two-seater crewmen KiA
8th and 9th of 63 total two-seater crewmen KiA

All:
4th and 5th of 6 total November crewmen KiA
12th and 13th of 17 1916 crewmen KiA
12th and 13th of 66 total pre-wound crewmen KiA
12th and 13th of 84 total crewmen KiA

Etc.
4th of 5 total November victories
10th of 15 total 1916 victories
10th of 57 total pre-wound victories
10th victory in Albatros D.I or D.II
5th of 5 two-seater victories flying from Lagnicourt
8th of 9 total victories flying from Lagnicourt
4th of 4 two-seater victories flying under Kirmaier
4th of 4 total victories flying under Kirmaier

Statistics Notes
* *All pusher victories are pre-wound*
Δ *All FE.2 victories are pre-wound*
± *All No.18 Squadron victories are pre-wound*
† *All No.18 Squadron deaths are pre-wound; KiA includes DoW*
†† *All pusher deaths are pre-wound*

Victory No.11

Above: This Aircraft Manufacturing Company DH.2 5084, photographed at School of Military Aeronautics, Langley Field, Virginia, provides an outstanding general view of the first true British fighter type Richthofen ever faced. In all he shot down five of them, although his first—during a classic "duel" with No.24 Squadron commanding officer Major Lanoe Hawker—remains the most known.

23 November
DH.2 5964
No.24 Squadron RFC

Day/Date: Thursday, 23 November
Time: 3PM
Weather: High pressure (30.12" – 30.42" Hg) dominated the region. Winds below 6000 feet SSW 15mph; winds above 6000 feet SW 30 mph. Forecast high temperature 50°F.
Attack Location: Above Bapaume
Crash Location: Near Ligny-Thilloy, ca. 230 metres east of Luisenhof Farm
Side of Lines: Friendly

RFC Communique No. 64: "Missing – L. G. Hawker, V.C., D.S.O., No. 24 Squadron. On November 23rd, a defensive patrol of No. 24 Squadron, consisting of Major Hawker, Capt. Andrews, and Lieut. Saundby, engaged two hostile machines near Bapaume and drove them East. They then saw two strong hostile patrols approaching high up. The patrol was about to retire when Major Hawker dived and continued the pursuit of the first hostile machines. The de Havillands were at once attacked by the two strong hostile patrols, one of the enemy's machines diving on to the tail of Major Hawker's de Havilland. This machine was driven off by Capt. Andrews, who was then attacked in the rear and having his engine damaged was forced to break off the combat. Lieut. Saundby drove off one hostile machine which was attacking Capt. Andrews and then engaged a second and drove it down out of control. Major Hawker was last seen engaging a hostile machine at about 3,000 feet."

KOFL 1. Armee Weekly Activity Report: "Erfolge im Luftkampf: am 23.11. nachm.3.00 bei Bapaume durch Lt. Frhr.v. Richthofen, Jagdstaffel 2.

Successes in Aerial Combat: On 23 November 16, 3:00PM near Bapaume by *Lt.Frhr.* v.Richthofen, *Jagdstaffel* 2."

MvR Combat Report: "Vickers one-seater, plane lying near Bapaume. Inmate: Major Hawker, dead.

I attacked together with 2 planes a Vickers one-seater in 3000 meters alt. After a long curve fight of 3 – 5 minutes I had pressed down adversary to 500 meters. He now tried to escape flying to the front, I pursued and brought him down after 900 shot.

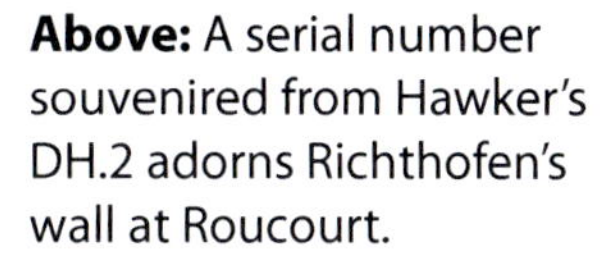
Above: A serial number souvenired from Hawker's DH.2 adorns Richthofen's wall at Roucourt.

Above Right: A Mark II Lewis machine gun, souvenired from Lanoe Hawker's DH.2, mounted on the wall of the Richthofen Museum.

Witnesses: *Leutnant* Wortmann. *Leutnant*. Collin, etc.

Frhr. v. Richthofen
Leutnant"

RFC Combat Casualties Report: "Left aerodrome 1.00PM. With two other de Havillands engaged 8 hostile machines over ACHIET. Not seen after this encounter."

RFC A/C:

Make/model/serial number: AMC DH.2 5964
Manufacturer: Aircraft Manufacturing Co., Ltd., Hendon, London, N.W.
Unit: No.24 Squadron RFC
Commander: Maj Lanoe Hawker, VC, DSO
Aerodrome: Bertangles
Sortie: Defensive Patrol
Colors/markings: u/k. Typically for the make/model, PC10 nacelle/wings/ailerons/horizontal stabilizers/elevator uppersurfaces; CDL undersurfaces; nacelle fabric either CDL or PC10, with white undersides; r/w/b rudder with black serial numbers, could be bordered in white when atop the red and blue. No.24 Squadron used red/white, blue/white, black/white markings on the interplane struts to identify individual machines within A, B, and C flights. Flights were further identified with red, white, and blue wheel covers, respectively. Nacelle undersides bordered with gray or PC10 "saw-tooth" pattern for all flights.
Engine Number: 6138 B 540
Gun: One Lewis, no. 14563
Manner of Victory: Unchecked shallow descent from ca. 30 meters until terrain impact.
Parts Souvenired: a. "5964" from the port side of the rudder, with borderless black numbers on b/w/r striped background. Photographed on display at Roucourt, Richthofen's home bedroom in Schweidnitz, and the post-war Richthofen Museum.
b. Mk.II Lewis machine gun. Photographed on display at Roucourt, Richthofen's home bedroom in Schweidnitz, and the post-war Richthofen Museum.

Above: Lanoe Hawker early in his pilot career.

RFC Crew:

Pilot: Major Lanoe Hawker (VC, DSO), 25 (DoB 30 December 1890), KiA
Pilot Victories: 7 (3 destroyed*, 2 forced to land**, 1 out of control‡, 1 captured ±)

Victory #	Date (1915)	Type
1	21 June	DFW.C‡
2	25 July	Alb.C**
3	25 July	Alb.C±
†4	2 Aug	C-type**
††5	11 Aug	Aviatik C*
††6	11 Aug	Fokker E*
7	7 Sept	Biplane*

†Attained in FE.2b, shared with observer Lt. Payne, No.6 Squadron ("Payze" in RFC Communique No.

Above: Hawker's stained glass memorial window in St. Nicholas Church, Longparish, England. (Mary Jo Darrah)

3.)
††Attained in FE.2b, shared with observer Lt. Noel Clifton, No.6 Squadron
Cause of Death: Shot in head by machine gun
Burial: In-field; grave lost during turmoil of war.

MvR A/C:
Make/model/serial number: Albatros D.II, possibly 481/16
Unit: *Jasta* 2
Commander: *Oblt.* Karl Bodenschatz (acting)
Airfield: Lagnicourt
Colors/markings: See Vic. #7.
Damage: u/k, believed to be none.
Initial Attack Altitude: 3000 meters.
Gunnery Range: u/k
Rounds Fired: 900
Known Staffel Participants: Ltn. Wortmann, *Ltn.* Collin

Notes:

1. Usually Hawker's DoB is given as 31 December 1890, the origin of which is his brother Tyrell's biography, *Hawker V.C.* However, birth records secured by the author from the District of Andover General Register Office reveal Hawker's actual DoB was 30 December 1890.
2. See page 71 for a detailed aeronautical analysis of this combat.

Above: Closeup of Hawker's memorial window. (Mary Jo Darrah)

Below: The memorial window displays Hawker's correct date of birth.

Victory No.11 Statistics:

Pushers*
3rd of 3 November total pusher victories
1st and only November single-seater pusher victory
1st of 4 1916 single-seater pusher victories
1st of 6 total single-seater pusher victories
5th of 18 total pusher victories

Single-Seaters
1st and only November single-seater victory
5th of 8 1916 single-seater victories
5th of 18 pre-wound single-seater victories
5th of 35 total single-seater victories

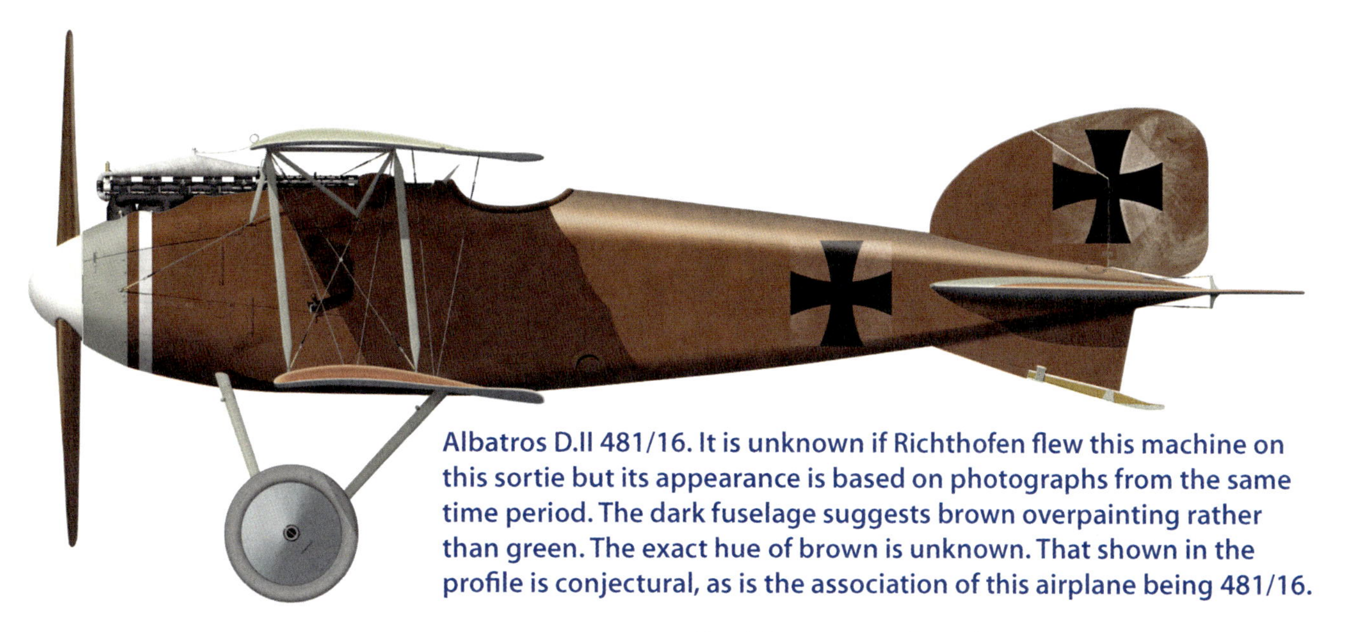
Albatros D.II 481/16. It is unknown if Richthofen flew this machine on this sortie but its appearance is based on photographs from the same time period. The dark fuselage suggests brown overpainting rather than green. The exact hue of brown is unknown. That shown in the profile is conjectural, as is the association of this airplane being 481/16.

DH.2 Δ
1st and only November DH.2 victory
1st of 4 1916 DH.2 victories
1st of 5 total DH.2 victories

No.24 Squadron ±
1st and only No.24 Squadron victory

Deaths†
No.24 Squadron:
1st and only No.24 Squadron crewman KiA

Pushers:
1st and only November single-seater pusher crewman KiA
1st of 2 1916 single-seater pusher crewmen KiA
1st of 4 total single-seater pusher crewmen KiA

Single-seaters:
1st and only November single-seater crewman KiA
5th of 6 1916 single-seater crewmen KiA
5th of 12 pre-wound single-seater crewmen KiA
5th of 21 total single-seater crewmen KiA

All:
6th of 6 total November crewmen KiA
14th of 17 1916 crewmen KiA
14th of 66 total pre-wound crewmen KiA
14th of 84 total crewmen KiA

Etc.
5th of 5 total November victories
11th of 15 total 1916 victories
11th of 57 total pre-wound victories
11th victory in Albatros D.I or D.II
4th of 4 single-seater victories flying from Lagnicourt
9th of 9 total victories flying from Lagnicourt
1st and only victory flying under Bodenschatz

Statistics Notes
* *All pusher victories are pre-wound*
Δ *All DH.2 victories are pre-wound*
± *All No.24 Squadron victories are pre-wound*
† *All No.24 Squadron deaths are pre-wound; KiA includes DoW*

Below: Lineup of three *Jasta* 2 Albatros D.Is and one D.II (left), believed to be Richthofen's, who is visible standing left of the camera tripod.

Victory No.12

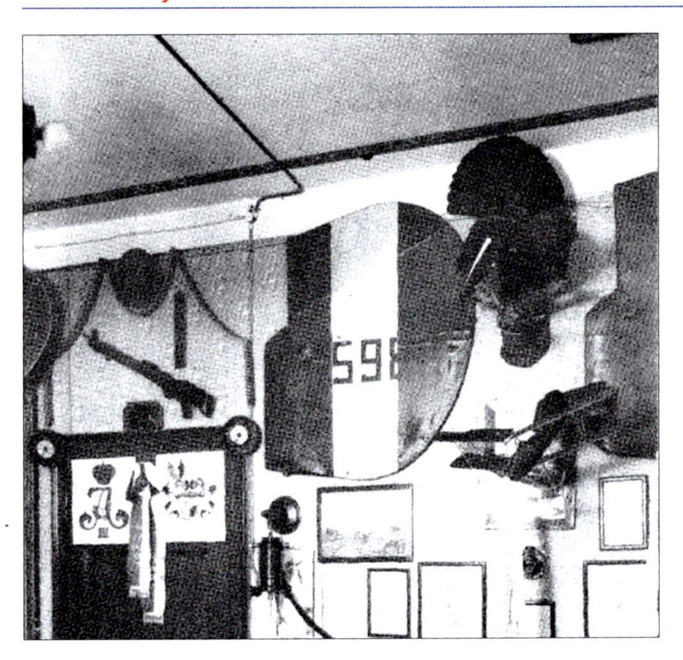

Left: DH.2 5986's entire rudder on display in Richthofen's home bedroom in Schweidnitz.

Above: A swatch of fabric from the starboard side of 5986's rudder, on display in Schweidnitz.

11 December
DH.2 5986
No.32 Squadron RFC

Day/Date: Monday, 11 December
Time: 11.55AM
Weather: Fine morning with some mist; rain later.
Attack Location: Above Mercatel, near Arras
Crash Location: Henin-s-Cojeul
Side of Lines: Friendly

RFC Communique No. 66: "Missing – Lieut. B. P. G. Hunt, No. 32 Squadron. The dumps and railway sidings east and north-east of Mory were attacked by six machines of No. 23 Squadron. Nineteen bombs were actually seen to hit the dumps, from which large clouds of red and black smoke issued. Lieut. B. P. G. Hunt, No. 32 Squadron, who was part of the escort, failed to return. His machine was seen to make a forced landing, apparently due to engine trouble, in the German lines."

KOFL 1. Armee Weekly Activity Report: "*Erfolge im Luftkampf: Ein Vickers-Einsitzer am 11.12.16 vorm. 11.55 bei Mercatel s.Arras Sieger Lt.Frhr.v.Richthofen, Jagdstaffel 2.*

Successes in Aerial Combat: one Vickers single-seater on 11 December [19]16 [at] 11.55AM near Mercatel south Arras. Victor *Lt.Frhr.* v.Richthofen, *Jagdstaffel* 2."

MvR Combat Report: "Vickers One-seater, Nr. 5986. Rotary Motor 30372. Inmate: made prisoner, wounded, Lieut. Hund [*sic*].

About 11.45 I attacked with Lieut.Wortmann in 2800 alt. and south of Arras enemy one-seater Vickers squad of 8 machines. I singled out one machine and after a short curve fight I ruined the adversary's motor and forced him to land behind our lines near Mercatel. Inmate not seriously wounded.

Frhr. v. Richthofen
Leutnant"

RFC Combat Casualties Report: "Left aerodrome 9.20AM."

RFC A/C:

Make/model/serial number: AMC DH.2 5986
Manufacturer: Aircraft Manufacturing Co., Ltd., Hendon, London, N.W.
Unit: No.32 Squadron RFC
Aerodrome: Léalvillers
Sortie: Escort
Colors/markings: u/k. Typically for the make/model, PC10 nacelle/wings/ailerons/horizontal stabilizers/elevator uppersurfaces; CDL undersurfaces; nacelle fabric either CDL or PC10 with white undersurface; r/w/b rudder with black serial number. No.32 Squadron used three variations of black/white concentric rings on the wheel covers to identify A, B, and C flights. White flight letters and individual numbers were painted

on the upper surfaces of the wings, repeated in black on nacelle undersurface.
Engine Number: 30372
Guns: Two Lewis. Nos. 13980, 11924
Manner of Victory: Machine gun bullets damaged engine, precipitating dead-stick forced landing.
Parts Souvenired: a. Entire b/w/r striped rudder with borderless black numbers across w/r stripes. Photographed on display (port side) in home bedroom in Schweidnitz.
b. Black 5986 on r/w/b background, from stbd side of rudder. Photographed in the home bedroom in Schweidnitz.
c. Black 5986 on r/w/b background, from stbd side of rudder. Photographed in the post-war Richthofen Museum, where it was mounted upside-down and displayed as "9865".

RFC Crew:
Pilot: Lt. Benedict Philip Gerald Hunt, 26, WiA/PoW
Manner of Injury: Wounded in liver.
RFC Combat Casualties Report: "A newspaper cutting from *The Times* forwarded by 5th Brigade states 'Captain Philip Hunt, Yeomanry, attached R.F.C., who was previously reported missing, is now reported to be wounded and a prisoner of war in Germany.'"
Imprisonment: u/k
Repatriation: Exchanged to Holland April 1918, due to illness. Interned there until 18 November 1918.

MvR A/C:
Make/model/Serial Number: Albatros D.II, possibly 481/16
Unit: *Jasta* 2
Commander: *Hptm*. Franz Walz
Airfield: Pronville
Colors/markings: See Vic. #7.
Damage: u/k
Initial Attack Altitude: 2800 meters.
Gunnery Range: u/k
Rounds Fired: u/k
Known Staffel *Participants: Ltn.* Wortmann

Note: This sortie reportedly was Hunt's first with a DH.2 equipped with two guns.

Victory No.12 Statistics:
Pushers*
1st of 4 December total pusher victories
1st of 3 December single-seater pusher victories
2nd of 4 1916 single-seater pusher victories
2nd of 6 total single-seater pusher victories

Above: Eventually the serial number from 5986's rudder was affixed to a rigid backing and inadvertently mounted upside-down in the Richthofen Museum, where it erroneously appeared as "9865."

6th of 18 total pusher victories

Single-Seaters
1st of 3 December single-seater victories
6th of 8 1916 single-seater victories
6th of 18 pre-wound single-seater victories
6th of 35 total single-seater victories

DH.2 Δ
1st of 3 December DH.2 victories
2nd of 4 1916 DH.2 victories
2nd of 5 total DH.2 victories

No. 32 Squadron ±
1st and only No.32 Squadron victory

Wounded †
No.32 Squadron:
1st and only No.32 Squadron crewman WiA

Single-Seaters:
1st and only December single-seater crewman WiA
1st and only 1916 single-seater crewman WiA
1st of 2 pre-wound single-seater crewmen WiA
1st of 8 total single-seater crewmen WiA

All:
1st and only December crewman WiA
2nd of 2 1916 crewmen WiA
2nd of 18 total pre-wound crewmen WiA
2nd of 26 total crewmen WiA

Etc.
1st of 4 total December victories
12th of 15 total 1916 victories

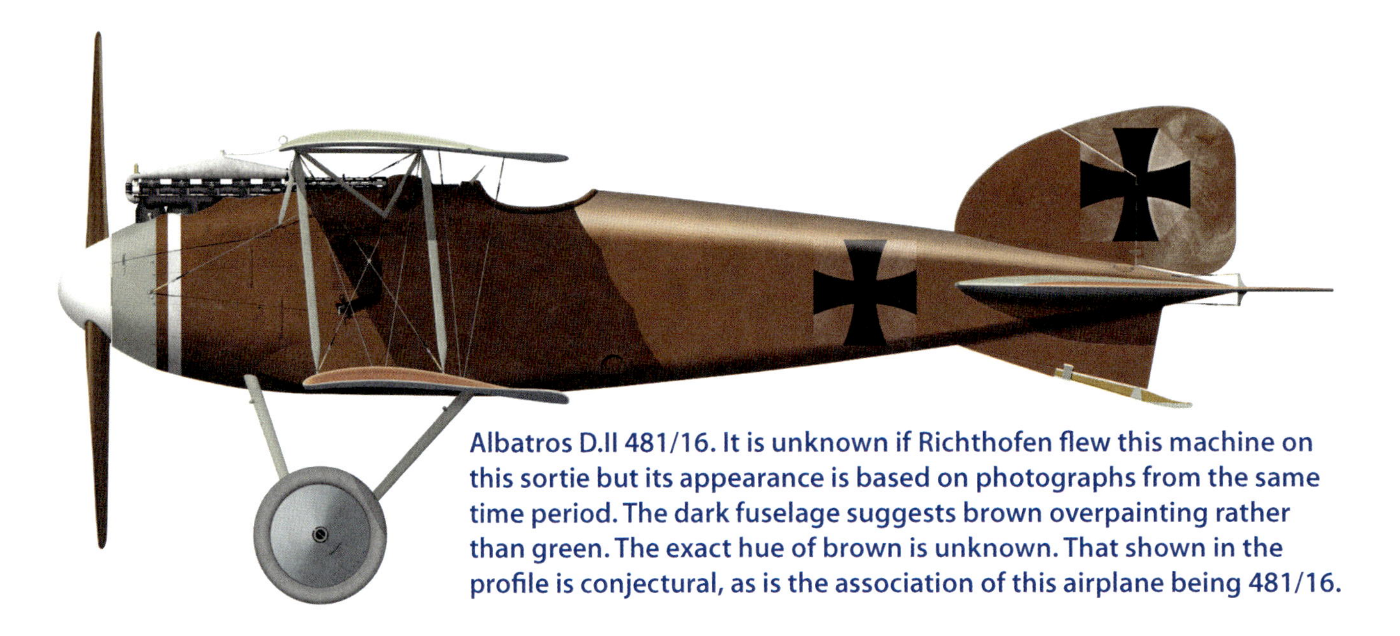

Albatros D.II 481/16. It is unknown if Richthofen flew this machine on this sortie but its appearance is based on photographs from the same time period. The dark fuselage suggests brown overpainting rather than green. The exact hue of brown is unknown. That shown in the profile is conjectural, as is the association of this airplane being 481/16.

12th of 57 total pre-wound victories
12th victory in Albatros D.I or D.II
1st of 4 single-seater victories flying from Pronville
1st of 5 total victories flying from Pronville
1st of 4 single-seater victories flying under Walz
1st of 5 total victories flying under Walz

Statistics Notes

* *All pusher victories are pre-wound*
Δ *All DH.2 victories are pre-wound*
± *All No.32 Squadron victories are pre-wound*
† *All No.32 Squadron WiA are pre-wound*

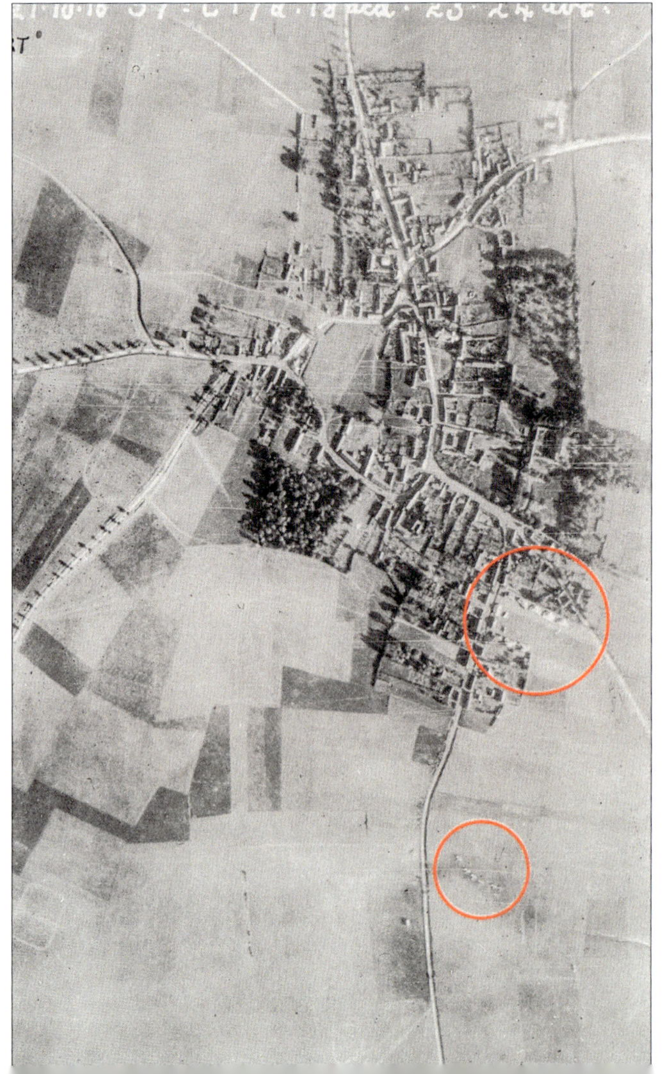

Above & Right: Aerial views of *Jasta* 2's airfield in Lagnicourt, France, taken 21 October 1916. The large red circle denotes the location of tent hangars, while the smaller circle reveals a line of six Albatros D.Is and/or D.IIs. (Detlev Mahlo)

Victory No.13

Above Left: 7927 fabric swatch on Richthofen's wall in Schweidnitz.

Above Right: 7927's souvenired Gnome engine placard.

20 December
DH.2 7927
No.29 Squadron RFC

Day/Date: Wednesday, 20 December
Time: 11.30AM
Weather: Fine all day.
Location: Above Monchy-au-Bois
Crash Location: East of Adinfer Wood
Side of Lines: Friendly

MvR Combat Report: "Vickers One-seater, Nr. 7929. Motor: Gnome, 30413. Inmate: Arthur Gerald Knight, Lieut. R.F.C. killed. Valuables enclosed. One machine gun taken.

About 11.30AM I attacked together with 4 planes and in 3000 meters altitude enemy one-seater above Menchy [*sic*]. After some curve fighting I managed to press adversary down to 1500 meters, where I attacked him at closest range. (Plane length) I saw immediately that I had hit enemy; first he went down in curves, then he dashed to the ground. I pursued him until 100 meters above the ground.

This plane had been only attacked by me.

Frhr. v. Richthofen
Leutnant"

RFC Combat Casualties Report: "Left aerodrome 9.45AM. No.3 A.A. report that a De Havilland was brought down in a spinning nose dive E. of ADINFER WOOD in Sq.X.28."

RFC A/C:

Make/model/serial number: AMC DH.2 7927
Manufacturer: Aircraft Manufacturing Co., Ltd., Hendon, London, N.W.
Unit: No.29 Squadron RFC

Above: Capt. Arthur Gerald Knight's headstone, 2011.

Aerodrome: Le Hameau
Sortie: Offensive Patrol
Colors/markings: u/k. Typically for the make/model, PC10 nacelle/wings/ailerons/horizontal stabilizers/elevator uppersurfaces; CDL undersurfaces; nacelle fabric either CDL or PC10 with white undersurface; r/w/b rudder with black serial numbers sometimes bordered in white on the red/blue. No.29 Squadron used colored numerals to identify flights: A Flight used red numeral with white shadow or border; B Flight used white numeral with blue shadow or border; and C Flight used blue numeral with white shadow or border.
Engine Number: Gnôme 30413 WD 4134
Gun: One Lewis. No. 19234
Manner of Victory: Descending spiral fight and then spinning nose dive until terrain impact.

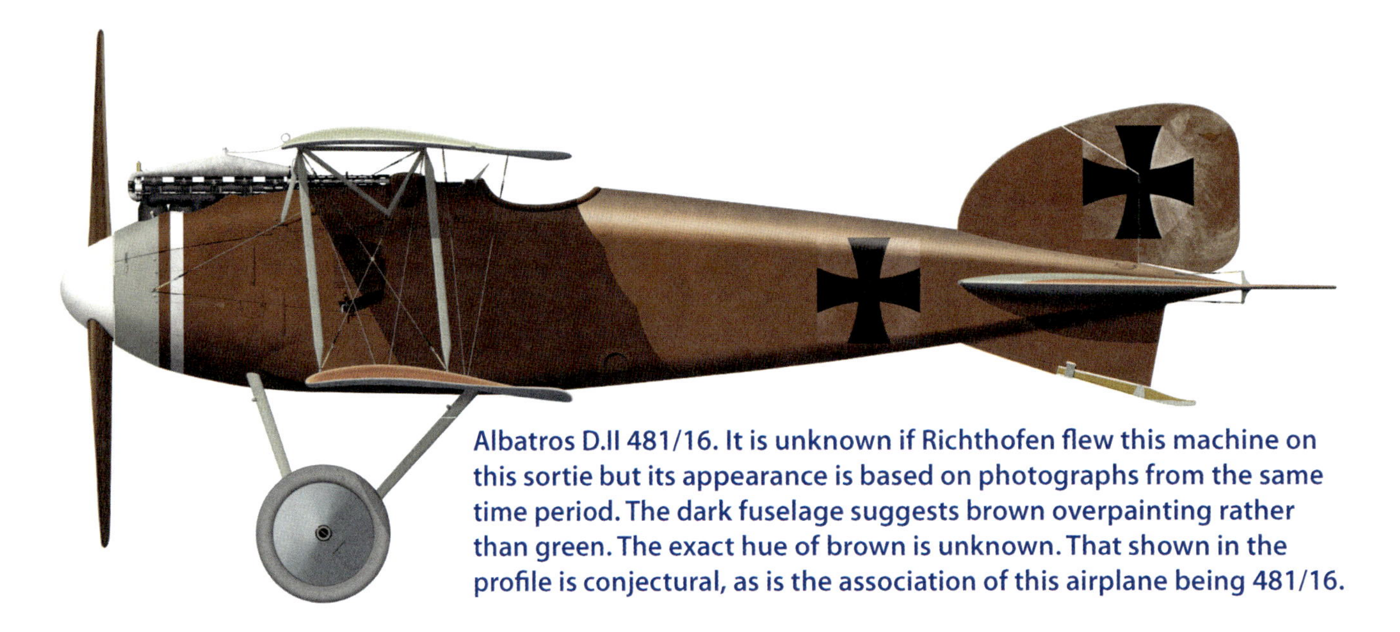

Albatros D.II 481/16. It is unknown if Richthofen flew this machine on this sortie but its appearance is based on photographs from the same time period. The dark fuselage suggests brown overpainting rather than green. The exact hue of brown is unknown. That shown in the profile is conjectural, as is the association of this airplane being 481/16.

Parts Souvenired: a. Black borderless "7927" on r/w background, taken from starboard side of rudder. Photographed on display in Richthofen's home bedroom in Schweidnitz.
b. Gnôme Engine Company placard. Photographed at Roucourt and the post-war Richthofen Museum.

RFC Crew:

Pilot: Capt. Arthur Gerald Knight (DSO, MC), 21, KiA

Pilot Victories: 8 (2 destroyed*, 2 shared destroyed**, 4 out of control‡)

Victory #	Date (1916)	Type
1	22 June	L.V.G. C*
2	19 July	Fokker E-type‡
3	2 Sept	unident. D-type‡
4	14 Sept	Fokker D-type**
5	15 Sept	unident. D-type**
6	17 Oct	LFG Roland C.II‡
7	9 Nov	unident. D-type*
8	16 Dec	unident. D-type‡

RFC Combat Casualties Report: "Reported died 20/12/16—German source. Death officially accepted as having occurred on 20/12/16."
Cause of Death: u/k. Either gunshot wound(s) or trauma associated with airplane crash.
Burial: Douchy-les-Ayette British Cemetery, Grave III. C. 11.

MvR A/C:

Make/model/serial number: Albatros D.II, possibly 481/16.
Unit: *Jasta* 2
Commander: *Hptm.* Franz Walz
Airfield: Pronville
Colors/markings: See Vic. #7.
Damage: u/k

Initial Attack Altitude: 3000 meters.
Final Attack Altitude: 100 meters
Gunnery Range: "Plane length"
Rounds Fired: u/k
Known Staffel *Participants:* u/k

Notes:

1. Knight had been the pilot pursued by *Hptm.* Boelcke and *Ltn.d.R.* Erwin Böhme when they collided 28 October, killing Boelcke.
2. *Jasta* 2 became *Jasta* Boelcke on 17 December. However, Richthofen had used "*Jagdstaffel* Boelcke" in the headings of his combat reports since his 12th victory on 11 December.

Victory No.13 Statistics:

Pushers*
2nd of 4 December total pusher victories
2nd of 3 December single-seater pusher victories
3rd of 4 1916 single-seater pusher victories
3rd of 6 total single-seater pusher victories
7th of 18 total pusher victories

Single-Seaters
2nd of 3 December single-seater victories
7th of 8 1916 single-seater victories
7th of 18 pre-wound single-seater victories
7th of 35 total single-seater victories

DH.2 Δ
2nd of 3 December DH.2 victories

3rd of 4 1916 DH.2 victories
3rd of 5 total DH.2 victories

No.29 Squadron
1st of 3 No.29 Squadron DH.2 victories
1st of 2 December No.29 Squadron victories
1st of 2 1916 No.29 Squadron victories
1st of 4 pre-wound No.29 Squadron victories
1st of 5 total No.29 Squadron victories

Deaths ±
No.29 Squadron:
1st and only December No.29 Squadron crewman KiA
1st and only 1916 No.29 Squadron crewman KiA
1st of 2 pre-wound No.29 Squadron crewmen KiA
1st of 3 total No.29 Squadron crewmen KiA

Pushers:†
1st and only December single-seater pusher crewman KiA
2nd of 2 1916 single-seater pusher crewmen KiA
2nd of 4 total single-seater pusher crewmen KiA

Single-Seaters:
1st and only December single-seater crewman KiA
6th of 6 1916 single-seater crewmen KiA
6th of 12 pre-wound single-seater crewmen KiA
6th of 21 total single-seater crewmen KiA

All:
1st of 3 total December crewmen KiA
15th of 17 1916 crewmen KiA
15th of 66 total pre-wound crewmen KiA
15th of 84 total crewmen KiA

Etc.
2nd of 4 total December victories
13th of 15 total 1916 victories
13th of 57 total pre-wound victories
13th victory in Albatros D.I or D.II
2nd of 4 single-seater victories flying from Pronville
2nd of 5 total victories flying from Pronville
2nd of 4 single-seater victories flying under Walz
2nd of 5 total victories flying under Walz

Statistics Notes
* *All pusher victories are pre-wound*
Δ *All DH.2 victories are pre-wound*
± *KiA includes DoW*
† *All pusher deaths are pre-wound*

Above: View of what is suspected to be Richthofen's Albatros D.II, parked near the tent hangars of Lagnicourt. Ltn. Jürgen Sandel's Albatros D.I 431/16 is at left.

Victory No.14

Left: A5446's entire rudder on display in Richthofen's bedroom, Schweidnitz.

20 December
FE.2b A5446
No.18 Squadron RFC

Day/Date: Wednesday, 20 December
Time: 1.45PM
Weather: Fine all day.
Attack Location: Moreuil
Crash Location: Between Queant and Lagnicourt
Side of Lines: Friendly

MvR Combat Report: "Vickers two-seater; A.5446. Motor: Beardmore, nr. 791. Inmates: Pilot Lieut. D'Arcy, Observer, unknown, had no identification disc. Inmates dead, plane smashed, one machine gun taken, valuables please find enclosed.

About 1.45PM I attacked together with 4 planes of our *Staffel* in 3000 meters altitude enemy squad above Noreuil [*sic*]. The English squad had thus far not been attacked by Germans and was flying somewhat apart. I had therefore opportunity to attack the last machine. I was foremost of our own people and other German planes were not to be seen. Already after the first attack the enemy motor began to smoke; the observer had been wounded. The plane went down in large curves. I followed and fired at closest range. I had killed, as was ascertained later on, also the pilot. Finally the plane crashed on the ground.

Plane is lying between Queant and Lagnicourt.

Frhr. v.Richthofen
Leutnant"

RFC Combat Casualties Report: "Left aerodrome 11.15AM. Was last seen recrossing to our side of the line over LE TRANSLOY at about 1.15PM [British time, one hour behind German time], apparently O.K., though going southwards."

RFC A/C:
Make/model/serial number: Royal Aircraft Factory FE.2b A5446
Manufacturer: Subcontracted to and built by Bolton & Paul, Ltd., Norwich
Unit: No.18 Squadron RFC
Aerodrome: St. Leger-les-Authie
Sortie: Offensive Patrol
Colors/markings: Presentation a/c *Malaya No.11*. Generally, for the make/model, PC10 nacelle/wings/ailerons/horizontal stabilizers/elevator uppersurfaces; CDL under-surfaces; r/w/b rudder with black serial numbers, usually bordered in white when atop the red and blue.
Engine: 160 hp Beardmore
Engine Number: Beardmore No.791
Guns: Two Lewis. Nos. 16021, 17924
Manner of Victory: Continuous descending spiral under continual attack until terrain impact.
Parts Souvenired: Machine gun; entire b/w/r striped rudder with black numbers across all stripes, with "A," "5," and "6" bordered in white. Rudder photographed on display (port side) in home bedroom in Schweidnitz.

RFC Crew:
Pilot: Lt. Lionel George D'Arcy, 28, KiA
RFC Combat Casualties Report: "2/Lt. L.G.D'Arcy. Death accepted by Army Council as having occurred on 20.12.16 on the evidence of a letter from the Central committee of the German Red Cross and lapse of time."
Cause of Death: Likely gunshot wound(s), based on Richthofen's combat report.
Burial: Either in-field or laid where fallen; location lost in the turmoil of war.
Obs: Sub-Lt. Reginald Cuthbert Whiteside, 21, KiA
Cause of Death: u/k. Either gunshot wound(s) or trauma associated with airplane crash.
Burial: Either in-field or laid where fallen; location lost in the turmoil of war.

MvR A/C:
Make/model/serial number: Albatros D.II, possibly 481/16 or 501/16

Above: Albatros D.II 501/16. A *Jasta* 2 machine occasionally flown later by *Oblt.* Adolf Tutschek, a handwritten caption on the reverse of this photograph indicates on this occasion 501/16 had been flown by Richthofen, who landed near a balloon company after suffering engine failure. Note the airplane's overall high sheen and various airfoil patches. (Reinhard Kastner)

Unit: *Jasta* Boelcke
Commander: *Hptm*. Franz Walz
Airfield: Pronville
Colors/markings: Albatros D.II 481/16: See Vic. #7. Albatros D.II 501/16: Factory finish shellacked and varnished "warm straw yellow" wooden fuselage with pale greenish-gray spinner/cowl/fittings/vents/hatches/struts. Upper surfaces of wings/ailerons and horizontal stabilizers/elevator were factory three-color camouflage of pale green, olive green and Venetian red. Wheel covers/undersurface of wings/ailerons and horizontal stabilizers/elevators light blue. Rudder was one of the dark camouflage colors. All crosses were black on square white cross fields.
Damage: u/k

Initial Attack Altitude: 3000 meters.
Gunnery Range: "Closest"
Rounds Fired: u/k
Known Staffel *Participants: Ltn.d.R.* Hans Imelmann, *Ltn.d.R.* Hans Wortmann

Notes:

1. This combat report and others has a line that reads "Witnessess," but features no names. Ostensibly, Richthofen compiled and listed witnesses to help secure confirmation of his victory claims, but mostly these were ignored during translation. The great exception is Victory #61, which includes a detailed list of witnesses and their corroborative testimonies.

Victory No.14 Statistics:

Pushers *

3rd of 4 December total pusher victories
1st and only December two-seater pusher victory
5th of 5 1916 two-seater pusher victories
5th of 12 total two-seater pusher victories
8th of 18 total pusher victories

Two-Seaters

1st and only December two-seater victory
7th of 7 1916 two-seater victories
7th of 39 pre-wound two-seater victories

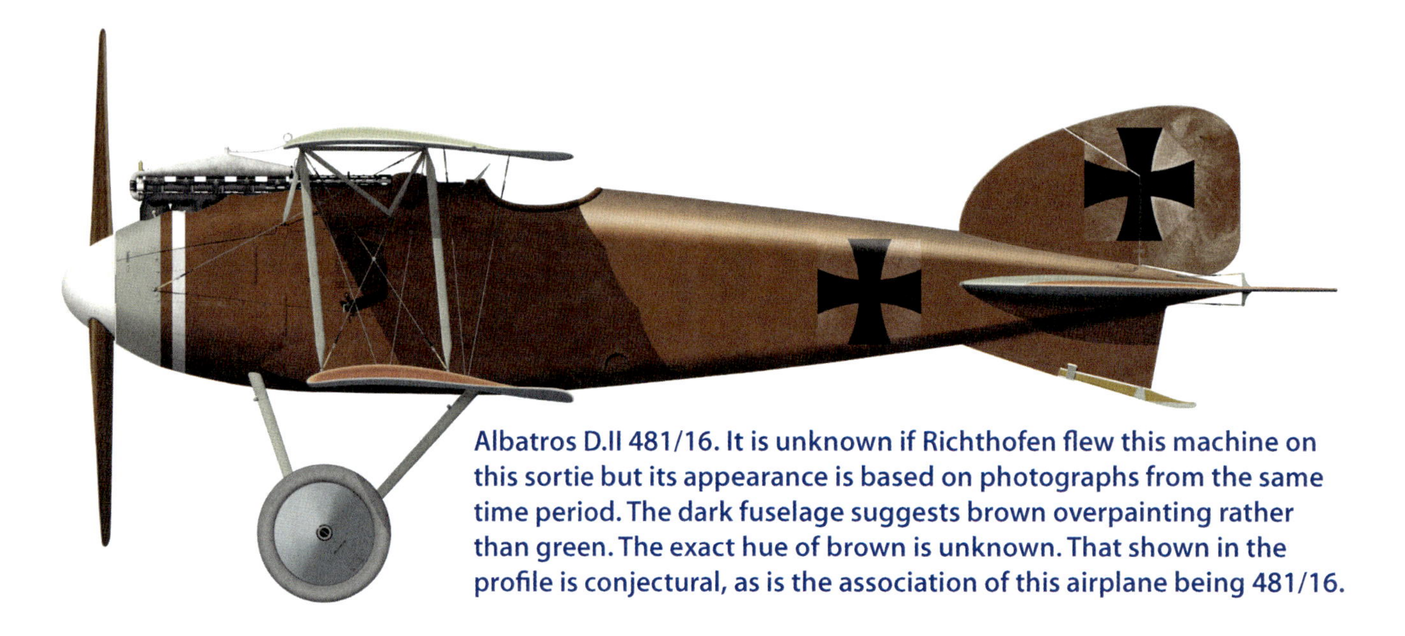

Albatros D.II 481/16. It is unknown if Richthofen flew this machine on this sortie but its appearance is based on photographs from the same time period. The dark fuselage suggests brown overpainting rather than green. The exact hue of brown is unknown. That shown in the profile is conjectural, as is the association of this airplane being 481/16.

7th of 45 total two-seater victories

FE.2 Δ
1st and only December FE.2 victory
5th of 5 1916 FE.2 victories
5th of 12 total FE.2 victories

No.18 Squadron ±
3rd of 4 No.18 Squadron FE.2 victories
1st and only December No.18 Squadron victories
3rd of 3 1916 No.18 Squadron victories
3rd of 4 total No.18 Squadron victories

Deaths†
No.18 Squadron:
1st and 2nd of 2 December No.18 Squadron crewmen KiA
5th and 6th of 6 1916 No.18 Squadron crewmen KiA
5th and 6th of 8 total No.18 Squadron crewmen KiA

Pushers: ††
1st and 2nd of 2 December two-seater pusher crewmen KiA
9th and 10th of 10 1916 two-seater pusher crewmen KiA
9th and 10th of 13 total two-seater pusher crewmen KiA

Two-Seaters:
1st and 2nd of 2 December two-seater crewmen KiA
10th and 11th of 11 1916 two-seater crewmen KiA
10th and 11th of 54 pre-wound two-seater crewmen KiA
10th and 11th of 63 total two-seater crewmen KiA

All:
2nd and 3rd of 3 total December crewmen KiA
16th and 17th of 17 1916 crewmen KiA
16th and 17th of 66 total pre-wound crewmen KiA
16th and 17th of 84 total crewmen KiA

Etc.
3rd of 4 total December victories
14th of 15 total 1916 victories
14th of 57 total pre-wound victories
14th victory in Albatros D.I or D.II
1st and only two-seater victory flying from Pronville
3rd of 5 total victories flying from Pronville
1st and only two-seater victory flying under Walz
3rd of 5 total victories flying under Walz

Statistics Notes
* *All pusher victories are pre-wound*
Δ *All FE.2 victories are pre-wound*
± *All No.18 Squadron victories are pre-wound*
† *All No.18 Squadron deaths are pre-wound; KiA includes DoW*
†† *All pusher deaths are pre-wound*

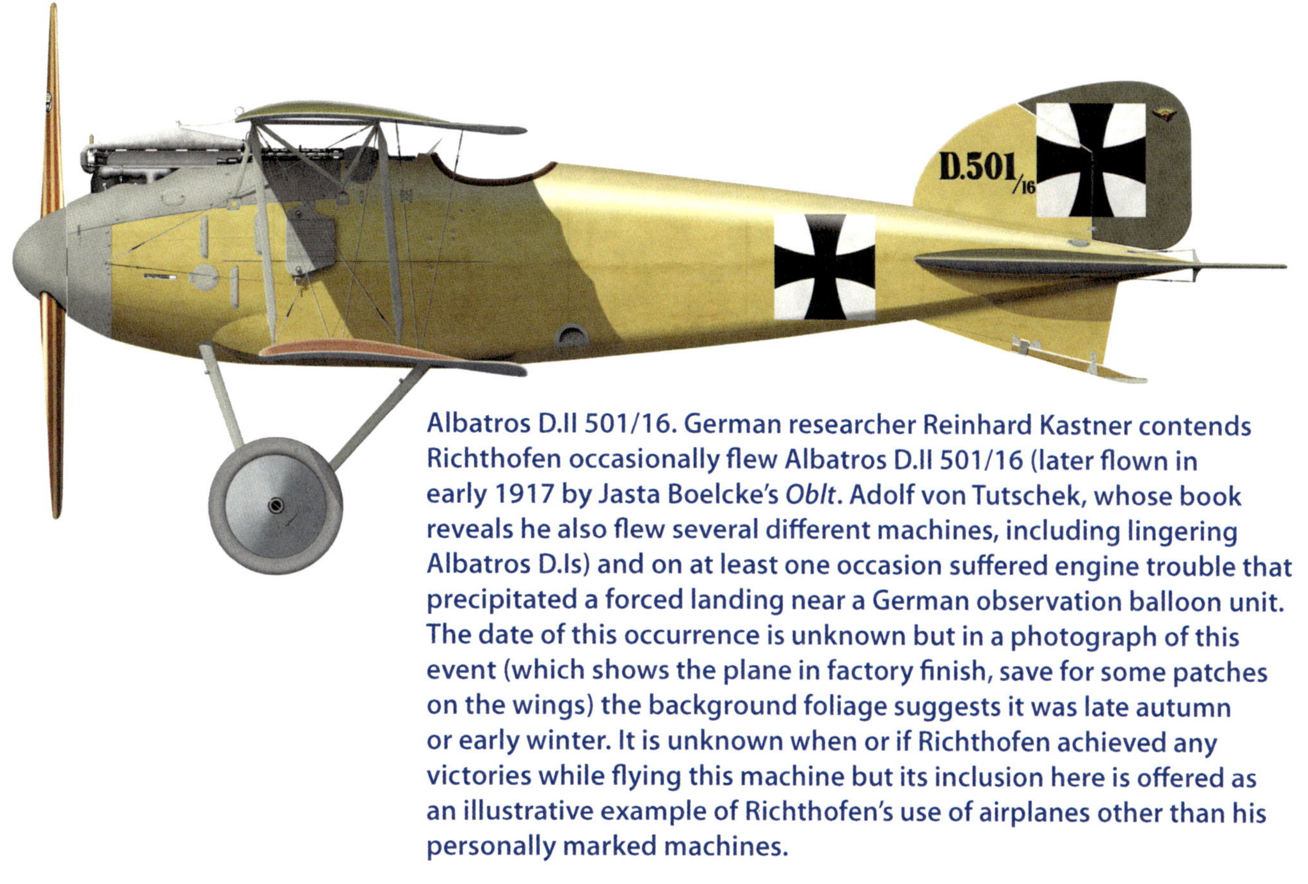

Albatros D.II 501/16. German researcher Reinhard Kastner contends Richthofen occasionally flew Albatros D.II 501/16 (later flown in early 1917 by Jasta Boelcke's *Oblt*. Adolf von Tutschek, whose book reveals he also flew several different machines, including lingering Albatros D.Is) and on at least one occasion suffered engine trouble that precipitated a forced landing near a German observation balloon unit. The date of this occurrence is unknown but in a photograph of this event (which shows the plane in factory finish, save for some patches on the wings) the background foliage suggests it was late autumn or early winter. It is unknown when or if Richthofen achieved any victories while flying this machine but its inclusion here is offered as an illustrative example of Richthofen's use of airplanes other than his personally marked machines.

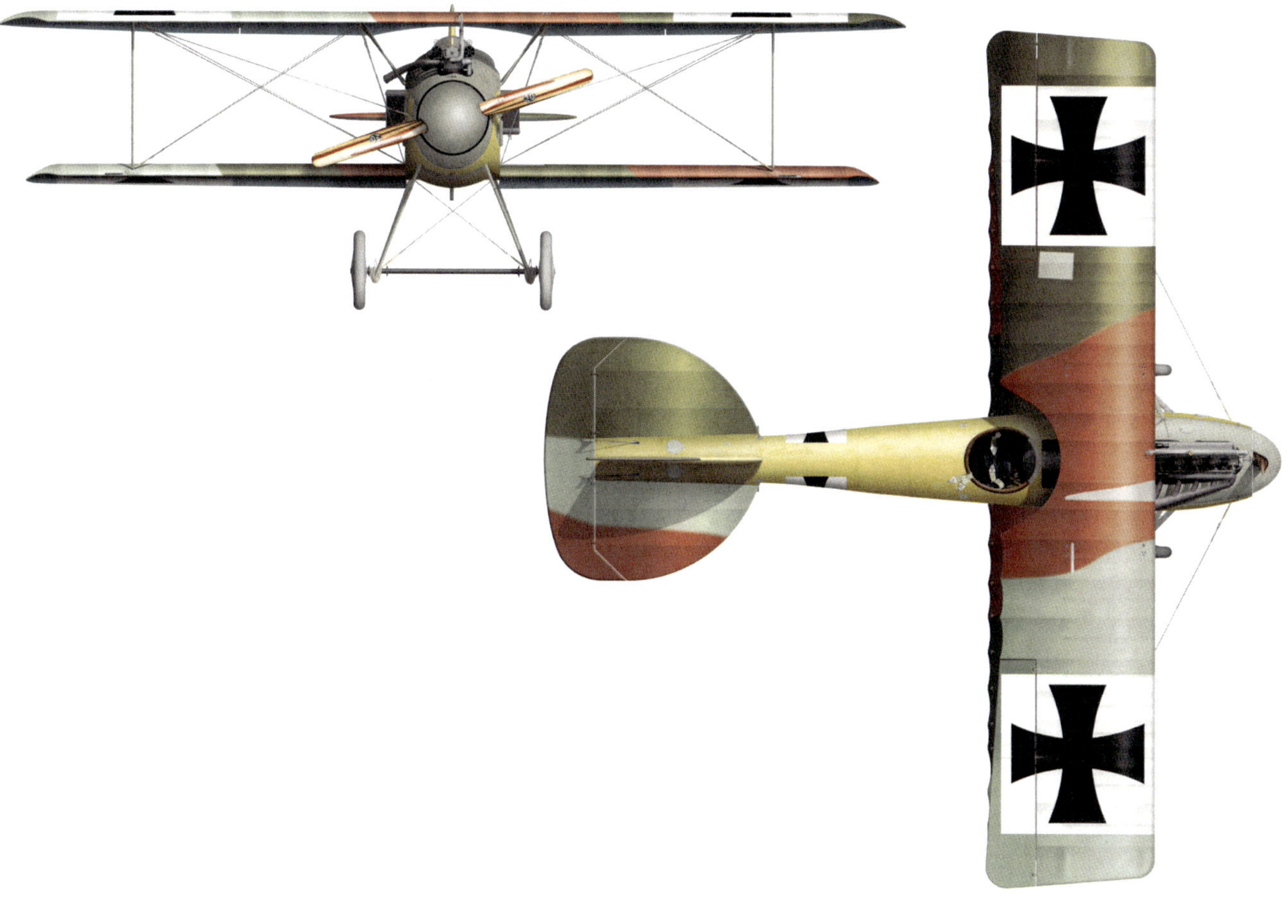

Victory No.15

Above: Five Albatros D.Is of *Jasta* 2 lined up at Lagnicourt airfield, autumn 1916.

27 December
DH.2 5985
No.27 Squadron RFC

Day/Date: Wednesday, 27 December
Time: 4.25PM
Weather: Mist in morning, clearing later.
Attack Location: Ficheux (MvR), "...east of Adinfer Wood..." (McCudden)
Presumed Crash Location: "One kilometer behind trenches near Ficheux"
Side of Lines: n/a

MvR Combat Report: "Albatros D.481. F.E. two-seater, was smashed, number etc. not recognizable.

At 4.15PM 5 planes of our *Staffel* attacked enemy squadron south of Arras. The enemy approached our lines, but was thrown back. After some fighting I managed to attack a very courageously flying Vickers two-seater. After 300 shots enemy plane began dropping, uncontrolled. I pursued the plane up to a 1000 meters above the ground. Enemy plane crashed to ground on enemy side, one kilometer behind trenches near Ficheux.

Only our Staffel planes took part in the fight.
Witnesses: Air Battery 47.
" " 13.

Frhr. v. Richthofen
Was acknowledge[d]"

RFC A/C:
Make/model/serial number: AMC DH.2 5985
Manufacturer: Aircraft Manufacturing Co., Ltd., Hendon, London, N.W.
Unit: No.29 Squadron RFC
Aerodrome: Le Hameau
Colors/markings: u/k. Typically for the make/model, PC10 nacelle/wings/ailerons/horizontal stabilizers/elevator uppersurfaces; CDL undersurfaces; nacelle fabric either CDL or PC10 with white undersurface; r/w/b rudder with black serial numbers sometimes bordered in white on the red/blue. No. 29 Squadron used colored numerals to identify flights: A Flight used red numeral with white shadow or border; B Flight used white numeral with blue shadow or border; and C Flight used blue numeral with white shadow or border.
Engine Number: u/k
Guns: One Lewis.

Above: Richthofen standing next to what is presumed to have been his Albatros D.II, Lagnicourt, November 1916. Another Albatros D.II is visible beyond the struts at far right. Although grainy, that Albatros appears to have light-colored tailskid keel and a black/white/black horizontal streamer between its interplane struts in a manner similar to that seen on Kirmaier's machine. These features tentatively identify the distant machine as his. (Lance Bronnenkant)

Manner of Victory: Spiraling or spinning decent. "I had been chased absolutely out of the sky from 10,000 feet to 800 by a Boche..."
Parts Souvenired: n/a

RFC Crew:
Pilot: Sgt. James Thomas Byford McCudden, 21
Pilot Victories: 1 (1 destroyed) (At the time)

Victory #	Date (1916)	Type
1	6 Sept	unident. C-type

MvR A/C:
Make/model/serial number: Albatros D.II 481/16 (specifically identified in combat report)
Unit: *Jasta* Boelcke
Commander: *Hptm*. Franz Walz
Airfield: Pronville
Colors/markings: See Vic. #7.
Damage: u/k
Initial Attack Altitude: 3000 meters.
Final Attack Altitude: "1000 meters" (MvR), "800 feet [244 meters]" (McCudden)
Gunnery Range: "Closest"
Rounds Fired: 300
Known Staffel *Participants:* u/k

Notes:

1. Although Richthofen claimed (corroborated by Air Batteries 13 and 47) and received credit for this DH.2 flown by future 57-victory ace and RFC luminary James McCudden, it did not crash. Later McCudden wrote that he and Richthofen engaged in a head-on firing pass initiated at 100 yards, during which McCudden's gun jammed. Richthofen continued to engage, forcing McCudden to "turn on [his] back and [dive] vertically" to escape. At 800 feet (Richthofen wrote 1000 meters) Richthofen disengaged via a climbing egress to the east amidst antiaircraft fire, allowing McCudden the freedom to rectify the jam and pursue, but he was unable to catch the Albatros due to the latter's superior climb performance. McCudden noted that his machine received no hits during this encounter.
2. From McCudden's combat report: "Going east of Arras I saw five HA. Lt Jennings attacked an HA and another JA was approaching from behind. I fired about 15 shots and drove him off. He turned and came towards me, firing. I opened fire at 100 yards and after about eight shots my gun stopped, due to a cross feed. As the hostile machine was engaging me at close range, I turned on my back

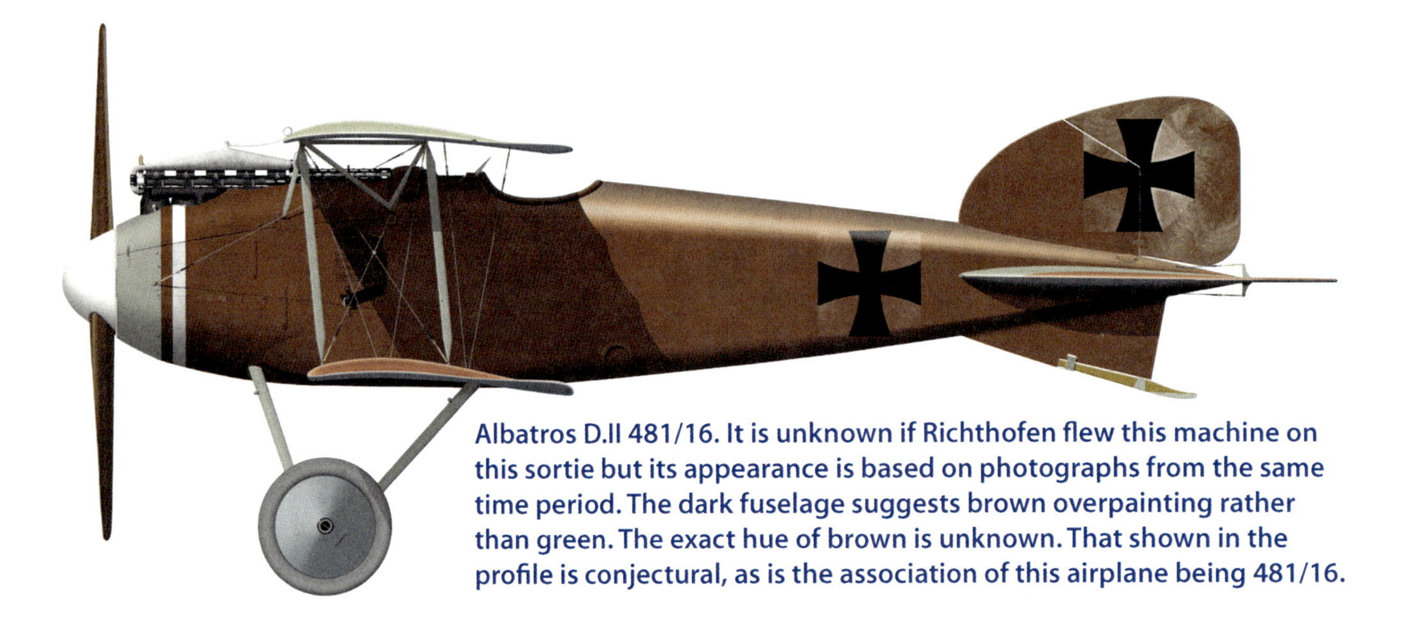

Albatros D.II 481/16. It is unknown if Richthofen flew this machine on this sortie but its appearance is based on photographs from the same time period. The dark fuselage suggests brown overpainting rather than green. The exact hue of brown is unknown. That shown in the profile is conjectural, as is the association of this airplane being 481/16.

and dived vertically, in a slow spin and in this way regained our lines. At 800 feet over Basseux the HA left me. I quickly rectified the stoppage and followed the HA across the trenches at 2,000 feet. Owing to his superior speed and climb he out distanced me and rejoined his patrol at about 5,000 feet. The hostile patrol then withdrew."

3. Although in his 1918 *Air Combat Operations Manual* Richthofen opined that head-on attacks against two-seaters (which he thought McCudden was flying) were 'very dangerous,' in late 1916 he was still refining the lessons learned from Boelcke and ostensibly had not compiled enough experience to conclude the tactical ineffectiveness of this attack methodology.

Victory No.15 Statistics:

Pushers ⋆
4rd of 4 December total pusher victories
3nd of 3 December single-seater pusher victories
4rd of 4 1916 single-seater pusher victories
4th of 6 total single-seater pusher victories
9th of 18 total pusher victories

Single-Seaters
3rd of 3 December single-seater victories
8th of 8 1916 single-seater victories
8th of 18 pre-wound single-seater victories
8th of 35 total single-seater victories

DH.2 Δ
3rd of 3 December DH.2 victories
4th of 4 1916 DH.2 victories
4th of 5 total DH.2 victories

No.29 Squadron
2nd of 3 No.29 Squadron DH.2 victories
2nd of 2 December No.29 Squadron victories
2nd of 2 1916 No.29 Squadron victories
2nd of 4 pre-wound No.29 Squadron victories
2nd of 5 total No.29 Squadron victories

Etc.
4th of 4 total December victories
15th of 15 total 1916 victories
15th of 57 total pre-wound victories
15th victory in Albatros D.I or D.II
3rd of 4 single-seater victories flying from Pronville
4th of 5 total victories flying from Pronville
3rd of 4 single-seater victories flying under Walz
4th of 5 total victories flying under Walz

Statistics Notes
⋆ *All pusher victories are pre-wound*
Δ *All DH.2 victories are pre-wound*

Aeronautical Analysis of Richthofen vs. Hawker, 23 November 1916

Persistent myths still surround Richthofen's famous battle with No.24 Squadron commanding officer Major Lanoe Hawker, VC, DSO (Richthofen's 11th victory). Particularly, who attacked who, the length of the battle, the particulars of the battle, and the state of Hawker's engine. Although the author has addressed these myths elsewhere, their persistent and infectious damage to history requires reexamination via retrograde event extrapolation and calculative aeronautical analysis to reveal the factual historical perspective. Therefore, setting the events in motion:

At 1300 hours (British time, one hour behind German time) on Thursday, 23 November 1916, four No. 24 Squadron 'A' Flight scouts took off from Bertangles aerodrome on a defensive patrol of the British 4th Army front near Bapaume, France. Leading this quartet of Aircraft Manufacturing Company (AMC) DH.2s was 'A' Flight Commander

Right: Colorized image of Lanoe Hawker. His "attack everything" mantra epitomized the RFC's spirit of *Per Ardua ad Astra* but target fixation and lack of combat recency begat a catastrophic event chain that led to his death. His loss was a great blow to the entire RFC, No.24 Squadron's Robert Saundby later regarding his commanding officer with a quote from Shakespeare: "His life was gentle, and the elements so mixed in him that Nature might stand up and say to all the world, 'This was a man.'"

Below: Eighteen No.24 Squadron AMC DH.2 fighters and one Morane "Bullet" lined up at Bertangles aerodrome. The DH.2 nacelles are finished in either gray or PC10 and wheel covers are painted in representative flight colors. The various identifying strut markings can also be seen.

Above: Fitters and riggers pose with Andrews' DH.2 5983. The red painted wheel covers and red/white striped struts are apparent, as is the starboard ammunition tray, Lewis machine gun, and No.24's sawtooth pattern on the belly of the nacelle.

Captain John Oliver Andrews, an experienced seven-victory ace, who the previous day shot down and killed *Jagdstaffel* 2 *Staffelführer Oberleutnant* Stefan Kirmaier.[1] Accompanying Andrews were Captain Robert Henry Magnus Spencer Saundby; 2nd Lieutenant John Henry Crutch; and—as a last-minute replacement—No.24 Squadron Commander and Royal Flying Corps luminary Major Lanoe George Hawker, VC, DSO.

Hawker was an accomplished pilot of considerable and varied experience. Having flown reconnaissance sorties during the early war, on 19 April 1915 Hawker flew a solo Royal Aircraft Factory BE.2c bombing mission against the Gontrode Zeppelin sheds near Ghent, Belgium, an action for which he was awarded the Distinguished Service Order (DSO). As the war progressed he transitioned to FE.2s and single-seat Bristol Scouts, to the latter of which he rigged a fuselage-mounted machine gun to fire obliquely outside the propeller arc. With his airplane so equipped, on 25 July 1915 Hawker attacked and received credit for downing two Albatros C-type two-seaters, one of which "burst into flames and turned upside down, the observer falling out. The machine and pilot crashed to earth..."[2] Hawker received the Victoria Cross (VC) for these two victories, the first RFC scout pilot so decorated. During the following month Hawker claimed three additional victories and upon returning to England 20 September his victory tally was seven.

Eight days later Hawker assumed command of No.24 Squadron, a weeks-old unit with which he spent the next four months training pilots on various two-seaters. Eventually re-equipped with DH.2's, the squadron departed England 7 February 1916 and "crossed the Great Divide" to St. Omer, France—becoming the first solely-equipped single-seater squadron to proceed to a combat zone[3] — and on 10 February established residence at Bertangles, whereupon "war-flying proper immediately commenced."[4] Although command responsibilities precluded Hawker's mission participation—which eroded his combat recency—he nevertheless made efforts to accompany his pilots whenever the pilot pool was low or there was a "young pilot about to go on leave"[5] in whose slot the altruistic Hawker occupied lest misfortune befall a pilot so close to respite.

Such was the reason for Hawker's participation with 'A' Flight's sortie 23 November. Climbing

Above: One *Jasta* 2 Albatros D.II and two D.Is at Lagnicourt, likely around the time of Richthofen's fight with Hawker. The overpainted and white-striped D.II at left is believed to be Richthofen's machine. Note that all three airplanes have been overpainted, even to the extent that the fuselage and wing crosses/cross fields have been covered entirely. (Lance Bronnenkant)

through clear skies dominated by high pressure,[6] 'A' Flight's ingress was uneventful until 1330 hours, when they spotted a "rough house going on over Grandcourt" between No.60 Squadron Nieuports and unidentified "H.A." (Hostile Aircraft).[7] 'A' Flight "went down with engines on"[8] to join the fight but arrived after the Nieuports had driven away most of the German airplanes. Crutch saw Hawker "go N[orth] after [a] single seater H.A."[9] and began following him but his engine—already running roughly for at least part, if not all, of the flight—"cut out on 2 cylinders and started to knock,"[10] necessitating his precautionary landing at No.9 Squadron's airfield[11] near Morlancourt.[12] There, Crutch discovered "2 plugs damaged and tappet rods out of adjustment"[13]—maladies all-too-common with DH.2 7919, an airplane plagued with sundry engine malfunctions.[14]

Andrews, Saundby and Hawker continued inbound until shortly after 1350 hours, when near Achiet[15] Andrews spotted "two H.A."[16] flying at 6,000 feet northeast of Bapaume, prompting him to lead 'A' Flight in a diving attack that "drove them [the Germans] east."[17] During this chase Andrews still scanned his immediate airspace and while doing so discovered "two strong patrols of H.A. scouts above me."[18] Realizing a long pursuit of the eastward-fleeing Germans was imprudent in the face of such an immediate threat above him, Andrews decided to break off his attack but "a D.H. Scout, [flown by] Maj. Hawker, dived past me and continued to pursue."[19] Not wanting to abandon Hawker—apparently unaware of the German scouts above him—Andrews and Saundby followed him eastward and "were at once attacked by the H.A., one of which dived on to Maj. Hawker's tail."[20]

These attacking scouts were *Jasta* 2 Albatros D.IIs (and likely also D.Is) led by Richthofen, at that time a ten-victory ace and de facto *Jasta* 2 *Staffelführer* in the wake of Kirmaier's death the previous day. As *Jasta 2* dived on 'A' Flight the English pilots found themselves involved in what Saundby's logbook described as a "violent fight".[21] Two Albatrosses attacked his DH.2, forcing him to spiral "two or three times"[22] before the Germans disengaged and "zoomed off."[23] Andrews went after the Albatros attacking Hawker and "drove him off firing about 25 rounds at close range"[24] but in the process was attacked by a fourth Albatros whose accurate gunfire struck his engine, incapacitating it and forcing a glide with which Andrews was "obliged to try to regain our lines."[25] The pursuing German fired continually, his astute gunnery reflected by Andrew's logbook: "Machine badly shot about. Large hole in cambox of engine, also one cylinder shot thro', smashed piston, etc."[26] Fortunately for Andrews, Saundby found himself in good position to attack this belligerent and fired "three-quarter double drum into him at about 20 yards range,"[27] after which the Albatros "wobbled"[28] and then power-dived away. Saundby went after him but the Albatros out-dived his DH.2 and he could not follow.

Pulling level, Saundby scanned his immediate airspace and saw that the German airplanes "appeared to have moved away east."[29] Although he could see Andrews nearby, there was no sign of Hawker. Andrews last saw Hawker "at about 3,000 feet near Bapaume, fighting with an H.A. apparently quite under control but going down"[30] but had to break visual contact to concentrate on his dead-stick approach and landing in Guillemont, which he executed successfully at 1410 hours.[31]

What Andrews glimpsed was the approximate midpoint of the now legendary combat between

Hawker and Richthofen, which by Andrews' eyewitness account had developed into a continuous series of tight descending spirals. As described in Richthofen's autobiography:

"Thus we both turned like madmen in circles, with engines running full-throttle at 3,500 meters height. First 20 times left, then 30 times right, each mindful of getting above and behind the other."[32]

At 500 meters Hawker ceased spiraling and began evasive aerobatics that involved "looping and such tricks,"[33] after which he broke for the lines in a zig-zagging descent from 100 meters. Richthofen pursued, firing steadily as his faster Albatros gained on the jinking DH.2 but at 30 meters altitude his guns jammed and "almost cost me success".[34] Clearing his weapons, Richthofen resumed firing until "fifty meters behind our lines"[35] he saw Hawker's DH.2 begin an unchecked shallow descent and then impact the artillery-ravaged terrain near Ligny-Thilloy, one mile (1.6 km) south of Bapaume. German soldiers later inspecting the wreck determined Hawker had been shot once in the back of the head. He was buried next to his destroyed airplane, his grave being ultimately lost in the turmoil of war.

Despite its familiarity among students of World War I aviation history, descriptions of this dogfight have been and continue to be the geneses of several continuously perpetuated myths, largely fueled by misrepresented translation and discrepancies between Richthofen's combat report and his autobiographical account.

Myth #1: 'A' Flight Dived On and Attacked *Jasta* 2

This myth's cornerstone is Richthofen's 23 November autobiographical account, the beginning of which differs from every other participant's account:

"I was on patrol that day and observed three Englishmen who had nothing else in mind than to hunt. I noticed how they ogled me, and since I felt ready for battle, I let them come. I was lower than the Englishmen; consequently, I had to wait until they came down to me. It did not take long before one dove for me, trying to catch me from behind. After a burst of five shots the sly fellow had to stop, for I was already in a sharp left curve. The Englishman attempted to get behind me while I attempted to get behind him."[36]

This account indicates 'A' Flight—specifically, Hawker—attacked an altitudinally disadvantaged Richthofen ("...I was lower than the Englishmen") and suggests Richthofen was flying alone ("...I was on patrol that day..."; "I noticed how they ogled me..."; "I was lower than the Englishmen..."). These recollections contradict those of 'A' Flight's survivors ("...I observed two strong patrols of H.A. scouts above me..."; "...We were at once attacked by the H.A., one of which dived on to Maj. Hawker's tail..."; "...We were dived on by a patrol of seven or eight *Walfisch* [*sic*][37]..."). Similarly, they contradict Royal Flying Corps Communiqué No. 64 ("The de Havillands were at once attacked by the two strong hostile patrols, one of the enemy's machines diving on to the tail of major Hawker's de Havilland")[38] as well as Richthofen's own combat report ("I attacked with two planes a Vickers one-seater in 3,000 meters altitude...").

So which account accurately describes *Jasta* 2's initial encounter with 'A' Flight — Richthofen's autobiography or his combat report? Irrefutable direct evidence supporting one more than the other does not exist but the preponderance of circumstantial evidence suggests the combat report is more accurate. It was written immediately after fighting Hawker when the event details were still fresh in his mind — details which match 'A' Flight's contention that they were attacked ("... we were dived on..."; "...attacked by two lots of H.A...."; "...we were at once attacked...") and the Germans did the attacking ("...I attacked with two planes..."). Contrastingly, Richthofen dictated his autobiography nearly six-months and 41 victories after fighting Hawker—suggesting possible event-detail cross-contamination with so many subsequent victories— during a period when Richthofen's accomplishments, rising fame and status had bolstered confidence and pride to such a degree that Richthofen later described his spirit as "insolent"—an opinion with which he also regarded his book ("When I read my book, I smile at the insolence of it. I no longer possess such an insolent spirit").[39]

Did this insolence foster event embellishment, or was Richthofen's autobiographical account simply propagandist? It is known portions of the dictated autobiography were further "edited" prior to publication, so editorial spin designed to ultra-elevate Richthofen's lofty pedestal is not implausible—i.e., England's finest pilot attacked an outnumbered and altitudinally disadvantaged Richthofen yet the latter *still* prevailed. Perhaps Richthofen's autobiographical account *does* coincide with his combat report—i.e., Richthofen was flying the Albatros that Andrews attacked ("...one...dived on to Maj. Hawker's tail...I drove him off firing about 25 rounds at close range...") and *that* attack was the one Richthofen evaded with a "sharp left curve". If so, after Andrews was hit and rendered lame, Richthofen could have freely reengaged

Above: DH.2 5925. Flown by Saundby during No.24 Squadron's fight with *Jasta* 2, it is shown here in 1917 as a training machine at Brooklands. The machine is unarmed but still carries traces of its striped outboard interplane struts. (H. Kilmer)

Hawker—by then aware of the attacking airplanes—who then evaded with his own sharp left curve, precipitating the kind of turning dogfight Andrews and Richthofen described.

In any event, Hawker and Richthofen's exact maneuvers prior to their turning dogfight will never be known. Regardless, based on all combat reports it is a reasonable certainty that *Jasta* 2 attacked 'A' Flight from above and not the other way around.

Myth #2: Richthofen and Hawker Fought for One-Half Hour

Upon review of Richthofen's 23 November combat report it is easy to see how the dogfight timeline has become inflated:

"Nov. 23. 1916, 3PM south of Bapaume. Vickers One-seater, plane lying near Bapaume.
Occupant: Major Hawker, dead.
I attacked together with two planes a Vickers one-seater in 3,000 meters altitude. After a long curving fight of 3–5 minutes I had pressed down adversary to 500 meters. He now tried to escape flying to the Front, I pursued and brought him down after 900 shots.
Witnesses: Lieut. Wortmann. Lieut. Collin etc.
Frhr. v. Richthofen.
Lieut."[40]

"3 – 5" (three-to-five) is often misinterpreted as "35," although sometimes it is rounded down to the generic "half-hour". For many, Richthofen's use of the adjective "long" ("...after a long curving fight...") is enough suggestion to conclude he *must* have meant 35 minutes, since 35 minutes is obviously much longer than three-to-five minutes. Yet, three-to-five minutes of one-on-one turning aerial combat *is* long, especially if steeply-banked with increased and persistent stick pressures, load factors, and need for broad situational awareness.

Furthermore, "long" is relevant to experience. Most of Richthofen's previous encounters had been with either two-seaters[41] (four FE.2bs and two BE.2cs) or two-seaters converted into single-seaters (three BE.12s, which were more powerful BE.2cs with faired forward cockpits and a machine gun added to the port fuselage or top wing), which retained two-seater maneuverability inadequate for one-on-one dogfighting. Richthofen's combat reports indicate he made short work of these airplanes, revealing the total round expenditures needed to bring down seven of his first ten victories:

n/a, FE.2b
300 rounds, Martinsyde G.100
200 rounds, FE.2b
400 rounds, BE.12
350 rounds, BE.12
200 rounds, BE.12
400 rounds, FE.2b
n/a, BE.2c

n/a, BE.2c
300 rounds, FE.2b
(307-round average for victories No. 2–7 & 10.)

Comparatively, it took 900 rounds to overcome the DH.2's superior performance and Hawker's experience, superb stick-and-rudder airmanship, and his "attack everything" mantra. As Richthofen wrote in a letter dated 25 November 1916: "It was the most difficult battle I have had."[42] (See Volume 2, page 434, for a comprehensive examination of Richthofen's round expenditure.)

The Two Scenarios

Calculating Richthofen and Hawker's combat timeline can be accomplished by employing Richthofen's "20 times left, then 30 times right" as an illustrative reference datum against which the DH.2 and Albatros D.II's performance data can be compared within the "3–5" and "35" minute time frames. Of course, without radar plots, onboard cameras, and flight data recorders, an exact determination of each airplane's performance is impossible—precise information regarding weights, airspeeds, power, and pilot techniques used 23 November will never be known. Neither will the exact number of circles flown nor their precise radii—Richthofen's "twenty left, thirty right" was likely illustrative approximation and not representative of actual numbers, since there would have been no practical use for either man to count his circles during their dogfight.

Regardless, a comparative performance and timeline determination is possible based on the knowledge that an airplane's horizontal lift component (which turns banked airplanes) equals centrifugal force in steady, turning flight, creating the following turning performance relationships:[43]

$$\text{Turn Radius: } r = \frac{V^2}{11.26 \tan \varnothing}$$

$$\text{Turn Rate: } ROT = \frac{1{,}091 \tan \varnothing}{V}$$

Where:

r = turn radius, feet
ROT = rate of turn, degrees/second
V = velocity, knots
ø = bank angle, degrees

These relationships define radius and turn rate as trigonometric functions[44] of bank angle and velocity. Therefore, when an airplane is flown in a steady, coordinated[45] turn at a specific bank angle and velocity, the turn rate and radius are fixed and independent of the airplane type. This assumes all airspeeds and bank angles are constant and that all turns are coordinated and flown at a constant altitude—unlikely during any dogfight and certainly not the case with Richthofen and Hawker—but the results are applicable for climbing and descending flight when the angles of each are relatively small, such as would occur while descending during a 35-minute dogfight.

By combining these trigonometric functions with the different timelines and 50-circle reference datum (hereafter referred to as *50/35*, *50/3* and *50/5*, circles and minutes, respectively), performance relationships can be plotted across a wide airspeed range. When these results are compared with various combat reports and the local meteorological data for 23 November, it can be demonstrated which dogfight timeline—"3–5" or "35" minutes — is the more likely.

50/35

Via the calculative methodology described above, the 35-minute scenario was examined first:

The results are conclusive. Even at the fastest airspeeds—which require the largest bank angles to sustain the fixed 8.6°/second turn rate—Figure One's bank angles are not representative of the type Richthofen described ("…both of us flying like madmen in a circle…"). Pilots holding 32° banks to fly one 528-meter (1,732 feet) diameter circle every 42 seconds can never be described accurately as "flying like madmen". These flight parameters are more descriptive of student pilots in Cessna 152s practicing constant altitude turns than they are of two fighter pilots trying to kill each other.

Additionally, a 528-meter diameter circle is 530–660% larger than what Richthofen described ("the circles which we made around each other were so close that I estimated them to be not further than 80 to 100 meters").[46] If Richthofen and Hawker were on opposite sides of the circle—a distance of nearly six American football fields placed end-to-end—it is unlikely Richthofen "could observe every movement of his [Hawker's] head… If he had not had his flying helmet on, I could have seen what kind of face he made". Even at 50/35's smallest radius of 145 meters (477 feet)—a 290 meter (954 feet) diameter, still over three football fields distant and 290–360% larger than Richthofen's estimated 80–100 meters—it is doubtful Richthofen could have recognized *his brother* in a DH.2 cockpit, let alone discern the facial expressions of any goggled pilot—or notice his quick wave from 528 meters away ("my opponent waved to me quite cheerfully as we were at a thousand meters altitude as if to say: 'Well, well, how do you do?'"[47]). The radii lengths in Figure One are just too great to warrant the metaphor "could have seen what kind of face he made", ostensibly

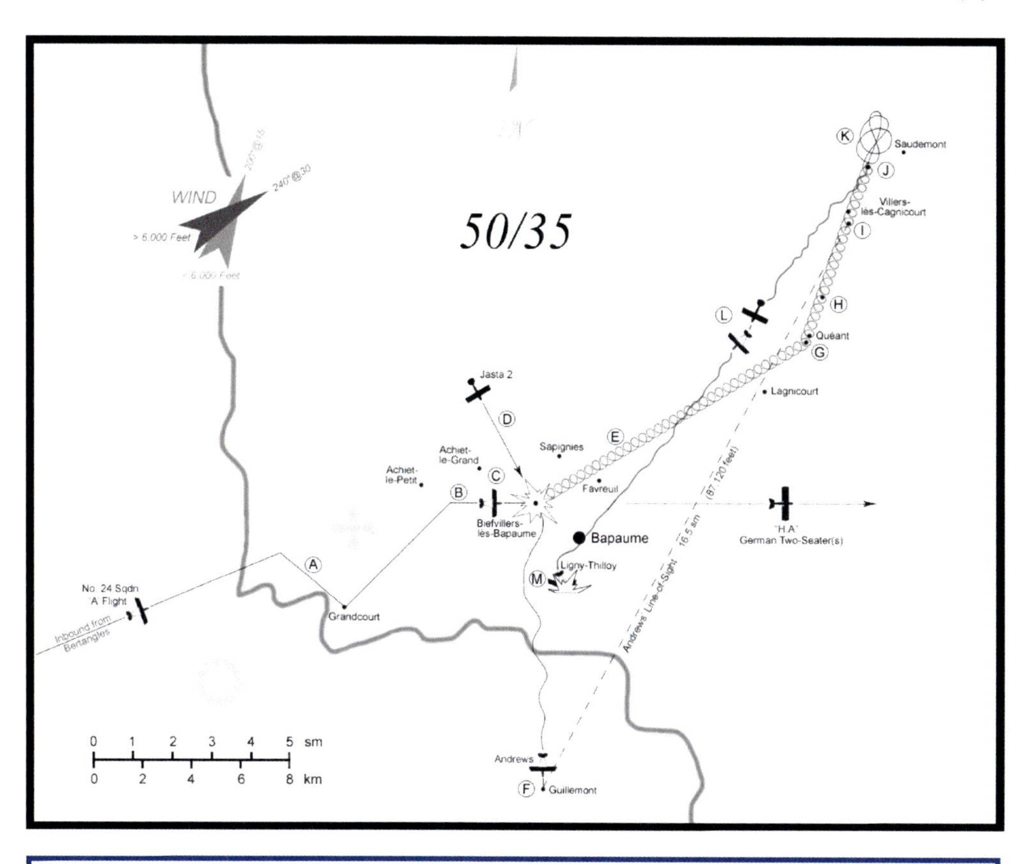

Right: Figure 1 illustrates the 50/35, or 35-minute dogfight scenario.
A. 'A' Flight power-dives on "rough house going on over Grandcourt."
B. 'A' Flight spots "H.A." two-seaters northeast of Bapaume and pursues same.
C. Andrews detects "two strong patrols of H.A. scouts" above him.
D. Jasta 2 dives on 'A' Flight.
E. Richthofen and Hawker's circling dogfight drifts northeast.
F. Andrews glides damaged DH.2 across the lines and dead-sticks in Guillemont.
G. 6,000 feet (1,829 meters).
H. 1,500 meters (4,921 feet).
I. 3,000 feet (914 meters). Hawker's location upon Andrews' last sighting.
J. 500 meters (1,640 feet).
K. Hawker's "looping and such tricks."
L. Richthofen and Hawker's ten minute-plus low-altitude chase.
M. Hawker KiA, crashes near Ligny-Thilloy.

If 50 Circles in 35 Minutes (50/35), Then:

Duration of each circle: 42 seconds & Turn Rate: 8.6°/second

MPH	Kts	Km/h	Turn Rate	Bank Angle	Turn Radius	Load Factor	Stall Speed Increase
90	78	145	8.6°/sec	32°	865'/264 m	1.18	9%
80	70	129	8.6°/sec	29°	785'/239 m	1.14	7%
70	61	113	8.6°/sec	26°	678'/207 m	1.11	5%
60	52	96	8.6°/sec	22°	594'/181 m	1.08	4%
50	43	80	8.6°/sec	19°	477'/145 m	1.06	3%

Right: Figure 2.

meant to illustrate close proximity.

In any event, Richthofen and Hawker descended continuously—a common occurrence during dogfights. Richthofen's combat report noted this descent lasted 2,500 meters, from 3,000 meters to 500 meters. Combining these figures with the 50/35 reference datum, basic mathematics was used to determine the following descent rates:

Figure Two illustrates that with 50/35 a 2,500 meter descent translates into a 71 meter/minute (mpm, or 233 foot/minute [fpm]) descent rate, the resultant descent angle of which would be less shallow than the gentle 3° glideslope modern airliners use during precision instrument approaches. Cessna pilots typically reduce altitude with 1,000 fpm descent rates and *those* are considered docile—a 233 fpm decent would not even be noticed until at very low altitudes. Yet neither Richthofen nor Hawker could overcome such a gentle altitude loss, even while continuously trying to climb above each other?

Regardless, that a continuous descent occurred is unquestionable. Therefore, based on 50/35, the above table lists salient descent altitudes and the times necessary to reach them from 3,000 meters:

Recall that Andrews last saw Hawker "at about 3,000 feet [914 meters] near Bapaume"—i.e., after Hawker and Richthofen had descended 6,844 feet (2,086 meters). Figure 3 illustrates that a 2,086-meter descent required 29 minutes, 24 seconds. Presuming *Jasta* 2 attacked 'A' Flight at 1355,[48] 50/35 stipulates that as Andrews glided towards friendly territory his last Hawker sighting must have been at 1424 (1355 + 29 min). However, according to Andrews'

Circling Descent from 3,000 m @ 71mpm		
Based on 50/35 Scenario Descent Rate		
From 3,000m to... (From 9,843' to...)	Total Descent	Descent Duration
2,500 m (8,202')	500 m (1,640')	7 min. 3 sec.
2,000 m (6,562')	1,000 m (3,281')	14 min. 6 sec.
1,500 m (4,921')	1,500 m (4,921')	21 min. 9 sec.
1,000 m (3,281')	2,000 m (6,562')	28 min. 12 sec.
914 m (3,000')	2,086 m (6,844')	29 min. 24 sec.
500 m (1,640')	2,500 m (8,202')	35 min. 12 sec.

Above: Figure 3.

logbook, at 1424 hours he had already dead-sticked his DH.2 in Guillemont and had been on the ground 14 minutes. 50/35 does not jibe.

Further discrepancies arise when cross-checking Figure Three's descent durations against recorded meteorological data for 23 November. Popular presumption is the winds that day were strong and westerly but synoptic weather charts reveal high pressure (30.42" Hg) was centered just south of Bapaume, the isobars of which indicate the winds were actually light and south-southwesterly. These charts dovetail with forecast amendments issued throughout that day:

9AM: Wind light—SW probably increasing to 25 mph later from south. At 6000 ft. west 20 mph changing to SW 40 m.p.h.

11AM: Wind SW or south 10 to 15 mph.

4PM: Wind SSW 15 mph changing to south or SSE and increasing to 30 mph.[49]

Airplanes in flight are directly influenced by wind because, when free of surface friction, they become part of the moving stream of air. Pilots normally counter this influence by varying the airplane's flightpath and airspeed. However, dogfighting airplanes are more subject to wind influence because pilots do nothing to temper its effects—altering flightpath to counter wind may provide the enemy an ideal firing opportunity. Thus, in calm winds a circling airplane remains stationary above a constant geographical location but with the slightest breeze it drifts across the ground in the same direction and speed as the wind. This uncorrected wind influence is known as *drift.*

Richthofen and Hawker's drift can be calculated using forecast and charted wind velocities, 50/35's 42-second circles and Figure 3's descent durations. Winds above 6,000 feet (1,829 meters) are presumed to have been west-southwesterly (240°) at 30 mph—an interpolation of the 9AM forecast—while below 6,000 feet they are presumed to have been south-southwesterly (200°) at 15 mph.[50]

Knowing 'A' Flight was "near Achiet" when they spotted the "two H.A." northeast of Bapaume; and knowing that the DH.2s flew east in pursuit of same; then *Jasta* 2's attack is estimated to have occurred near Biefvillers-lès-Bapaume, a small village less than one mile northwest of Bapaume itself. Since it is likely that Richthofen and Hawker began their engagement within one minute of *Jasta* 2's attack, Biefvillers-lès-Bapaume is presumed also to be the start location of their circling dogfight. Thence:

Start of Richthofen/Hawker's Dogfight: Above Biefvillers-lès-Bapaume
Circling Descent from 3,000 meters to 1,829 meters (6,000 feet) @ 71 mpm: 16 minutes, 30 seconds
Wind above 6,000 feet: 240° @ 30 mph
Resultant Wind Drift during Descent to 6,000 feet: 16:30 @ 30 mph = 8.3 sm (13 km)
Location when at 6,000 feet: Above Quéant, 8 sm northeast Bapaume
Wind below 6,000 feet: 200° @ 15 mph
Circling Descent from 1,829 meters to 500 meters @ 71 mpm: 18 minutes, 42 seconds
Resultant Wind Drift during Descent to 500 meters: 18:42 @ 15 mph = 4.7 sm (7.6 km)
Richthofen/Hawker's Location when at 500 meters: Near Saudemont, 12 sm northeast Bapaume

This location near Saudemont contrasts with Richthofen's combat report ("...I had forced down my opponent to 500 meters near Bapaume..."). At 12 miles distant, Saudemont cannot be considered "near Bapaume". It is much nearer *Jasta* 2's Lagnicourt airfield, within a half-mile of which the pair would have drifted as they fought—a noteworthy detail if so, curious in its absence. Richthofen's autobiography indicates "we were circling more and more over our positions until finally we were nearly over Bapaume, about a kilometer behind our Front." He does not state the altitude but the surrounding text suggests it was near 1,500 meters—i.e., at 1,500 meters the combatants were approximately 1 km (0.62 miles) behind the front, near Bapaume. However, when calculated using the same 50/35 methodology as before, Richthofen and Hawker would have taken 21 minutes to descend to 1,500 meters, during which

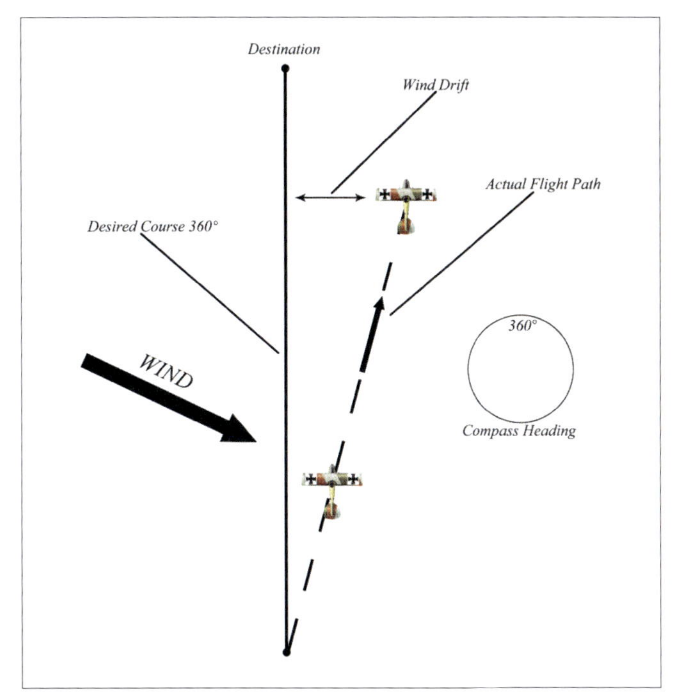

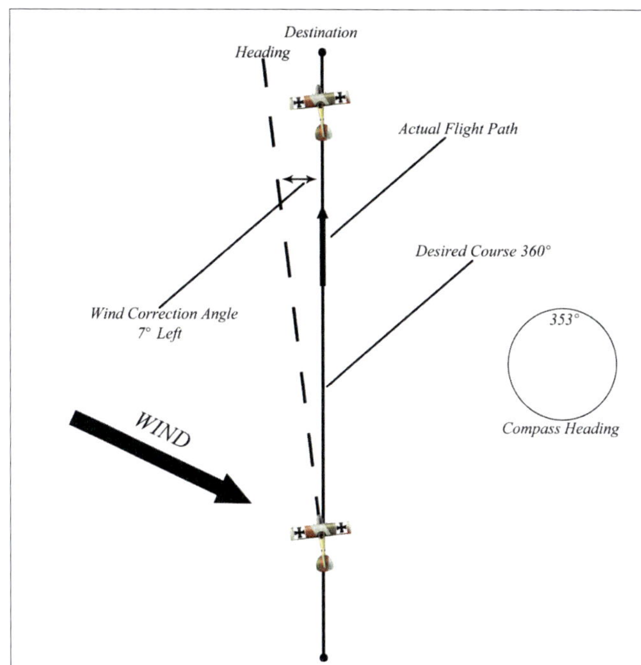

Above: Example of wind drift encountered by a pilot trying to fly a 360 course between the north/south points. Although the pilot consistently holds a compass heading of 360 (magnetic deviation and compass errors omitted for clarity) the wind blows the airplane sideways as it flies, resulting (in this example) in a flight path 15 degrees from the one desired. The higher the wind speed and slower the airplane, the greater the drift.

Above Right: To combat drift, the pilot must fly the airplane into the wind as much as the wind is blowing it off course. In this example the 353 compass heading is west of the desired course—i.e., into the wind—but the resultant flight path matches the desired course. Wind correction angles are calculated via predetermined wind speeds and known airplane performance, although trial-and-error is often and at times wholly employed.

Right: Circling airplanes in a no-wind situation remain stationary above the ground (assuming constant bank angles); i.e., the circling flight path and ground track are the same. The influence of wind is such that the resultant ground speeds on one half of the circle are higher than those on the other half. The pilots are still flying a circular flight path, but the varying wind-driven groundspeeds result in an uneven ground track (looping blue line) as the airplanes drift in the direction of the wind.

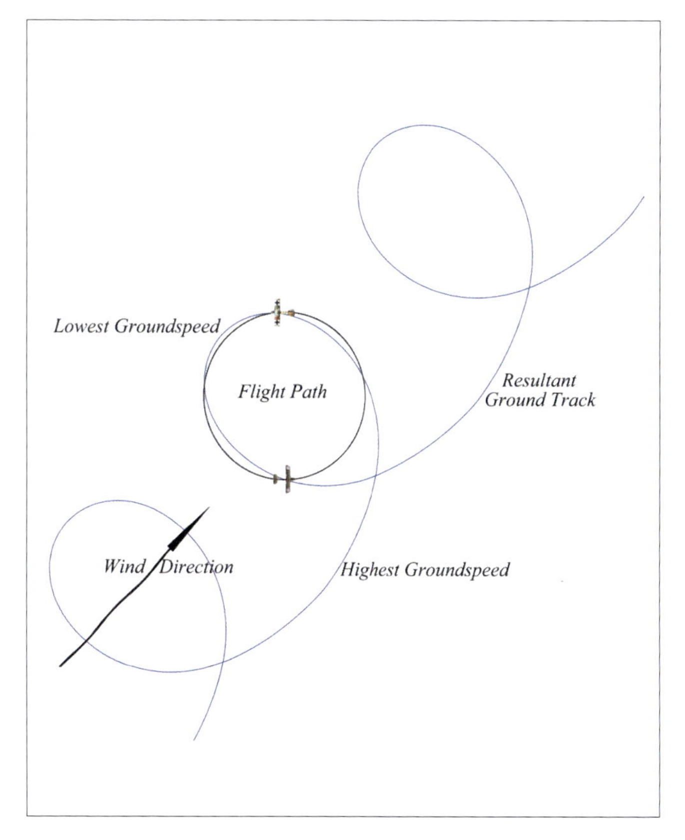

time they would have drifted until just northeast of Quéant, a location approximately nine miles from Bapaume and over ten miles from the front:

Start of Richthofen/Hawker's Dogfight: Above Biefvillers-lès-Bapaume
Circling Descent from 3,000 meters to 1,829 meters (6,000 feet) @ 71 mpm: 16 minutes, 30 seconds
Wind above 6,000 feet: 240° @ 30 mph
Resultant Wind Drift during Descent to 6,000 feet: 16:30 @ 30 mph = 8.3 sm (13 km)
Location when at 6,000 feet: Above Quéant, 8 sm northeast Bapaume
Wind below 6,000 feet: 200° @ 15 mph
Circling Descent from 1,829 meters to 1,500 meters

@ 71 mpm: 4 minutes, 36 seconds
Resultant Wind Drift during Descent to 1,500 meters: 4:36 @ 15 mph = 1.15 sm (1.9 km)
Richthofen/Hawker's Location when at 1,500 meters: 1 sm northeast Quéant, 10.5 sm northeast Bapaume

Andrews' combat report ("...I last saw Maj. Hawker at about 3,000 feet near Bapaume...") matches Richthofen's recollections with respect to Hawker's Bapaume proximity ("...I had forced down my opponent to 500 meters near Bapaume...") and both of their given altitudes reveal Hawker descended continuously "near Bapaume"—i.e., somewhat geographically stationary. Yet 50/35's calculated wind drift would have blown the descending combatants 11 statute miles from Biefvillers-lès-Bapaume to near Villers-lès-Cagnicourt, northeast of Bapaume:
Start of Richthofen/Hawker's Dogfight: Above Biefvillers-lès-Bapaume
Circling Descent from 3,000 meters to 1,829 meters (6,000 feet) @ 71 mpm: 16 minutes, 30 seconds
Wind above 6,000 feet: 240° @ 30 mph
Resultant Wind Drift during Descent to 6,000 feet: 16:30 @ 30 mph = 8.3 sm (13 km)
Location when at 6,000 feet: Above Quéant, 8 sm northeast Bapaume
Wind below 6,000 feet: 200° @ 15 mph
Circling Descent from 1,829 meters to 914 meters (3,000 feet) @ 71 mpm: 12 minutes, 48 seconds
Resultant Wind Drift during Descent to 3,000 feet: 12:48 @ 15 mph = 3.2 sm (5.2 km)
Richthofen/Hawker's Location when at 3,000 feet: Villers-lès-Cagnicourt, 11 sm northeast Bapaume

Andrews' Last View of Hawker

Recall that 50/35's calculations indicate when Andrews last saw Hawker at 3,000 feet he could only have done so at 1424 hours, which is 14 minutes *after* his dead-stick landing in Guillemont. This is a significant timeline discrepancy concerning an observation that occurred presumably while Andrews was still airborne. Yet 50/35's wind drift also eliminates the remote possibility that it occurred after his landing. Villers-lès-Cagnicourt is 16.5 statute miles northeast of Guillemont, a distance at which a DH.2 would appear as a featureless dark speck—if it could be discerned at all. It is unlikely Andrews could identify a dark speck as a *specific* airplane; i.e., *this* dark speck is Hawker's DH.2 5964 and *that* dark speck *is not*.

To appreciate fully the visual challenges, consider 16.5 statute miles is 87,120 feet. Modern commercial airliners fly at altitudes approximately one-third that distance yet via the naked eye it is impossible to distinguish between a cruising Delta B-737 and a Continental B-737, or discern a B-737 from an Airbus A-320—airplanes as similar in appearance to each other as are the DH.2 and FE.8. Decreasing a B-737's 138-foot length to that of a 25-foot DH.2 while increasing the observation some 300% illustrates the unlikelihood of Andrews' airplane make-and-model recognition (DH.2? FE.8? FE.2b?) and individual recognition (Hawker? Saundby?) from over 16 miles away.

Furthermore, trigonometric calculations reveal an airplane 3,000 feet high and 87,120 feet distant has an 87,172-foot slant range distance between it and an observer. This creates a 1.97° elevation angle at which the airplane is visible. Thus, even if Andrews had used field glasses to find Hawker he would have had to train them *horizontally* to do so, rather than up in the sky as would be natural. Anything protruding higher than two degrees above horizontal would have blocked Andrews' line-of-sight—vehicles, trees, buildings, hangars, lay of the land, rubble, smoke, as well as the villages of Lesboefs, le Transloy, Villers-au-Flos, Recquuigny, Beaulencourt, Bancourt, Frémicourt, Beugny, all of which were located on or adjacent to Andrews' line-of-sight between him and Villers-lès-Cagnicourt.

Aerobatics and the Low-Altitude Pursuit

After Richthofen and Hawker spiraled down to 500 meters near Saudemont, Hawker obviously concluded no amount of circling would provide his tactical advantage or escape. Thus he began a series of aerobatics, about which Richthofen recalled: "...he finally had to decide whether to land on our side or fly back to his own lines. Naturally, he attempted the latter, after trying in vain to evade me through looping and such tricks."[51]

Richthofen's recollection clearly indicates these aerobatics were evasive, designed to extricate Hawker from the dogfight and provide as much "head start" separation as possible before dashing for the English lines. Some contend these aerobatics were actually offensive and enabled Hawker to open fire on Richthofen, his bullets "narrowly missing" or "almost killing" Richthofen, depending upon the source read. Richthofen noted no such firing, however, revealing only that after Hawker's aerobatics he "tried to escape flying to the front", not "he then opened fire on me."

The belief the opposite occurred is likely the product of erroneous or misunderstood translation of Richthofen's account. Such was featured in the 1918 English translation of Richthofen's autobiography and reprinted in 1990 as *The Red Air Fighter*:

"Of course he tried the latter, after having endeavored in vain to escape me by loopings and such tricks. At that time his first bullets were flying around me, for so far neither of us had been able to do any shooting."[52]

That translation clearly states Hawker fired on Richthofen. However, rather than stating Hawker's gunfire "narrowly missed" Richthofen, the 1917 and 1933 German editions of *Der Rote Kampfflieger* and 1920's *Ein Heldenleben* agree that, in fact, Richthofen fired at Hawker in the midst of the latter's escape attempt:

"Gradually the good sportsman realized things were too much, and he had to finally decide if he wanted to land near us, or fly back to his lines. Naturally he attempted the latter, after trying in vain to evade me through looping and such tricks. Then followed my first blue beans [*blauen Bohnen*, archaic slang for bullets] around his ears, as until now there was no shooting."

The key lies within the German text *Dabei flogen meine ersten blauen Bohnen* ("my first blue beans flew") *ihn um die Ohren* ("around the ears"). *My* first blue beans. If Hawker had fired on Richthofen at this juncture Richthofen would have used *seine ersten blauen Bohnen — his* first blue beans.

In any event, despite Hawker's aerobatic mastery he could not elude Richthofen. Upon reaching 100 meters Hawker discontinued his "looping and such tricks" and then flew south towards the lines, utilizing a zig-zagging flight path that frustrated the pursuing Richthofen's aim. As Richthofen got closer buffeting caused by the DH.2's downwash and wingtip vortices would have further hampered his accurate gunnery, as would have any low-altitude turbulence—the presence of which is unknown on 23 November, although common with high pressure systems featuring cool evenings followed by clear, sunny days.

Hawker's erratic flight path also eroded his airspeed (due to drag from displaced control surfaces and any yawing flight) and groundspeed (from reduced airspeed coupled with a less than straight flight path), which aided Richthofen's closure. Some contend they *permitted* his closure, coupled with fuel flow troubles and a poorly running engine that "robbed" the DH.2 of full power. Belief in this is certainly curious when one considers only Hawker could have reported such maladies. Yet the speculation is plausible—constant engine malfunctions *plagued* No.24 Squadron's DH.2s. Squadron record books indicate Hawker's DH.2 5964 repeatedly suffered reduced engine power or outright engine failure, most often stemming from fouled plugs, stuck valves and misaligned or broken tappet rods—during one forced landing 5964 was even flown into high tension wires.[53] Regardless, the possibility of 5964's engine problems is moot upon the realization that Hawker's engine was running well enough 23 November to enable his low-altitude aerobatics—and even if flying straight-and-level with a perfectly running engine a flown-hard DH.2 was not going to outrun an Albatros D.II.

Especially when one considers how far 50/35 stipulates Hawker had to fly. Recall that the point at which the combatants reached 500 meters is 12 miles northeast of Bapaume and 13 miles northeast of Ligny-Thilloy, the location of Hawker's crash. True course from near Saudemont to Ligny-Thilloy is 209°. The 15-mph wind was from 200°, almost a direct headwind. Actual airspeeds during the final low-altitude chase are unknown but presuming a DH.2 best-case figure of 92 mph, then flying in a 15-mph headwind resulted in a 77-mph groundspeed (flying 92-mph forward through an airmass moving 15-mph in the opposite direction nets a forward speed of 77-mph across the ground). Thence:

True Course from near Saudemont to Ligny-Thilloy: 209°
Airspeed: 92 mph
Wind: 200° @ 15 mph
Groundspeed: 77 mph
Distance: 13 sm
Time to fly from near Saudemont to Ligny-Thilloy @ 77mph: 10 minutes, 6 seconds. (Applicable to straight-and-level flight only. Zigzagging would increase flight time.)

Fascinatingly, history has ignored these ten minutes, inasmuch as Richthofen vs. Hawker is always regarded as a 35-minute battle. This presumed 35 minutes defines the spiraling portion *only*, after which Hawker initiated his aerobatics. What about the subsequent ten-minute pursuit to Ligny-Thilloy? The recorded weather conditions dictate the 35-minute scenario's 12-mile drift could not occur without it. Thus, a 45 minute fight should have been supposed.

Hawker Under Fire

Since Richthofen did not match Hawker's aerobatics it is likely that after their cessation he required some interval to gain firing position behind the southward fleeing Hawker. Hawker's zig-zagging suggests this interval was brief—if not in position already (i.e. most likely six o'clock, since Richthofen was not a deflection shooter) Richthofen must have been in firing range or on the immediate precipice thereof, otherwise there was no need to initiate aim-spoiling maneuvers. Additionally, during the pursuit

Richthofen suffered gun jams that also required time to remedy. The time necessary to clear these jams was undocumented and thus will never be known but presuming each gun required 15 seconds to clear, then combined with a 15-second interval to gain firing position behind Hawker suggests the ten-minute low-altitude flight exposed Hawker to a nine minute, 16-second firing window (10:06 – 00:45 [3 x 15] = 09:21).

During this window Richthofen recalled "incessantly firing", estimating he shot 900 bullets before killing Hawker. Two guns firing 900 total rounds in nine minutes, 21 seconds would average 96 rounds per minute (rpm) or 1.6 rounds per second (rps), with each individual gun averaging 48 rpm or 0.8 rps. Such a firing rate is well below that of the Maxim lMG 08/15,[54] standard armament of the Albatros D.II, which employed a set synchronized firing rate of 450 rounds per minute. Although this rate was a "set" optimum meant to reduce overheating and jams, this did not mean the LMG 08/15 *always* fired 450 rpm. Propeller revolutions dictated the actual firing rate, which required commensurate rate changes to compensate for various propeller speeds in front of the gun muzzles, thus ensuring the guns fired only when the propeller blades were clear.

The 0.8 rps average firing rate does not illustrate Richthofen's attack accurately—i.e., it is extremely unlikely he could employ the necessary trigger management to fire such an abbreviated burst every second, from weaponry designed and calibrated to fire 900% faster. Even if he could, *why?* It would defeat the weapon's intended use. Rather, Richthofen grouped his shots into "bursts" in order to prevent gun overheating and jams and to provide shot assessment and time enough to re-aim if necessary. Hawker's jinking and zig-zagging suggest Richthofen's bursts durations were three seconds or less but because Richthofen's engine rpms are unknown, the actual synchronized firing rate cannot be calculated. However, illustrative representations of Richthofen's attack intensity can be determined based on these burst durations and the lMG 08/15 450 rpm firing rate:

900 Rounds Fired via Three-Second Bursts

Number of bullets per burst: 45 (two guns), 22.5 (each gun)
Number of 3-sec bursts available if 900 rounds fired: 20 (twenty 45-round bursts = 900 rounds)
Firing Window: 09:21 (561 seconds)
Firing Cadence during 561 seconds: 1 burst commencing every 28 seconds
(20 bursts x 28-seconds = 561 seconds)
Firing status during each 28-second burst cycle:
Firing: 3 seconds (11% of cycle)
Non-firing: 25 seconds (89% of cycle)
Average firing rate: 96 rpm (two guns), 48 rpm (each gun)

The same firing/non-firing percentages (11–89%, respectively) are found within one-second bursts:

900 Rounds Fired via One-Second Bursts

Number of bullets per burst: 15 (two guns), 7.5 (each gun)
Number of 1-sec bursts available in 900 rounds fired: 60 (sixty 15-round bursts = 900 rounds)
Firing Window: 09:21 (561 seconds)
Firing Cadence during 561 seconds: 1 burst commencing every 9.35 seconds
(60 bursts x 9.35 seconds = 561 sec)
Firing status during each 9.35-second burst cycle:
Firing: 1 second (11% of cycle)
Non-firing: 8.35 seconds (89% of cycle)
Average firing rate: 96 rpm (two guns), 48 rpm (each gun)

These figures are illustrative of attack intensity and do not imply Richthofen rhythmically fired a three-second burst and then paused 25 seconds; fired a three-second burst and then paused 25 seconds; etc. The actual firing cadence was certainly less consistent and instead relied upon his ability to counter Hawker's erratic flight path and fire his guns in the brief moments he was able to bring them to bear. Regardless, neither a 96-rpm firing rate nor an 89% non-firing status can be considered "incessant"— i.e., ceaseless; continuing without interruption.

50/3; 50/5

After analyzing the 35-minute scenario, the same methodology was used to analyze the three and five minute scenarios to determine if *they* more accurately represented the events of 23 November. As before, aircraft-turning performance was combined with trigonometric functions and the 50-circle reference datum and then plotted across various airspeeds:

Figure 4's bank angles are much more representative of Richthofen's "both of us flying like madmen in a circle", a description absolutely applicable to pilots holding 82° banks while flying a 46-meter-diameter circle every four seconds. However, its largest radii are only half that of Richthofen's estimated 80 to 100 meters and most if not all its load factors would have exceeded the physiological thresholds of both pilots—they could not have endured +7G's for three consecutive minutes, even if their airplanes could.

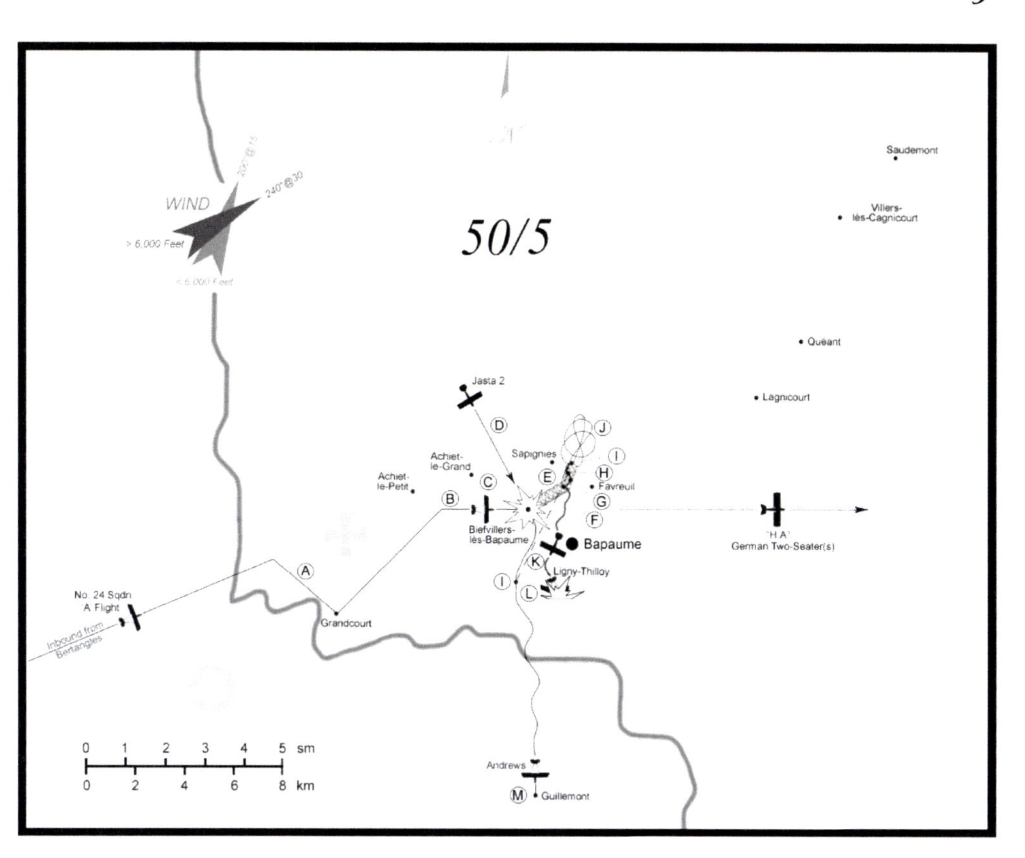

Right: 50/5 Map
A. 'A' Flight power-dives on "rough house going on over Grandcourt."
B. 'A' Flight spots "H.A." two-seaters northeast of Bapaume and pursues same.
C. Andrews detects "two strong patrols of H.A. Scouts" above him.
D. Jasta 2 dives on 'A' Flight
E. Richthofen and Hawker's circling dogfight drifts northeast.
F. 6,000 feet (1,829 meters).
G. 1,500 meters (4,921 feet).
H. Andrews and Hawker's relative positions when Andrews last saw Hawker "at about 3,000 feet (914 meters) near Bapaume.
I. 500 meters (1,640 feet).
J. Hawker's "looping and such tricks."
K. Richthofen and Hawker's two minute-plus low-altitude chase.
L. Hawker KiA, crashes near Ligny-Thilloy.
M. Andrews glides damaged DH.2 across the lines and dead-sticks in Guillemont.

If 50 Circles in 3 Minutes (50/3), Then:

Duration of each circle: 3.6 seconds & Turn Rate: 100°/second

MPH	Kts	Km/h	Turn Rate	Bank Angle	Turn Radius	Load Factor	Stall Speed Increase
90	78	145	100°/sec	82°	76′/23 m	7.18	9%
80	70	129	100°/sec	81°	69′/21 m	6.39	7%
70	61	113	100°/sec	80°	58′/18 m	5.76	5%
60	52	96	100°/sec	78°	51′/16 m	4.81	4%
50	43	80	100°/sec	76°	41′/12 m	4.13	3%

Right: Figure 4.

Comparatively, the bank angles in Figure Five are also representative of Richthofen's "flying like madmen" reference but while its radii are still smaller than those Richthofen estimated, its load factors are much more physiologically tolerable than those found in 50/3. This is significant when one recalls Richthofen was not an aerobatic pilot and therefore without load tolerance beyond that experienced briefly during turns and pullouts, while Hawker's load tolerance eroded due to reduced flight time—he had sortied only 5.6 hours since 1 October and not at all since 20 October.[55] Thus, 50/3's stipulation that Richthofen and Hawker sustained above-average grayout thresholds for at least three minutes is unlikely. Load factors in banked turns (commonly referred to as "Gs" or "G-forces") feature head-to-foot accelerations, which impel blood toward the lower portions of the body and reduce the heart's blood output to the eyes and brain. Diminishing this blood supply disrupts normal function and promotes *grayout* (partial vision loss characterized by reduced color perception and decreased peripheral vision); *blackout* (complete vision loss); muscular uncoordination and unconsciousness. Figure 4's maximum load factor of 7.18Gs far eclipses the average unconsciousness threshold of 5.4Gs, which at a load application of 1G per second would render both men unconscious in less than six seconds and therefore incapable of circling each other for an additional two minutes, 54 seconds. It also exceeds the aircraft structural limits.

Yet the increased *induced drag* (an unavoidable lift byproduct, which during banked turns results from increasing angle of attack to generate more

If 50 Circles in 5 Minutes (50/5), Then:							
Duration of each circle: 6 seconds & Turn Rate: 60°/second							
MPH	Kts	Km/h	Turn Rate	Bank Angle	Turn Radius	Load Factor	Stall Speed Increase
90	78	145	60°/sec	77°	125′/23 m	4.44	110%
80	70	129	60°/sec	75.5°	112′/21 m	4.0	100%
70	61	113	60°/sec	73.5°	98′/18 m	3.52	88%
60	52	96	60°/sec	71°	83′/16 m	3.07	75%
50	43	80	60°/sec	67°	70′/12 m	2.56	60%

Left: Figure 5.

vertical lift to maintain altitude) at these bank angles would have eroded airspeed significantly, making it likely Richthofen and Hawker flew their circles at airspeeds considerably less than 90 mph. However, Figure 4 illustrates if induced drag reduced airspeed to 50 mph, the commensurate load reduction to 4.13Gs would *still* be right at the average physiological grayout threshold of 4.1Gs. Yet Richthofen recalled he "could observe every movement of his [Hawker's] head... If he had not had his flying helmet on, I could have seen what kind of face he made." If so, it is a curious observation from one visually impaired during sustained grayout.

Pilots can avoid these various negative physiological effects by reducing the bank angle. This lessens angle of attack required to maintain altitude, reduces induced drag, decreases stall speed and diminishes load factor — yet places the airplane closer to a straight and level flight attitude best avoided during a dogfight. A more practical solution would be to reduce back-pressure while in a turn. This also reduces the angle of attack and lift sufficient to maintain altitude but allows the airplane to descend while still turning in a banked attitude without the excessive loads—i.e., such as occurred with Richthofen and Hawker's descending spiral.

50/5

While airplane performance and load factor avoidance are primary causes of Richthofen and Hawker's continuous altitude loss in the three *and* five minute scenarios, it is the author's contention that 50/3's too-small radii and physiologically-excessive load factors render the three-minute scenario too unlikely to warrant further consideration. Therefore, hereafter this work will focus on 50/5. Toward that end, descent rates for the five-minute scenario only have been calculated below:

Figure 6 illustrates that a 2,500-meter descent translates into a 500 mpm (1,640 fpm) descent rate. This is much more in line with load-avoidance altitude loss than is 50/35's 71 mpm (273 fpm) rate. Employing this larger rate, the adjacent table illustrates salient descent altitudes and the times necessary to reach them from 3,000 meters:

Recall that Andrews last saw Hawker "at about 3,000 feet near Bapaume." Figure 7 illustrates that Richthofen and Hawker's 2,086 meter (6,844 feet) descent required four minutes, 12 seconds — markedly less (86%) than 50/35's 29 minutes, 24 seconds. Having established that *Jasta* 2 attacked 'A' Flight at 1355, 50/5 stipulates that Andrews must have last seen Hawker at 1359 (1355 + 4 min)—11 minutes *prior* to his dead-stick landing, while still gliding near enough to see Hawker "fighting with an H.A. apparently quite under control but going down."[56] The plausibility that Andrews was better able to see and identify Hawker from this much closer aerial proximity than 50/35's stipulated 16.5 sm (87,120 feet) ground proximity is undeniable.

Reduced Wind Drift

The half-hour difference between the five minute and thirty-five minute scenarios creates their largest discrepancy: wind drift. Recall that weather charts, forecasts and observations for 23 November indicate winds above 6,000 feet were 240° at 30 mph while below 6,000 feet they were 200° at 15 mph. Using this information in conjunction with the Biefvillers-lès-Bapaume start location, wind drift was calculated for a five-minute descent from 3,000 meters:

Start of Richthofen/Hawker's Dogfight: Above Biefvillers-lès-Bapaume
Circling Descent from 3,000 meters to 1,829 meters (6,000 feet) @ 500 mpm: 2 minutes, 18 seconds
Wind above 6,000 feet: 240° @ 30 mph
Resultant Wind Drift during Descent to 6,000 feet: 2:18 @ 30 mph = 1.0 sm (1.67 km)
Location when at 6,000 feet: 1 sm north Bapaume
Wind below 6,000 feet: 200° @ 15 mph
Circling Descent from 1,829 meters to 500 meters @ 500 mpm: 2 minutes, 42 seconds
Resultant Wind Drift during Descent to 500 meters:

Descent Rates if 50 Circles in 5 Minutes (50/5)					
Duration of Each Circle: 6 seconds					
Altitude Start/ Finish	Total Descent	Meters/Minute	Feet/Minute	Meters/Circle	Feet/Circle
3,000–500 m (9,843′–1,640′)	2,500 m (8,203′)	500	1,640	50	164

Above: Figure 6.

Circling Descent from 3,000 m @ 71mpm		
Based on 50/35 Scenario Descent Rate		
From 3,000m to... (From 9,843′ to...)	Total Descent	Descent Duration
2,500 m (8,202′)	500 m (1,640′)	1 min. 0 sec.
2,000 m (6,562′)	1,000 m (3,281′)	2 min. 0 sec.
1,500 m (4,921′)	1,500 m (4,921′)	3 min. 0 sec.
1,000 m (3,281′)	2,000 m (6,562′)	28 min. 0 sec.
914 m (3,000′)	2,086 m (6,844′)	4 min. 12 sec.
500 m (1,640′)	2,500 m (8,202′)	5 min. 0 sec.

Above: Figure 7.

2:42 @ 15 mph = 0.7 sm (1.1 km)
Richthofen/Hawker's Location when at 500 meters: Between Sapignies and Favreuil, 1.8 sm (2.9 km) north of Bapaume

These results mirror Richthofen's combat report ("...I had forced down my opponent to 500 meters near Bapaume..."). A 1.8-mile post-drift location absolutely can be considered "near" Bapaume—certainly more so than 50/35's calculated 12-mile drift to Saudemont. The 1.8-mile drift also dovetails with Richthofen's recollection "we were circling more and more over our positions until finally we were nearly over Bapaume, about a kilometer [0.62 sm] behind our Front." As presented earlier, it is presumed this occurred at 1,500 meters altitude. Based on this, 50/5 stipulates that when the combatants descended to that altitude they would have been approximately 1.25 miles north of Bapaume and 2.25 miles behind the front:

Start of Richthofen/Hawker's Dogfight: Above Biefvillers-lès-Bapaume
Circling Descent from 3,000 meters to 1,829 meters (6,000 feet) @ 500 mpm: 2 minutes, 18 seconds
Wind above 6,000 feet: 240° @ 30 mph
Resultant Wind Drift during Descent to 6,000 feet: 2:18 @ 30 mph = 1.0 sm (1.67 km)
Location when at 6,000 feet: 1 sm north Bapaume
Wind below 6,000 feet: 200° @ 15 mph
Circling Descent from 1,829 meters to 1,500 meters @ 500 mpm: 42 seconds
Resultant Wind Drift during Descent to 1,500 meters: 0:42 @ 15 mph = 0.25 sm (0.4 km)
Richthofen/Hawker's Location when at 1,500 meters: 1.25 sm (2 km) north Bapaume

This distance is further from the front than Richthofen's "about a kilometer" estimation but matches more closely than 50/35's contention that upon reaching 1,500 meters the combatants were nine miles northeast of Bapaume and over ten miles from the front. Nine miles cannot be considered as "near Bapaume" as 1.25 miles.

Recall Andrews' combat report supports Hawker's continuous near-Bapaume proximity ("...I last saw Hawker at about 3,000 feet [914 meters] near Bapaume..."), which matches Richthofen's recollection that "I had forced down my opponent to 500 meters [1,640 feet] near Bapaume". These various altitudes reveal Hawker descended somewhat geographically stationary, always "near Bapaume"—at 3,000 feet Hawker was "near Bapaume" and at 1,640 feet he was *still* "near Bapaume." This contrasts with 50/35's contention that the pair drifted 11 miles to Villers-lès-Cagnicourt by the time they reached 3,000 feet. In actuality, the pair only drifted about one mile north, to just west of Favreuil:

Start of Richthofen/Hawker's Dogfight: Above Biefvillers-lès-Bapaume
Circling Descent from 3,000 meters to 1,829 meters (6,000 feet) @ 500 mpm: 2 minutes, 18 seconds
Wind above 6,000 feet: 240° @ 30 mph
Resultant Wind Drift during Descent to 6,000 feet: 2:18 @ 30 mph = 1.0 sm (1.67 km)
Location when at 6,000 feet: 1 sm north Bapaume
Wind below 6,000 feet: 200° @ 15 mph
Circling Descent from 1,829 meters to 914 meters (3,000 feet) @ 500 mpm: 1 minute, 48 seconds
Resultant Wind Drift during Descent to (3,000 feet): 1:48 @ 15 mph = 0.5 sm (0.8 km)
Richthofen/Hawker's Location when at (3,000 feet):

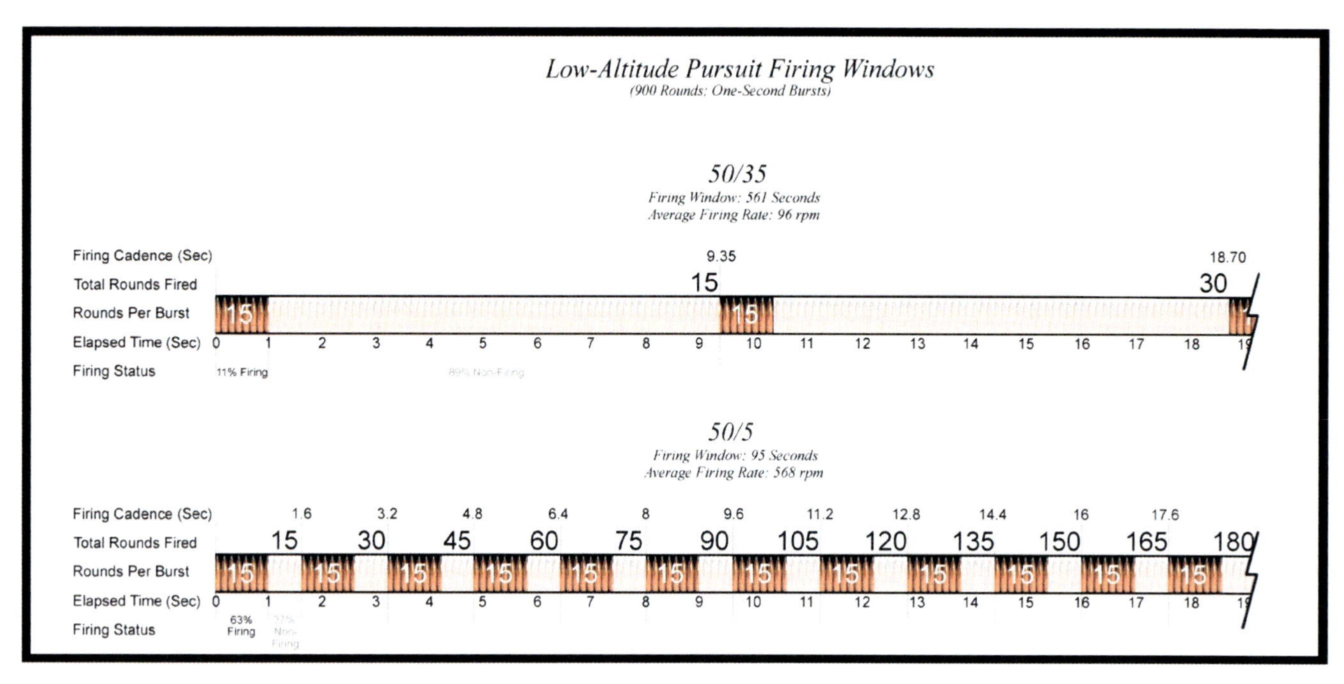

Above: Figure 8.

Just west of Favreuil, approx. 1 sm (1.6 km) north Bapaume

A More Likely Low-Altitude Pursuit

50/35's 29-minute drift placed the combatants 13 miles from Ligny-Thilloy, resulting in a ten-minute low-altitude pursuit. Contrastingly, 50/5's 1.8-mile drift creates a mere three-mile run to Ligny-Thilloy, resulting in a low-altitude pursuit of just under two and a half minutes:

True Course from Sapignies to Ligny-Thilloy: 189°
Airspeed: 92 mph
Wind: 200° @ 15 mph
Groundspeed: 77 mph
Distance: 3 sm
Time to fly from Sapignies to Ligny-Thilloy @ 77 mph: 2 minutes, 20 seconds

As was done with 50/35, 45-seconds had to be deducted from this flight time to illustrate Richthofen's initial pursuit closure and subsequent gun jams. Once so adjusted, the two-minute 20-second low-altitude pursuit sustains a one-minute 35-second firing window—*substantially* less than 50/35's nine-minute, 21-second firing window. The commensurate reduction in firing rate, cadence, and status is significant: two guns firing 900 total rounds in one minute, 35 seconds average 568 rpm or 9.5 rps, with each individual gun averaging 284 rpm or 4.75 rps. This rate is still below the Maxim's set optimum rate of 450 rpm yet significantly higher than 50/35's anemic 48 rpm, greatly increasing Richthofen's attack intensity:

900 Rounds Fired via Three-Second Bursts
Number of bullets per burst: 45 (two guns), 22.5 (each gun)
Number of 3-sec bursts available if 900 rounds fired: 20 (twenty 45-round bursts = 900 rounds)
Firing Window: 01:35 (95 seconds)
Firing Cadence during 95 seconds: 1 burst commencing every 4.75 seconds
(20 bursts · 4.75-seconds = 95 seconds)
Firing status during each 4.75-second burst cycle:
Firing: 3 seconds (63% of cycle)
Non-firing: 1.75 seconds (37% of cycle)
Average firing rate: 568 rpm (two guns), 284 rpm (each gun)

900 Rounds Fired via One-Second Bursts:
Number of bullets per burst: 15 (two guns), 7.5 (each gun)
Number of 1-sec bursts available in 900 rounds fired: 60 (sixty 15-round bursts = 900 rounds)
Firing Window: 01:35 (95 seconds)
Firing Cadence during 95 seconds: 1 burst commencing every 1.58 seconds
(60 bursts x 1.58 seconds = 95 sec)
Firing status during each 1.58-second burst cycle:
Firing: 1 seconds (63% of cycle)
Non-firing: 0.58 seconds (37% of cycle)
Average firing rate: 568 rpm (two guns), 284 rpm (each gun)

Although a 63% firing status is not "incessant" it is certainly more so than 50/35's anemic 11% firing status. The gulf between the two is graphically illustrated in Figure 8.

Even upon first glance the difference between Figure Eight's attack intensities is ostentatious: 50/35's gunnery is virtually non-existent when compared to 50/5's 0.6-second non-firing durations. At over eight seconds between bursts, 50/35's firing cadence creates an 89% non-firing status and limits its total rounds fired to one-fifth those of 50/5 — 20 to 100 rounds, respectively. As noted previously, these firing cadences illustrate attack intensity only and do not represent Richthofen's actual gunnery during the low-altitude pursuit, the determination of which is impossible. However, it is patently obvious that for 900 total rounds fired the smaller the firing window the greater the firing rate, and the greater the firing rate the greater the firing status.

Conclusion

The details of Richthofen and Hawker's epic fight will never be known with certainty. However, after careful evidence analysis and crosschecking weather, combat reports and timelines, it is this author's conclusion that the 50/5 scenario better represents the events of 23 November 1916 than does the 50/35 scenario.

Across the board, the 35-minute scenario matches none of the participant testimonials or timelines. It features docile flight performance parameters more indicative of banner-tow airplanes circling football stadiums than those of dogfighting airplanes. Its radii are too large. Its shallow decent rates easily could have been arrested. Its calculated wind drift greatly exceeds that observed and recorded by Richthofen and "A" Flight. Its 35-minute duration ignores the ten-minute low-altitude pursuit—never is it regarded the fight took *45* minutes—and its 48-rpm per gun firing rate is leisurely compared to Richthofen's self-described "incessant" gunnery. At its core, the scenario purports the premise that Richthofen and Hawker flew wide, lazy circles around each other for over half an hour with neither man able to shoot down the other—an unfathomable supposition.

Contrastingly, the five-minute scenario features steep bank angles and a commensurate altitude loss much more in line with dogfighting airplanes being flown by "madmen". Its radii more closely match Richthofen's estimates. Its wind drift validates eyewitness testimonials stating the combatants continuously remained in the vicinity of Bapaume. Its two-minute low-altitude pursuit supports a firing rate and intensity more in-line with one fighter pilot trying to kill another with machine guns. Its core analysis reveals that when the five-minute descent and two-minute pursuit are coupled with 30 seconds each for Richthofen's initial attack and Hawker's eventual aerobatics, then instead of a 45-minute dogfight drifting 12 miles to near Saudemont, the actual dogfight was more on the order of eight minutes near Bapaume.

Endnotes

1. Shores, Franks, Guest, *Above the Trenches*, (1990), p.52. Shared victory with Lt. Kelvin Crawford.
2. *RFC Communiqué No.1*, 27 July 1915.
3. Illingworth, *A History of 24 Squadron*, p.14.
4. *Ibid.*
5. O'Connor, *Airfields & Airmen, Somme*, (2002), p.87.
6. Meteorological Office Library, London. Diaries of forecasts issued to service units on the Western Front between 24 October 1916 and 30 March 1919.
7. *No.24 Squadron Record Book, Vol. IV*. Saundby's report, 23 November 1916, Air 1/170/15/160/9.
8. *Ibid.*, Saundby's report, 23 November 1916.
9. *Ibid.*, Crutch's report, 23 November 1916.
10. *Ibid.*
11. *Ibid.*
12. Jefford, *RAF Squadrons*, (2001), p.30.
13. *No. 24 Squadron, op.cit.*
14. *Ibid.*
15. Hawker, *Hawker VC*, (1965), p.250.
16. *No. 24 Squadron, op.cit.*, Andrews' report; Saundby's report indicates "1 H.A."
17. *Ibid.*
18. *Ibid.*
19. *Ibid.*
20. *Ibid.*
21. Hawker, *op.cit.*, p.249.
22. *No. 24 Squadron*, Saundby's report, *op.cit.*
23. *Ibid.*
24. *Ibid*, Andrews' report.
25. *Ibid.*
26. Hawker, *op.cit.*, p.249.
27. *No.24 Squadron, op.cit.*, Saundby's report.
28. *Ibid.*
29. *Ibid,*
30. *Ibid.*
31. *Ibid.*
32. Richthofen, *der Rote Kampfflieger*, (1917 ed.), pp.103, 104.
33. Richthofen, *The Red Baron*, (1969 ed.), p.61.
34. *Ibid.*, p.62.
35. Richthofen, (1917), *op.cit.*, p.105.
36. Richthofen, (1969), *op.cit.*, p.61.
37. *Walfisch* is more commonly associated with the LFG Roland C.II, a fast two-seater. However, Saundby's consistent use of the term indicates he was referring to the Albatros D.II, since *Jasta 2* did not fly the Roland C.II.
38. *RFC Communiqué No.64*, 2 December 1916.

39. Richthofen, (1969), *op.cit.*, p.119.
40. *Richthofen Combat Reports* (Translations), Public Records Office, London, (n.d.) (PRO File AIR 1/686/21/13/2250 #212534)
41. Lone single-seater was a Martinsyde G.100.
42. Richthofen, (1969), *op.cit.*, p.62
43. Hunt, Jr., *Aerodynamics for Naval Aviators*, (1960), pp.35, 37.
44. A *function* is a variable so related to another that for each value assumed by one there is a value determined for the other.
45. Enough rudder to overcome adverse yaw and prevent slipping or skidding.
46. Richthofen, (1917), *op.cit.*, p.104.
47. Richthofen, (1969),*op.cit.*, p.61.
48. Exact time is unknown. Times noted in the various combat reports vary by ten minutes; 1355 is the average time.
49. Meteorological Office Library, London. Diaries of forecasts issued to service units on the Western Front between 24 October 1916 and 30 March 1919.
50. Normally, there is a gradual shift in wind velocity with altitude but because of the enormous range of possibilities in that regard this work denotes only the given velocities above and below 6,000 feet.
51. Richthofen, (1969), *op.cit.*, p.62.
52. Richthofen, *The Red Air Fighter*, (1990), pp.101–102.
53. *No.24 Squadron, op.cit.*, Crawford's report, 10 November: "Tappet rod fallen from rocker arm. Plug gone on No. 3 [cylinder]. Landed No. 7 [aerodrome]. Left No. 7 on 12 Nov. 16. Engine conked over Allonville, didn't quite make aerodrome. High tension wire disconnected."
54. Williams, Gustin, *Flying Guns of World War* 1, (2003), p.36.
55. *No.24 Squadron, op.cit.*, 1 October, 1.7 hours; 10 October, 1.6 hours; 20 October, 2.3 hours.
56. *Ibid.*, Andrews' report.

Left: A Halberstadt D.II built under license by Hannover. The D.II was powered by 120 hp Mercedes D.II engine. The Halberstadt fighters were characterized by excellent maneuverability and handling qualities but their performance was inferior to the Albatros.

Below: A Halberstadt D.V, which differed from the D.II in its cabane strut design for better forward visibility and its use of a 120 hp Argus As.II engine.

Victory Details 1917 (Until WiA)

Victory No.16

Above: Sopwith Pup. After shooting down his first Pup, Richthofen wrote "...the enemy plane was superior to ours," no doubt a reference to the Pup's noted maneuverability.

4 January
Sopwith Pup N5193
No.8 Squadron RNAS

Day/Date: Thursday, 4 January
Time: 4.15PM
Weather: Low clouds and rain in the morning; bright in the afternoon.
Attack Location: Near Metz-en-Couture
Crash Location: Near Metz-en-Couture
Side of Lines: Friendly

RFC Communique No.69: "A patrol of the R.N.A.S. Squadron had several combats."

MvR Combat Report: "Sopwith One-seater (lying south of this place), Nr.L.R.T.5193. Motor: 80 H.P. Le Rhone No.5187. A new type plane, never seen before, but as wings broken, badly discernable. Inmate: Lieut. Todd killed, paper and valuables enclosed.

About 4.15PM just starting, we saw above us in 4000 meters altitude 4 planes, unmolested by our artillery. As the archies were not shooting, we took them for our own. Only when they were approaching we noticed they were English. One of the English planes attacked us and we saw immediately, that the enemy plane was superior to ours. Only because we were three against one, we detected the enemy's weak points. I managed to get behind him and shot him down. The plane broke, whilst falling, as under [*sic*—perhaps a typo for 'asunder'].

Frhr. v. Richthofen
Leutnant"

RFC Combat Casualties Report: "Left aerodrome 2.30PM. Last seen with formation during combat with 7 hostile aeroplanes near BAPAUME about 3.15PM."

RNAS A/C:

Make/model/serial number: Sopwith Pup N5193
Manufacturer: Built by Sopwith Aviation Co.,

Above: N5193's serial numbers hang on Richthofen's wall at Roucourt.

Ltd., Canbury Park Road, Kingston-on-Thames
Unit: No.8 Squadron RNAS
Aerodrome: Vert Galant
Sortie: Offensive Patrol
Colors/markings: Generally, PC10 fuselage, with engine cowl and metal fuselage panels unpainted engine-turned metal; wing uppersurfaces PC10 with CDL undersurfaces; black serial number on white rectangle, located near fuselage/empennage junction; likely CDL vertical stab with black Sopwith company markings; r/w/b rudder.
Engine: 80 hp Le Rhone
Engine Number: 5187
Gun: One Vickers, no. L5178
Manner of Victory: Gunfire caused in-flight structural breakup and uncontrolled descent until terrain impact.
Parts Souvenired: a. Fabric with port/stbd serial numbers. Photographed on display at Roucourt, Richthofen's home bedroom in Schweidnitz, and in the post-war Richthofen Museum.
b. Sopwith Aviation Co., Ltd. manufacture placard. Photographed on display at Roucourt and in the post-war Richthofen Museum.

RNAS Crew:
Pilot: Flt. Lt. Allan Switzer Todd, 30, KiA
Cause of Death: u/k. Either gunshot wound(s) or trauma associated with airplane crash.
Burial: Either in-field or laid where fallen; location either unrecorded or lost during turmoil of war.

MvR A/C:
Make/model/serial number: Albatros D.II, possibly 481/16.
Unit: *Jasta* Boelcke
Commander: *Hptm.* Franz Walz
Airfield: Pronville
Colors/markings: See Vic. #7.
Damage: u/k
Initial Attack Altitude: ca. 4,000 meters
Gunnery Range: u/k

Above: N5193's souvenired manufacture placard.

Rounds Fired: u/k
Known Staffel Participants: u/k. At least two others.

Notes:
1. Last victory with *Jasta* Boelcke.
2. Presumed to be last victory flying an Albatros D.II.
3. "Archie" was an allied term for anti-aircraft. Its use in Richthofen's combat report likely was slang-influenced translation and not literally used in Richthofen's original combat report.
4. From the *KOFL 6. Armee Weekly Activity Report* of 14–21 January 1917:
   a. "Organisatorisch Veraenderungen: Die Fuehrung von Jagdstaffel 11 uebernahm am 20.1.17 Lt. Frh. von Richthofen (Jagdstaffel Boelcke). Bisheriger Fuehrer der Jagdstaffel 11, Oberlt. Lang, uebernahm am gleichen Tage die Fuehrung von Jagdst. 28 bei 4. Armee.
   Organizational Changes: The leadership of *Jagdstaffel* 11 was taken over by *Lt. Frh.* von Richthofen (*Jagdstaffel* Boelcke) on 20 January [19]17. Former leader of *Jagdstaffel* 11, *Oblt.* Lang, took over the leadership of *Jagdst.* 28 at 4th army."
   b. *"Besonderes: Dem Lt. Frh. von Richthofen, Jagdstaffel 11 wurde am 21.1.17 der* Orden Pour le Mérite *verliehen fuer erfolgreichen anerkannten Abschuss 16 feindlicher Flugzeuge.*

   Special [note]: *Lt. Frh.* von Richthofen, *Jagdstaffel* 11 was awarded [with] the order *Pour le Mérite* on 21 January [19]17 for successful [and] confirmed downing of 16 enemy aircraft."

Victory No.16 Statistics:
Single-Seaters
1st of 2 January single-seater victories
1st of 15 1917 single-seater victories

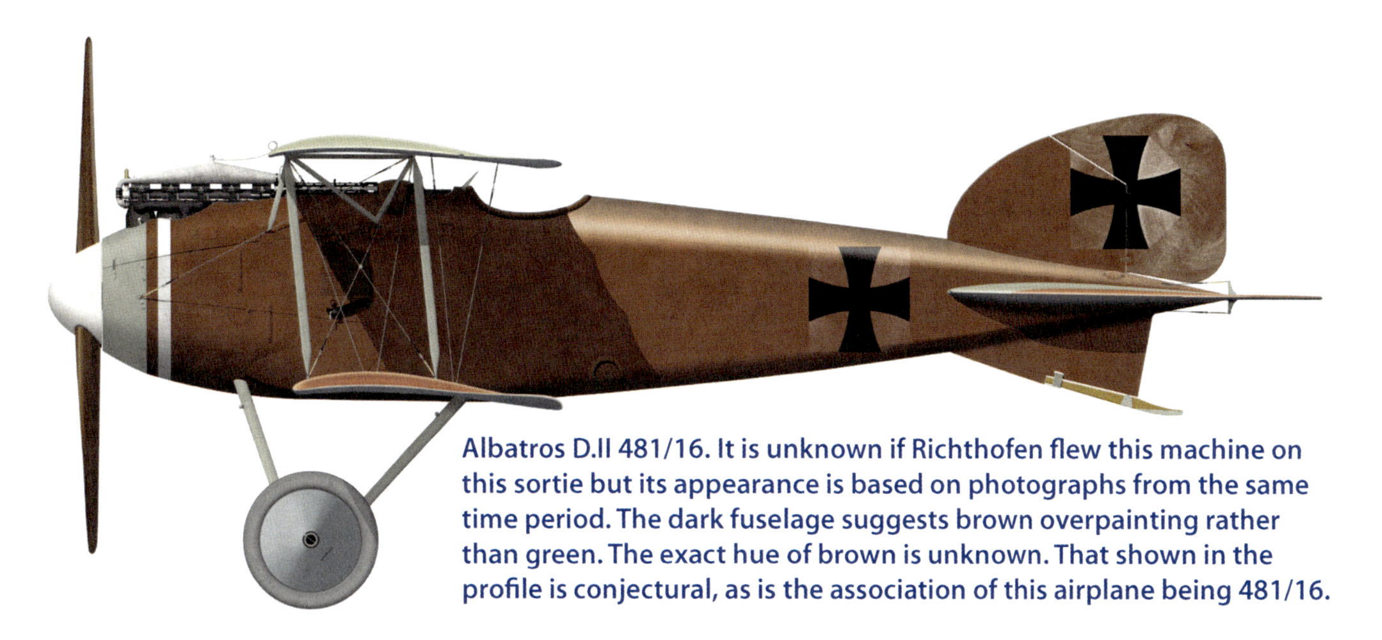

Albatros D.II 481/16. It is unknown if Richthofen flew this machine on this sortie but its appearance is based on photographs from the same time period. The dark fuselage suggests brown overpainting rather than green. The exact hue of brown is unknown. That shown in the profile is conjectural, as is the association of this airplane being 481/16.

9th of 18 pre-wound single-seater victories
9th of 35 total single-seater victories

Sopwith Pup
1st and only January Sopwith Pup victory
1st of 2 1917 Sopwith Pup victories
1st of 2 total Sopwith Pup victories

No.8 Squadron
1st of 2 No. 8 Squadron victories

Deaths
No.8 Squadron:
1st and only January No.8 Squadron crewman KiA
1st of 2 1917 No.8 Squadron crewmen KiA
1st of 2 total No.8 Squadron crewmen KiA

Single-Seaters:
1st of 2 January single-seater crewmen KiA
1st of 9 1917 single-seater crewmen KiA
7th of 12 pre-wound single-seater crewmen KiA
7th of 21 total single-seater crewmen KiA

All:
1st of 2 total January 1917 crewmen KiA
1st of 53 total 1917 crewmen KiA
18th of 66 pre-wound crewmen KiA
18th of 84 total crewmen KiA

Etc.
1st of 3 total January victories
1st of 48 total 1917 victories
16th of 57 total pre-wound victories
16th victory in Albatros D.I or D.II
4th of 4 single-seater victories flying from Pronville
5th of 5 total victories flying from Pronville
4th of 4 single-seater victories flying under Walz
5th of 5 total victories flying under Walz

Above: Richthofen and Wolff (right) next to a Rumpler C.I.

Victory No.17

Left: Royal Aircraft Factory FE.8. Similar in design to the AMC DH.2, a salient difference between them is FE.8 booms tapered longitudinally to join the horizontal stabilizer, while DH.2 booms tapered laterally to join the horizontal stabilizer (upper booms) and rudder post (lower booms, which met in a vee).

23 January
FE.8 6388
No.40 Squadron RFC

Day/Date: Tuesday, 23 January
Time: 4.10PM
Weather: Fine all day
Attack Location: West of Lens
Crash Location: Near Lens
Side of Lines: Enemy

RFC Communique No.72: "2nd Lieut. Hay, No.40 Squadron, engaged a hostile machine over La Bassée. The wings were seen to break away from the German machine before it crashed. 2nd Lieut. Hay then attacked and destroyed the leading machine of a formation of eight. He was then brought down himself and killed."

KOFL 6. Armee Weekly Activity Report: "Erfolge im Luftkampf: Am 23.1. 1 engl. Flugzeug durch Lt. Frh. von Richthofen, J. St. 11, ueber Lens in Brand geschossen um jenseits der Linie zum Absturz gebracht. (17. Flugzg.)

Successes in Aerial Combat: On 23 January 1 English aircraft shot ablaze over Lens by *Ltn. Frh.* von Richthofen, *Jagdstaffel* 11, and brought down to crash across the lines (17th aircraft)."

MvR Combat Report: "Trenches above southwest of Lens. No details, plane dropped on enemy's side.

About 4.10PM. I attacked together with seven of my planes enemy squad. west of Lens. The plane I had singled out caught fire after 150 shots discharged from a distance of 50 meters. The plane fell, burning. Occupants [*sic*] fell out of plane in 500 meters height.

Immediately after the plane had crashed on ground, I could see a heavy black smoke cloud arising. The plane burnt for quite a while with frequent flares of flame.

Frhr. v. Richthofen.
Acknowledged."

RFC Combat Casualties Report: "Left aerodrome 1.12PM. Brought down near LENS at 3.50pm in combat with 5 hostile machines. Report received from "C" A.A. Batt. that at 3.50PM an F.E.8 was attacked by three Germans and seen to go down in flames the other side of LENS."

RFC A/C:
Make/model/serial number: Royal Aircraft Factory FE.8 6388
Manufacturer: Subcontracted to and built by the Darracq Motor Engineering Co., Ltd., Townmead Road, Fulham, London, S.W.6
Unit: No.40 Squadron RFC
Aerodrome: Treizennes
Sortie: Line Patrol & Photographic Escort
Colors/markings: PC10 nacelle with large white "4" forward the cockpit, with borderless roundel adjacent cockpit; wing/aileron/horizontal

Above: John Hay in FE.8 6388. Note the flexible Lewis machine gun and attached bag to catch spent cartridge casings.

stabilizer/elevator uppersurfaces PC10 with CDL undersurfaces; battleship gray tail booms with unpainted wooden struts; PC10 vertical stabilizer; r/w/b rudder with black serial number on a black-bordered white background. Description based on photograph of the machine.
Engine: 100 hp Gnome Monosoupape
Engine Number: 30445 WD 4166
Gun: One Lewis, no. 14510
Manner of Victory: Uncontrolled decent in flames until terrain impact. Pilot fell/jumped at 500 meters and was found by Canadian troops two miles southeast of Aix Noulette.
Parts Souvenired: none

RFC Crew:
Pilot: 2nd Lt. John Hay, 28, KiA

Above: John Hay.

Pilot Victories: 3 (destroyed)

1916

Victory#	Date	Details
1	22 Oct. 1916	*Oblt.* Linck, *Staffelführer, Jasta* 10
2	23 Jan. 1917	*FA* 240, crew Mohs/Riehl, at Provin-La Bassee
3	23 Jan. 1917	*FA* 13, crew Wasselewski/Schwarz

Cause of Death: u/c. Likely blunt force trauma from post free-fall terrain impact.
Burial: Aire Communal Cemetery, Grave I. H. 7.

MvR A/C:
Make/model/serial number: Albatros D.III, likely *Le Petit Rouge*, serial number u/k. (See notes.)
Unit: *Jasta* 11
Commander: *Ltn.* Manfred von Richthofen
Airfield: La Brayelle
Colors/markings: *Le Petit Rouge* – Red spinner, cowl, fuselage, empennage, wheel covers, and cabane/interplane/landing gear struts. The exact shade of red is unknown but likely the same or

Above: Closeup of the nameplate on Hay's propeller cross. (Australian War Memorial)

Left: Propeller cross erected on Hay's first grave. (Australian War Memorial)

Above: Hay's headstone. (Chris and Jean Cosgrove)

similar to that applied to *Jasta* 11 Ltn. Georg Simon's Albatros D.III 2015/16, forced down with a damaged engine 4 June 1917 and captured by the British, who described its mostly red paint as "bright red, either vermillion with crimson lake or geranium lake only." The uppersurfaces of the upper wings were factory camouflaged Venetian red, pale green, and olive green, sloped left. Undersurfaces were light blue. Crosses were bordered white without cross fields; those on bottom wing undersurface were not bordered. All fuselage and empennage insignia were overpainted with a red "wash" through which the crosses remained discernable, although the Albatros

Damage: u/k

Initial Attack Altitude: u/k

Gunnery Range: 50 meters

Rounds Fired: 150

Known Staffel Participants: u/c

Notes:

1. No.40 Squadron Lt. Benbow reported being engaged head-on by a red machine during this engagement.
2. Richthofen's first victory in an Albatros D.III.
3. Richthofen's first victory with *Jasta* 11.
4. A cornucopia of theories abounds regarding *when* Richthofen painted his Albatros D.III red. *Why* he painted it red ties in with what Lothar said about Manfred's reason to overpaint his Albatros D.I and D.II while with *Jasta* 2: "When Manfred began to gain his first successes with *Jagdstaffel* Boelcke, he was annoyed because he felt he was much too visible to his enemies in aerial combat and that they saw him much too early. He tried using a variety of colors to make himself invisible. At first he emphasized the earth colors. From above one would not detect these colors if there were no movement, which is of course impossible in a plane. To his sorrow, Manfred found that no one color was useful in the air. There is no camouflage for the flier with which he can make himself invisible. Then, in order to at least be recognized as the leader by his comrades in the air, he chose the color bright red. Later the red machine also became known to the English as *'Le petit rouge'* and the other names that accompanied it."

Some researchers indicate that *Jasta* Boelcke received the first Albatros D.IIIs in late December/early January and that Richthofen had

Albatros D.III, *Le Petit Rouge*. Its appearance is based on descriptions, anecdotes, and photographs from the time period.

been assigned one, which he subsequently took with him when he transferred to *Jasta* 11; a unit equipped primarily with the Halberstadt D.V. If so, did Richthofen paint his new Albatros D.III red while still with *Jasta* Boelcke? One theory suggests Richthofen had painted his D.II red even *before* the arrival of the D.III. However, if so painted for leader recognition why would a red airplane also require pennants for recognition, as McCudden witnessed on the Albatros he fought 27 December? This would have been redundant and something Richthofen did not do with any of his Albatrosses in *Jasta* 11, judging by their absence in every photograph of his machines—although (as will be seen) at least one of his partially-red Fokker triplanes used streamers for a brief period, though more for enemy taunting than friendly recognition. In any event, McCudden said nothing about the airplane he fought being red, which would have been a noteworthy detail at that stage of the war. Thus, is seems unlikely Richthofen's Albatros D.II 481/16 was painted red, at least as of 27 December 1916.

Regardless, a supporting theory for Richthofen's flying a red D.III with *Jasta* Boelcke is that Richthofen needed time for his red machine to become well known to the enemy by the time he flew with *Jasta* 11, because after Victory No.18 on 24 January (as will be seen), the captured British crew stated they were familiar with the red airplane (Richthofen wrote, "...my red painted plane is not unknown to them..."). Furthermore, they were the ones who coined the nickname "Le Petit Rouge." If Richthofen did not bring an Albatros D.III with him to *Jasta* 11, then the first D.III he could have procured arrived at that unit on 21 January. This would have given him only three days to have the machine painted, have the oil-based paint dry, and then let him fly it around frequently enough to become "known" along the front for having a red airplane (although Note 1 above indicates No.40 Squadron reported being attacked by at least one red machine 23 January). Some believe there was not enough time to become so known.

However, this belief is based on supposition that Richthofen's bright red Albatros D.III was well known by **all** the RFC units in that area, which is a bit of a stretch. According to the crew of FE.2b 6997 downed on 24 January (see Victory No. 18), they recalled, "we had previously seen it [the red Albatros D.III], but we did not know who it was." I.e, they *personally* had seen it before and were *personally* familiar with it. "We" being the FE.2b crew in question, not the whole of the RFC. At that stage of the war, even one sighting of a red airplane easily could have created a lasting impression of familiarity with one crew.

Therefore, this work asserts Richthofen painted his Albatros D.III after his arrival with *Jasta* 11 on 16 January 1917, whether he brought a D.III with him from *Jasta* Boelcke or procured one with his new unit. This ties in well with the reason that Richthofen chose red was "to be recognized as the leader by his comrades in the air." In *Jasta* 11 he was no longer just the *de facto* leader in the air as he had been with *Jasta* 2. He was now the *Staffelführer*—the appointed commander, leader in the air and on the ground—of a lackluster unit that needed much polishing and training to meet Richthofen's standards. This was to be done hands-on, and the importance of leader recognition is ostensible. However, while well plausible that Richthofen painted his Albatros D.III red after joining *Jasta* 11, be advised this conclusion remains conjectural.

Victory No.17 Statistics:

Pushers*

1st of 2 January pusher victories
1st and only January single-seater pusher victory
1st of 2 1917 single-seater pusher victories
5th of 6 total single-seater pusher victories
10th of 18 total pusher victories

Single-Seaters

2nd of 2 January single-seater victories
2nd of 15 1917 single-seater victories
10th of 18 pre-wound single-seater victories
10th of 35 total single-seater victories

FE.8 Δ

1st and only January FE.8 victory
1st and only 1917 FE.8 victory
1st and only total FE.8 victories

No.40 Squadron ±

1st and only No.40 Squadron victory

Deaths †

No.40 Squadron:
1st and only January No.40 Squadron crewman KiA
1st and only 1917 No.40 Squadron crewman KiA
1st and only total No.40 Squadron crewmen KiA

Single-Seaters:

2nd of 2 January single-seater crewmen KiA
2nd of 9 1917 single-seater crewmen KiA
8th of 12 pre-wound single-seater crewmen KiA
8th of 21 total single-seater crewmen KiA

All:

2nd of 2 total January 1917 crewmen KiA
2nd of 53 total 1917 crewmen KiA
19th of 66 pre-wound crewmen KiA
19th of 84 total crewmen KiA

Etc.

2nd of 14 victories behind enemy lines
2nd of 3 total January victories
2nd of 48 total 1917 victories
17th of 57 total pre-wound victories
1st victory in an Albatros D.III
1st of 6 single-seater victories flying from La Brayelle
1st of 28 total victories flying from La Brayelle
1st of 9 single-seater victories as *Staffelführer*
1st of 39 total victories flying as *Staffelführer*

Statistics Notes

* *All pusher victories are pre-wound*
Δ *All FE.8 victories are pre-wound*
± *All No.40 Squadron victories are pre-wound*
† *All No.40 Squadron deaths are pre-wound; KiA includes DoW*

Left: Albatros D.II cockpit. Dual-grip control column contains two machine gun triggers at center and the primary throttle on the left side; an auxiliary throttle was located on the port side of the cockpit forward, out of view in this photograph. Note the overall glossy sheen of the fuselage wood, the diminutive windshield, a rear-view mirror at top, and flares stored externally on the port side of the cockpit.

Above: Richthofen in front of an Albatros D.III(OAW). Although adorned in Jasta 11 markings, Richthofen has no known victories flying this model of Albatros.

Below: Another photo of Richthofen and the Albatros D.III(OAW), taken within moments of the previous photo. The white square on the fuselage is the rigging diagram. Note the man climbing down from the cockpit is doing so with neither ladder nor footstep and is kneeling on the cockpit rim, demonstrating the strength of the wood in this area.

Victory No.18

Above: FE.2b 6997's souvenired serial number on display in the Richthofen Museum.

Above: FE.2b 6997's souvenired Beardmore engine placard.

24 January
FE.2b 6997
No.25 Squadron RFC

Day/Date: Wednesday, 24 January
Time: 4.10PM
Weather: Fine all day
Attack Location: Near Vimy
Crash Location: Near Vimy
Side of Lines: Friendly

RFC Communique No. 72: "882 photographs were taken during the day. 2nd Lieut. J. L. Leith and 2nd Lieut. W.D. Matheson, No. 25 Squadron, engaged and drove down out of control, a hostile machine which had forced one of our aeroplanes to land in German territory."

KOFL 6. Armee Weekly Activity Report: "Am 23.1. 1 engl. F.e. DD [Doppeldecker]. *160 PS. Durch Lt. Frh. von Richthofen, J. St. 11, bei Vimy zur Landung gezwungen. (18 Flugzeug) Besatzung 2 engl. Offiziere gefangen, Flugzeug verbrannt.*

On 23 January 1 English. F.E. biplane. 160 PS. Forced to land near Vimy by Lt. Frh. von Richthofen, *Jasta* 11. (18th aircraft) Crew 2 English officers captured, aircraft burned."

MvR Combat Report: "Fixed motor. Plane No.6937 [*sic*]. Motor No.748. Occupants: Pilot, Capt. Craig [*sic*]. Lieut. McLennan.

Accompanied by Sergt.Howe I attacked about 12.15PM the commanding plane of an enemy formation. After a long fight I forced adversary to land near Vimy. The inmates burnt plane after landing. I myself had to land, as one wing had been cracked in 300 meters. I was flying Albatross [sic] D.III.

According to the English inmates my red painted plane was not unknown to them, as when being asked who had brought them down they answered: '*Le petit Rouge*'.

Two machine-guns have been seized by my *Staffel*. The plane was not worth being removed, as it was completely burned.

Frhr. v. Richthofen
Was acknowledged."

RFC Combat Casualties Report: "Left aerodrome at 9.50AM. Brought down by hostile scout near VIMY. Seen to make a good landing."

RFC A/C:
Make/model/serial number: Royal Aircraft Factory FE.2b 6997
Manufacturer: Subcontracted to and built by Boulton & Paul, Ltd., Norwich.
Unit: No.25 Squadron RFC
Aerodrome: Lozinghem
Sortie: Photography. ROUVROY.
Colors/markings: Presentation a/c *Punjab No.28*. Generally, PC10 nacelle with one, two, or three horizontal white-bordered black bands across nose (corresponding with A, B, or C Flight); likely white "PUNJAB 28" on sides of nacelle; wing uppersurfaces PC10 with CDL undersurfaces; likely upperwing roundels had white surrounds.
Engine: 160 hp Beardmore
Engine Number: 748
Guns: Two Lewis.
Manner of Victory: Damage to fuel tank/engine/propeller precipitated long controlled glide and successful forced landing, after which crew burned

Albatros D.III, *Le Petit Rouge*. Its appearance is based on descriptions, anecdotes, and photographs from the time period.

the machine.
Items Souvenired: a. 6997 port/starboard sides of rudder. Black numbers/white borders on r/w/b and b/w/r backgrounds. Photographed on display at Roucourt, Richthofen's home bedroom in Schweidnitz, and in the post-war Richthofen Museum.
b. Beardmore Co. Ltd engine placard, no. 748. Photographed on display at Roucourt and the post-war Richthofen Museum.
c. Two machine guns taken for *Jasta* 11.

RFC Crew:
Pilot: Capt. Oscar Greig, 27, WiA/PoW
Manner of Injury: Machine gun bullet to the right ankle
Repatriation: 12 December 1918. Was not repatriated but walked away from PoW camp post-armistice and made his way back to England, arriving 22 December.
Obs: Lt. John Eric MacLennan, 20, PoW
Repatriation: 2 January 1919.
RFC Combat Casualties Report: "Officially reported prisoners of war."

MvR A/C:
Make/model/serial number: Albatros D.III, likely *Le Petit Rouge*, serial number u/k. (See notes.)
Unit: *Jasta* 11
Commander: *Ltn.* Manfred von Richthofen
Airfield: La Brayelle
Colors/markings: See Vic. #17
Damage: In-flight structural compromise of lower wing which precipitated a cautionary/emergency landing: "As I shot down my eighteenth, one of my wings broke in two during the air battle at three-hundred-meter altitude. It was only through a miracle that I reached the ground without going kaput."
Initial Attack Altitude: u/k
Gunnery Range: u/k
Rounds Fired: u/k
Known Staffel Participants: Fw Hans Howe

Notes:
1. MvR wrote of this victory in *Der rote Kampfflieger* but there are several discrepancies between that account, his combat report, and other reports/recollections of the combatants. In his autobiography Richthofen wrote that after damaging the FE.2 ("two-seat Vickers") he felt the crew had been wounded and experienced "deep compassion for my opponent and decided not to send him plunging down" (he wrote the FE.2 eventually burst into flames before reaching the ground), after which he experienced "at about five-hundred-meter altitude, a malfunction in my machine during a normal glide [that] forced me to land before making another turn." He then described landing his Albatros D.III nearby amongst some barbed wire near the FE.2 and then overturning, after which he spoke with Greig and MacLennan personally ("I enjoyed talking with them") about this "careless" landing and learned from them that his red Albatros was known as "*Le Petit Rouge*." Richthofen's combat report (written immediately after the event and not dictated some four months later) states Greig and MacLennan actually set fire to their plane themselves on the ground ("the inmates burnt plane after landing") and that his Albatros wing "cracked" at 300 meters, an altitude which dovetails with what Richthofen wrote in a personal letter to his mother 27 January ("one

of my wings broke in two during the air battle at three-hundred-meter altitude"). His combat report also states "according to the English inmates my red painted plane was not unknown to them, as when being asked who had brought them down they answered: 'Le petit Rouge.'"

However, Floyd Gibbons's postwar interview with MacLennan indicates he never spoke with Richthofen: "As regards the red machine, we had previously seen it, but we did not know who it was. I am glad to hear that he had to land, as I did not know this." Unless MacLennan is lying it is almost without question that had such an event and subsequent conversation about that event occurred, one would retain the memory of speaking with the man who had just shot one down and then crash-landed nearby. Richthofen's combat report agrees that the Englishmen knew of his red plane but it does not state specifically that this knowledge was gleaned first hand via personal conversation. This suggests Richthofen learned of the name "*Le petit Rouge*" either through conversations with the soldiers who had captured and spoken with Greig and MacLennan and then relayed the information to Richthofen (perhaps as he later scavenged the wreckage of 6997 for souvenirs), or by similar second-hand means. Whether Richthofen actually up-ended his Albatros upon landing—which is mentioned neither in his combat report nor personal letters, as are other similar events—or the event was included as autobiographical embellishment is uncertain and open for conjecture.

2. MacLennan revealed Richthofen continually attacked the gliding 6997 (always from its six o'clock low) until "the machine was but a few hundred feet from the ground," corroborating the relatively low altitude at which Richthofen experienced structural failure and contradicting *Der Rote Kampfflieger's* claim of "deep compassion" for the enemy. Rather, it is another example of his no-quarter *modus operandi.*
3. Due to the structural problems of the Albatros D.III, which led to *Idflieg* grounding the make/model on 27 January, Richthofen began flying a Halberstadt D-type (see Victory No. 19).

Victory No.18 Statistics:

Pushers*

2nd of 2 January pusher victories
1st and only January two-seater pusher victory
1st of 7 1917 two-seater pusher victories
6th of 12 total two-seater pusher victories
10th of 18 total pusher victories

Two-Seaters

1st and only January two-seater victory
1st of 33 1917 two-seater victories
8th of 39 pre-wound two-seater victories
8th of 45 total two-seater victories

FE2 Δ

1st and only January FE.2 victory
1st of 7 1917 FE.2 victories
6th of 12 total FE.2 victories

No.25 Squadron ±

1st of 4 No.25 Squadron FE.2 victories
1st and only January No.25 Squadron victory
1st of 4 1917 No.25 Squadron victories
1st of 4 total No.25 Squadron victories

Wounded

No.25 Squadron:
1st and only January No.25 Squadron crewman WiA
1st and only 1917 No.25 Squadron crewman WiA
1st and only total No.25 Squadron crewmen WiA

Pushers:

1st and only January two-seater pusher crewman WiA
1st of 4 1917 two-seater pusher crewmen WiA
1st of 5 total two-seater pusher crewmen WiA

Two-Seaters:

1st and only January two-seater crewman WiA
1st of 11 1917 two-seater crewmen WiA
2nd of 16 pre-wound two-seater crewmen WiA
2nd of 18 total two-seater crewmen WiA

All:

1st and only total January crewmen WiA
1st of 19 total 1917 crewmen WiA
3rd of 18 total pre-wound crewmen WiA
3rd of 26 total crewmen WiA

Etc.

3rd of 3 total January victories
3rd of 48 total 1917 victories
18th of 57 total pre-wound victories
2nd victory in an Albatros D.III
1st of 22 two-seater victories flying from La Brayelle
2nd of 28 total victories flying from La Brayelle
1st of 30 two-seater victories as *Staffelführer*
2nd of 39 total victories flying as *Staffelführer*

Statistics Notes

* *All pusher victories are pre-wound*
Δ *All FE.2 victories are pre-wound*
± *All No.25 Squadron victories are pre-wound*

Victory No.19

Above: Halberstadt D.V.

1 February
BE.2d 6742
No.16 Squadron RFC

Day/Date: Thursday, 1 February
Time: 4PM
Weather: Overcast morning, but fine for the remainder of the day.
Attack Location: Over trenches 1 kilometer southwest of Thelus
Crash Location: "Barbed wire of our first lines"
Side of Lines: Friendly

KOFL 6. Armee Weekly Activity Report: "Am 1.2.17 ein engl. B.E. Zweisitzer 1 km s.w. Thelus durch Lt. Frh. v. Richthofen, Fuehrer von J. St. 11 abgeschossen. (als 19.). Besatzung: Fuehr. Captain Murray [sic], *Beobachter Lt. McBar* [sic] *tot. Zugehoerigkeit nicht festzustellen.*

On 1 February 17 one English B.E. two-seater 1 kilometer southwest of Thelus shot down by *Lt. Frh.* v. Richthofen, leader of *Jagdstaffel* 11 (as 19th). Crew: Pilot Captain Murray [*sic*], observer Lt. McBar [*sic*] dead. Affiliation [with a specific squadron] not detectable."

MvR Combat Report: "BE Two-Seater. Occupants: Lieut. Murray – Lieut. McBar [*sic*], both wounded and died on February 2nd. Plane No. 6742.

About 4PM I spotted, flying with Lieut.Allmenroeder, in 1800 meters height an artillery flyer. (BE two-seater). I managed to approach him within 50 yards [*sic*] apparently unnoticed, with my Halberstaedter machine.
From this distance up to only the length of a plane I fired 150 shots. The enemy plane then went down in large, uncontrolled right hand curves, pursued by Allmenroeder and myself. The plane crashed into the barbed wire of our first lines. The occupants were both wounded and were made prisoners by the infantry. It is impossible to remove the plane.

Frhr. v. Richthofen
Was acknowledged."

RFC Combat Casualties Report: "Left aerodrome at 2.30PM. Centre Group 3rd Can.Div1.Art. report one of our machines shot down and landed in (51b) A.10.d at 3.10PM as result of fight between 3 British and 4 hostile machines. Machine destroyed by shell fire 20 minutes later."

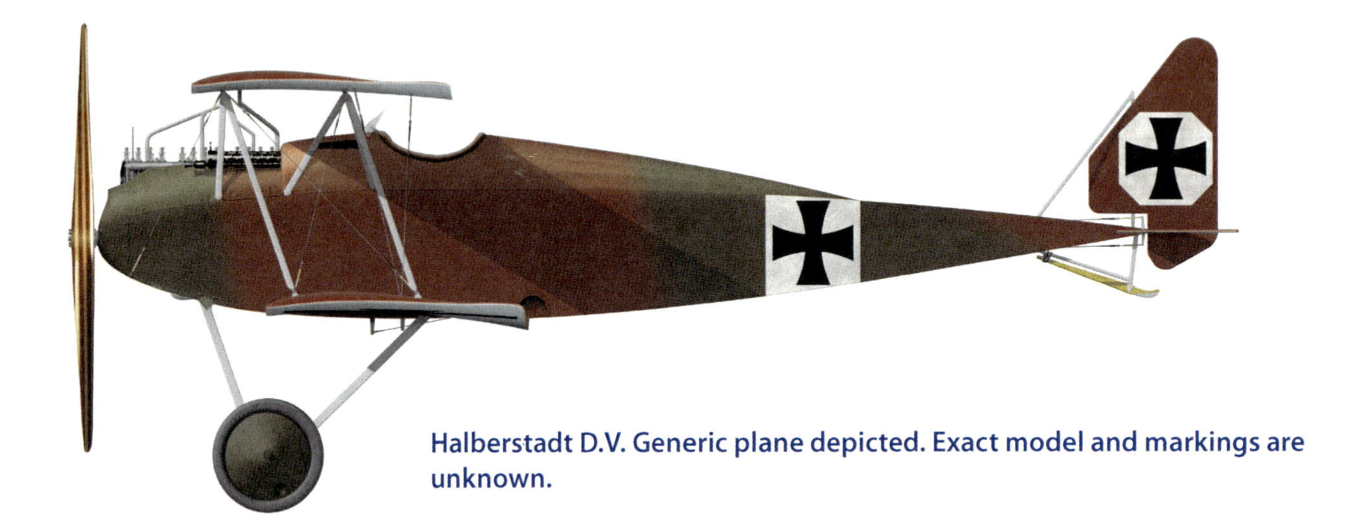

Halberstadt D.V. Generic plane depicted. Exact model and markings are unknown.

RFC A/C:

Make/model/serial number: Royal Aircraft Factory BE.2d 6742
Manufacturer: Subcontracted to and built by The Vulcan Motor & Engineering Co. (1906), Ltd., Crossens, Southport.
Unit: No.16 Squadron RFC
Aerodrome: Bruay
Sortie: Photography
Colors/markings: u/c. Generally, either overall CDL wings/ailerons/vertical and horizontal stabilizers/elevators or the same in PC10; battleship gray engine cowl; b/w/r rudder; black serial number on vertical stabilizer. No.16 Squadron markings were two vertical fuselage bands astride roundel (black on CDL machines, white on PC10).
Engine: 90 hp R.A.F. 1a
Engine Number: 521 WD 1884
Guns: E633; E17453
Manner of Victory: Clockwise spiral descent until crash landing.
Items Souvenired: Likely none due to shelling.

RFC Crew:

Pilot: Lt. Percival William Murray, 20, PoW/DoW 2.2.17
Cause of Death: u/c, likely gunshot wound(s)
RFC Combat Casualties Report: "An unconfirmed report in a German publication states that Lt. P.W. Murray was killed."
Burial: Bois-Carré British Cemetery (Thelus), France, Grave I. D.
Obs: Lt. Duncan John McRae, 24, PoW/DoW 2.2.17
Cause of Death: u/c, likely gunshot wound(s)
Burial: Bois-Carré British Cemetery (Thelus), France, Grave I. D.

MvR A/C:

Make/model/serial number: Halberstadt D-type, model and serial number u/k
Unit: *Jasta* 11
Commander: *Ltn.* Manfred von Richthofen
Airfield: La Brayelle, France
Colors/markings: Likely reddish-brown/olive green camouflaged fuselage, empennage, wings and horizontal stabilizer uppersurfaces. Clear-doped fabric or light blue undersurfaces. Light grey struts. Fuselage and upper-wing crosses on square white cross fields; rudder cross on octagonal white cross field.
Damage: n/k

Initial Attack Altitude: 1,800 meters
Gunnery Range: From 50 meters to a "plane length"
Rounds Fired: 150
Known Staffel Participants: Ltn. Karl Allmenröder

Notes:

1. First victory flying a Halberstadt D-type.
2. It is unknown which Halberstadt D-type Richthofen flew during this period. Popular consensus has been it was a D.III but more likely it was a D.V, based on that model's overwhelming numerical superiority in *Jasta* 11 during Jan/Feb 1917. However, since the *Staffel* also was equipped with one Halberstadt D.II and one D.III until 28 February, on any sortie he could have flown the Halberstadt D.II, D.III or D.V.

Victory No.19 Statistics:

Two-Seaters

1st of 3 February two-seater victories
2nd of 33 1917 two-seater victories
9th of 39 pre-wound two-seater victories
9th of 45 total two-seater victories

Above: Murray's headstone, 2011.

Above: McRae's headstone, 2011.

BE.2 ∆
1st of 3 February BE.2 victories
1st of 15 1917 BE.2 victories
3rd of 17 total BE.2 victories

No.16 Squadron ±
1st of 6 No.16 Squadron BE.2 victories
1st and only February No.16 Squadron victory
1st of 6 1917 No.16 Squadron victories
1st of 6 total No.16 Squadron victories

Deaths †
No.16 Squadron:
1st and 2nd of 2 February No.16 Squadron crewmen KiA
1st and 2nd of 12 1917 No.16 Squadron crewmen KiA
1st and 2nd of 12 total No.16 Squadron crewmen KiA

Two-Seaters:
1st and 2nd of 3 February two-seater crewman KiA
1st and 2nd of 44 1917 two-seater crewmen KiA
12th and 13th of 54 pre-wound two-seater crewmen KiA
12th and 13th of 63 total two-seater crewmen KiA

All:
1st and 2nd of 3 total February crewmen KiA
3rd and 4th of 53 total 1917 crewmen KiA
20th and 21st of 66 total pre-wound crewmen KiA
20th and 21st of 84 total crewmen KiA

Etc.
1st of 3 total February victories
4th of 48 total 1917 victories
19th of 57 total pre-wound victories
1st victory in a Halberstadt
2nd of 22 two-scater victories flying from La Brayelle
3rd of 28 total victories flying from La Brayelle
2nd of 30 two-seater victories as *Staffelführer*
3rd of 39 total victories flying as *Staffelführer*

Statistics Notes
∆ All BE.2 victories are pre-wound
± All No.16 Squadron victories are pre-wound
† All No.16 Squadron deaths are pre-wound; KiA includes DoW

Victory No.20

Above: 2nd Lt. Herbert Arthur Croft's headstone, 2011.

14 February
BE.2d 6231
No.2 Squadron RFC

Day/Date: Wednesday, 14 February
Time: 12 noon
Weather: Fine
Attack Location: West of Loos
Crash Location: "Wreckage landed in the fire zone."
Side of Lines: Friendly

RFC Communique No. 75: "Sixty-nine targets were dealt with by aeroplane observation."

KOFL 1. Armee Weekly Activity Report: "1 Uhr N [Nachmittag]. *ein engl. DD-Zweisitzer, östl. Loos innerhalb unserer Linien durch Lt. Frhr.v.Richthofen (als 20).*

1 o'clock in the afternoon an English biplane two-seater, [brought down] east of Loos within our lines by *Lt. Frh.* von Richthofen (as 20th)."

MvR Combat Report: "BE Two-seater. Occupants: one killed, the other heavily wounded. Name of pilot: Lieut. Bonnet [*sic*] (died). No details concerning plane, as wreckage landed in the fire zone

After flying back from conference with the *Staffel* Boelcke, I spotted an enemy artillery flyer at a height of 2,000 meters west of Loos. I attacked the enemy and approached him unnoticed by some 50 meters. After several hundred shots the plane dashed down falling into our trenches. The pilot was killed in the air, observer seriously injured when landing.

Frhr. v. Richthofen"

RFC Combat Casualties Report: "Left aerodrome at 9.45AM. Machine was observed by Lt. E.M Lugard to be attacked by H.A., and was last seen over CITE ST.AUGUSTE descending in a quick spiral."

RFC A/C:
Make/model/serial number: Royal Aircraft Factory BE.2d 6231
Manufacturer: Subcontracted to and built by Ruston, Proctor & Co., Ltd., Lincoln.
Unit: No.2 Squadron RFC
Aerodrome: Hesdigneul
Sortie: Artillery Observation
Colors/markings: u/c. Generally, either overall CDL wings/ailerons/vertical and horizontal stabilizers/elevators or the same in PC10; battleship gray engine cowl; b/w/r rudder; black serial number on vertical stabilizer. No.2 Sqn markings were a triangle aft of the roundel and sometimes on the turtledeck (black on CDL machines, white on PC10).
Engine: 90 hp R.A.F. 1a
Engine Number: 21996 WD 516
Guns: Two Lewis. Nos. 6098, 3913.
Manner of Victory: Rapid spiral descent until terrain impact.
Parts Souvenired: n/k

RFC Crew:
Pilot: 2nd Lt Cyril Douglas Bennett, 19, WiA/PoW
Manner of Injury: Hip injury and basilar skull fracture resulting from airplane crash.

Halberstadt D.V. Generic plane depicted. Exact model and markings are unknown.

Unconscious two or three days; no memory of the event.
RFC Combat Casualties Report: "Officially reported wounded prisoner of war (3rd Echelon Part II Orders, AFO 1810 dated 4/8/17.)"
Imprisonment: Ulm, Saarbrücken, Hannover, Holzminden.
Repatriation: 6 December 1918
Obs: 2nd Lt. Herbert Arthur Croft, KiA
Cause of Death: Gunshot wound(s), perhaps exacerbated by trauma associated with airplane crash. Pulled from wreckage alive but died shortly thereafter.
RFC Combat Casualties Report: "Officially accepted as having been shot dead during an aerial combat on 14/2/17 on the evidence of 2/ Lieut. C.D.Bennett, a prisoner of war in Germany. (3rd Echelon Part II Orders, A.F.O.1810, dated 23/6/17)."
Burial: Cabaret-Rouge British Cemetery (Souchez), France, Grave VII. H. 11.

MvR A/C:
Make/model/serial number: Either a Halberstadt D-type, mode/serial number u/k; or Albatros D.III, poss. *Le Petit Rouge*
Unit: *Jasta* 11
Commander: *Ltn.* Manfred von Richthofen
Airfield: La Brayelle
Colors/markings: Halberstadt D.V – See Vic. #19. *Le Petit Rouge* – See Vic. #17.
Damage: n/k
Initial Attack Altitude: 2,000 meters
Gunnery Range: 50 meters
Rounds Fired: "Several hundred."
Known Staffel Participants: n/a, lone sortie

Notes:

1. Richthofen does not state the make/model/ serial number of the airplane he flew during this sortie. Identification is tenuous because *Idflieg's* grounding of all Albatros D.IIIs was still in effect and remained so until 19 February, until wing inspections, reinforcements and load tests had demonstrated a more satisfactory level of structural integrity. However, it seems that once an Albatros was repaired or reinforced it was returned to flight status on an individual basis, *prior* to the official rescinding of the grounding. This is based on RFC records that reveal several Albatros D.IIIs were shot down and captured while the grounding was still in effect. Regarding Richthofen's Albatros D.III, by the 14 February sortie it had been three full weeks since *Le Petit Rouge* suffered its "cracked" wing. Although speculative it is well possible and even likely that during those 21 days its failed lower wing had been replaced and the other wing reinforced, or possibly also replaced.
   Support favoring Richthofen flying an Albatros is based on his description of firing "several hundred" shots, which if one considers "several hundred" to be at least 700 would exceed the 500-bullet capacity of the Halberstadt. Yet "several hundred" is an inexact figure dependent upon an individual's perception of the definition, so it cannot be used to exclude use of a Halberstadt. Regardless, this work leans toward Richthofen's use of an Albatros D.III.
2. Richthofen's comment that the BE.2d pilot had been killed and observer wounded, the reverse of what happened, was perhaps due to the BE.2 observer cockpit being situated forward of the pilot's.

Albatros D.III, *Le Petit Rouge*. Its appearance is based on descriptions, anecdotes, and photographs from the time period.

Victory No.20 Statistics:

Two-Seaters
2nd of 3 February two-seater victories
3rd of 33 1917 two-seater victories
10th of 39 pre-wound two-seater victories
10th of 45 total two-seater victories

BE.2 Δ
2nd of 3 February BE.2 victories
2nd of 15 1917 BE.2 victories
4th of 17 total BE.2 victories

No.2 Squadron
1st of 4 No.2 Squadron BE.2 victories
1st of 2 February No.2 Squadron victories
1st of 4 1917 No.2 Squadron victories
1st of 5 total No.2 Squadron victories

Deaths
No.2 Squadron:
1st and only February No.2 Squadron crewman KiA
1st of 3 1917 No.2 Squadron crewmen KiA
1st of 5 total No.2 Squadron crewmen KiA

Two-Seaters:
3rd of 3 February two-seater crewman KiA
3rd of 44 1917 two-seater crewmen KiA
14th of 54 pre-wound two-seater crewmen KiA
14th of 63 total two-seater crewmen KiA

All:
3rd of 3 total February crewmen KiA
5th of 53 total 1917 crewmen KiA
22nd of 66 total pre-wound crewmen KiA
22nd of 84 total crewmen KiA

Wounded
No.2 Squadron:
1st of 2 February No.2 Squadron crewmen WiA
1st of 3 1917 No.2 Squadron crewmen WiA
1st of 3 total No.2 Squadron crewmen WiA

Two-Seaters:
1st of 2 February two-seater crewman WiA
2nd of 11 1917 two-seater crewmen WiA
2nd of 16 pre-wound two-seater crewmen WiA
2nd of 18 total two-seater crewmen WiA

All:
1st of 2 total February crewmen WiA
2nd of 19 total 1917 crewmen WiA
4th of 18 total pre-wound crewmen WiA
4th of 26 total crewmen WiA

Etc.
2nd of 3 total February victories
5th of 48 total 1917 victories
20th of 57 total pre-wound victories
3rd victory in an Albatros D.III
3rd of 22 two-seater victories flying from La Brayelle
4th of 28 total victories flying from La Brayelle
3rd of 30 two-seater victories as *Staffelführer*
4th of 39 total victories flying as *Staffelführer*

Statistics Notes
Δ *All BE.2 victories are pre-wound*

Victory No.21

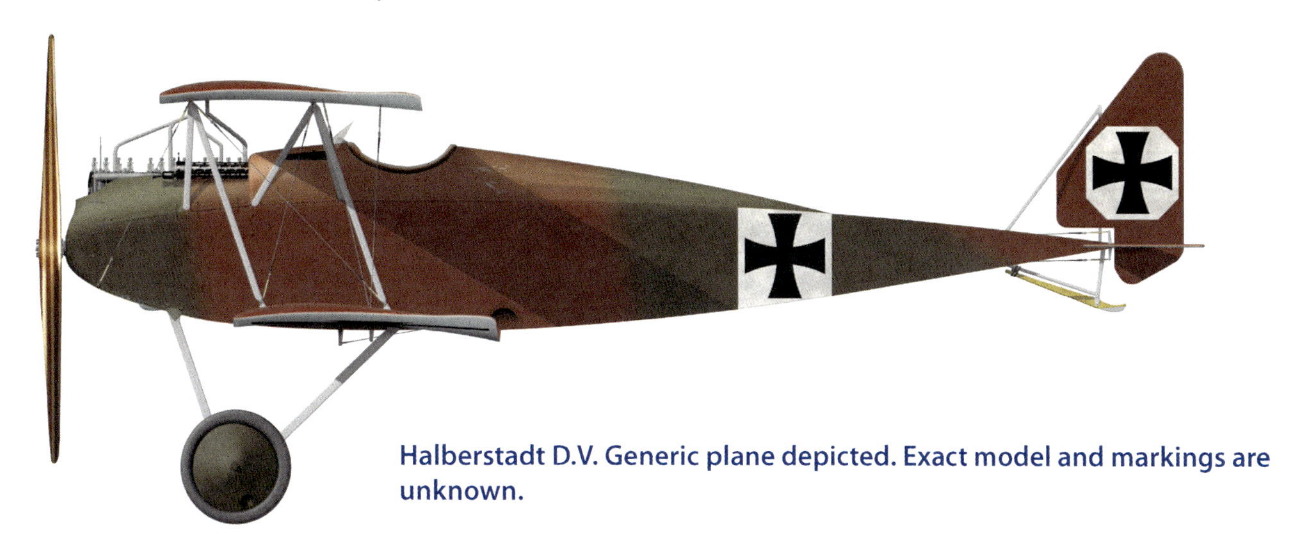

Halberstadt D.V. Generic plane depicted. Exact model and markings are unknown.

14 February
BE.2c 2543
No.2 Squadron RFC

Day/Date: Wednesday, 14 February
Time: 4.45PM
Weather: Fine, misty later.
Attack Location: Near Lens
Crash Location: 1,500 meters southwest of Mazingarbe.
Side of Lines: Enemy

KOFL 6. Armee Weekly Activity Report: *"Ausserdem 5 Uhr N. ein weiteres engl. Flugzeug im Loosbagen, jenseits der Linien (als 21) durch Lt. Frhr. von Richthofen abgeschossen.*

Also 5 o'clock in the afternoon another English aircraft shot down in the Loos-bend, across the lines (as 21st) by *Lt. Frh.* von Richthofen."

MvR Combat Report: "BE Two-seater. No details, as plane landed on enemy's side.

About 4.45PM I attacked with my *Staffel* of 5 planes artillery flyers in low altitude near Lens. Whilst my gentlemen attacked a second BE, I attacked the one flying nearest me. After the first 100 shots the observer stopped shooting. The plane began to smoke and twisted in uncontrolled curves to the right. As this result was not satisfactory to me, especially over the enemy's line, I shot at the falling plane until the left part of the wings came off. As the wind was blowing at a velocity of 20 yards a second [41 mph], I had been drifting far over to the enemy's side. Therefore I could observe that the enemy plane touched the ground southwest of Mazingarbe. I could see a heavy cloud of smoke in the snow arising from where the plane was lying.

As it was foggy and already rather dark I have not witnesses either from the air or from the earth.

Frhr. v. Richthofen
Was acknowledged."

RFC A/C:
Make/model/serial number: Royal Aircraft Factory BE.2c 2543
Manufacturer: Subcontracted to and built by Wolseley Motors, Ltd., Adderley Park, Birmingham.
Unit: No.2 Squadron RFC
Aerodrome: Hesdigneul
Sortie: Artillery registration
Colors/markings: u/c. Generally, either overall CDL wings/ailerons/vertical and horizontal stabilizers/elevators or the same in PC10; battleship gray engine cowl; b/w/r rudder; black serial number on vertical stabilizer. No.2 Sqn markings were a triangle aft of the roundel and sometimes on the turtledeck (black on CDL machines, white on PC10).
Engine: 90 H.P. R.A.F. 1a
Engine Number: u/k
Guns: u/k
Manner of Victory: Rapid clockwise spiraling descent.
Parts Souvenired: None.

RFC Crew:
Pilot: Capt. George Cyril Bailey, 26 (DoB 15 July

Albatros D.III, *Le Petit Rouge*. Its appearance is based on descriptions, anecdotes, and photographs from the time period.

1890), WiA
RFC Combat Casualties Report: "Slightly wounded in leg in aerial combat."
Manner of Injury: Gunshot wound in knee.
Obs: 2nd Lt. George William Betts Hampton, 30 or 31 (DoB 1886)

MvR A/C:
Make/model/serial number: Either a Halberstadt D-type, mode/serial number u/k; or Albatros D.III, poss. *Le Petit Rouge*
Unit: *Jasta* 11
Commander: *Ltn.* Manfred von Richthofen
Airfield: La Brayelle
Colors/markings: Halberstadt D.V – See Vic. #19. *Le Petit Rouge* – See Vic. #17.
Damage: n/k

Initial Attack Altitude: ca. 6,800 feet (according to Bailey and Hampton)
Gunnery Range: u/k
Rounds Fired: Over 100
Known Staffel Participants: "Five planes," identities unknown

Notes:

1. Although credited as Richthofen's 21st victory, 2543 was not shot down. During his descent Bailey noted the enemy had disengaged, allowing him to recover and return to base.
2. Upon being wounded Bailey had entered a spiral dive to escape, during which Richthofen's gunfire knocked the observer's rear gun overboard. This is believed to be what Richthofen perceived as pieces of the BE.2's wings being shot away and explains why the BE.2's return gunfire was limited ("...after the first 100 shots the observer stopped shooting..."). Although uncertain, it is likely falling darkness hampered Richthofen's visibility to discern clearly the events below, where an expectation of victory coupled with dark objects and smoke on the ground conspired to be confirmation of a crash.
3. *Idflieg's* grounding of the Albatros D.III was still in effect, yet make/model identification is tenuous. (See Vic. #20 Notes.)

Victory No.21 Statistics:

Two-Seaters
3rd of 3 February two-seater victories
4th of 33 1917 two-seater victories
11th of 39 pre-wound two-seater victories
11th of 45 total two-seater victories

BE.2Δ
3rd of 3 February B.E.2 victories
3rd of 15 1917 B.E.2 victories
5th of 17 total B.E.2 victories

No.2 Squadron
2nd of 4 No.2 Squadron B.E.2 victories
2nd of 2 February No.2 Squadron victories
2nd of 4 1917 No.2 Squadron victories
2nd of 5 total No.2 Squadron victories

Wounded
No.2 Squadron:
2nd of 2 February No.2 Squadron crewmen WiA
2nd of 3 1917 No.2 Squadron crewmen WiA
2nd of 3 total No.2 Squadron crewmen WiA

Two-Seaters:
2nd of 2 February two-seater crewman WiA
3rd of 11 1917 two-seater crewmen WiA
3rd of 16 pre-wound two-seater crewmen WiA

3rd of 18 total two-seater crewmen WiA

All:
2nd of 2 total February crewmen WiA
3rd of 19 total 1917 crewmen WiA
5th of 18 total pre-wound crewmen WiA
5th of 26 total crewmen WiA

Etc.
3rd of 14 victories behind enemy lines
3rd of 3 total February victories
6th of 48 total 1917 victories
21st of 57 total pre-wound victories
4th victory in an Albatros D.III
4th of 22 two-seater victories flying from La Brayelle
5th of 28 total victories flying from La Brayelle
4th of 30 two-seater victories as *Staffelführer*
5th of 39 total victories flying as *Staffelführer*

Statistics Notes
Δ *All BE.2 victories are pre-wound*

Above: Aerial view of La Brayelle aerodrome. This expansive pre-war airfield served many German units and was *Jasta* 11's home from October 1916 until mid-April 1917. After Richthofen's tenure there began in January 1917 he amassed 28 credited victories. (DEHLA Collection)

Victory No.22

Albatros D.III, *Le Petit Rouge*. Its appearance is based on descriptions, anecdotes, and photographs from the time period.

4 March
BE.2d 5785
No.2 Squadron RFC

Day/Date: Sunday, 4 March
Time: 12.50PM
Weather: Fine.
Attack Location: One kilometer north of Loos
Crash Location: n/a "In front of our trenches."
Side of Lines: n/a

RFC Communique No. 78: "1,187 photographs were taken during the day."

KOFL 6. Armee Weekly Activity Report: "12.50 Nach. 1. Fdl. B.E. DD noerdl. Loos durch Lt.Frh. v.Richthofen, Fuehrer J.St.11.

12.50 in the afternoon 1 enemy B.E. biplane [shot down] north [of] Loos by *Lt. Frh.* von Richthofen, leader of *Jagdstaffel* 11."

MvR Combat Report: "B.E. Two-seater. Details unknown, plane dropped on enemy's side.

I had started all by myself and was just looking for my squad when I spotted a single B.E. My first attack was apparently a failure as the adversary tried to escape by curves and dives. After having forced my adversary downwards from 2800 to 1200 meters, he imagined himself safe and flew once more straight on. I took advantage of this, put myself behind him and fired some 500 shots at him. My adversary dived, but in such a steep way that I could not follow.

According to our infantry observations the plane crashed to the ground in front of our trenches.

Frhr. v. Richthofen
Was acknowledged."

RFC A/C:

Make/model/serial number: Royal Aircraft Factory BE.2d 5785
Manufacturer: Subcontracted to and built by the British and Colonial Aeroplane Co., Ltd., Filton, Bristol.
Unit: No.2 Squadron RFC
Aerodrome: Hesdigneul
Sortie: Photo Op
Colors/markings: u/c. Generally, either overall CDL wings/ailerons/vertical and horizontal stabilizers/elevators or the same in PC10; battleship gray engine cowl; b/w/r rudder; black serial number on vertical stabilizer. No. 2 Sqn markings were a triangle aft of the roundel and sometimes on the turtledeck (black on CDL machines, white on PC10).
Engine: 90 H.P. R.A.F. 1a
Engine Number: u/k
Guns: u/k
Manner of Victory: Machine gun fire caused BE.2 to dive "in such a steep way that I could not follow."
Parts Souvenired: none

RFC Crew:
Pilot: Lt. James Benjamin Evelyn Crosbee, 20 (DoB 11 February 1897)
Obs: Sgt. John Edward Prance, 33 (DoB 3 November 1884), WiA
RFC Combat Casualties Report: "Wounded in leg by H.A. Struck off strength."
Manner of Injury: Gunshot wound to leg.

MvR A/C:
Make/model/serial number: Most likely an Albatros D.III, possibly *Le Petit Rouge*, serial number u/k
Unit: *Jasta* 11
Commander: *Ltn.* Manfred von Richthofen
Airfield: La Brayelle
Colors/markings: See Vic. #17.
Damage: n/k
Initial Attack Altitude: 2,800 meters
Gunnery Range: u/k
Rounds Fired: 500+
Known Staffel Participants: n/a, lone sortie.

Notes:
1. Richthofen's combat reports submit this claim as his 23nd victory.
2. Although he received confirmation of this victory, in actuality 5785 was able to escape and return to base. Richthofen's combat report does not state he saw the BE.2 hit the ground. Although speculative, he could have had a reasonable "might have" assumption of this having happened, based on 5785's steep dive. If such an assumption occurred it is unknown if he would have filed a claim based upon that. However, upon returning to base it is obvious inquiries were made and he was informed that "according to our infantry observations the plane crashed to the ground in front of our trenches." Therefore it seems these ground eyewitness observations provided the final basis for the claim.
3. Richthofen is often associated with flying a Halberstadt throughout March. Seemingly corroborating this at least in part is the crew of 5785 reported being attacked by a Halberstadt. But Richthofen's combat report states his first attack (which wounded Prance) had been "a failure" and that as he re-attacked he fired "some 500 shots." Although Richthofen's account estimates his rounds fired it suggests these combined attacks expended more than 500 bullets. This exceeds the 500 round capacity of the single gun Halberstadt Ds but falls well within the 1,000 round capacity of the twin gunned Albatros D.III.

Additionally, *Idflieg's* grounding of Albatros D.IIIs had been rescinded on 19 February, nearly two weeks before this encounter and 39 days after Richthofen's wing failure of 24 January. It is doubtful that with the renewed availability of the repaired Albatros D.IIIs that he would have continued flying a Halberstadt, a make and model he held in lesser standing than the Albatros. As Ferko's *Richthofen* reveals, on 22 December 1916 Richthofen attended a conference in Cambrai regarding the future of fighter development and was the lone voice supporting the development of higher powered machines rather than the 120 hp Halberstadt. In a report the following year he directly compared the Albatros and Halberstadt D-types and although he had criticism for the Albatros, clearly he preferred it to the Halberstadt: *"Aussicht: Aussicht oben, unten, seitlich muss einwandfrei sein. Albatros D III gut; Alb. D II besonders nach unten schlech. Alb. D I nach oben im Kurvenkampf nicht zu gebrauchen. Beim Halberstadter D ist oberes Tragdeck zu nahe an den Augen und genau in Augenhöhe. Etwas weiter ab vom Auge und etwas höher ist für der Luftkampf angenehm.*

View: View up, down, sideways must be impeccable. Albatros D.III good; Alb.D II especially downwards bad. Alb. D I [the view] up [is] useless in turn fights. On the Halberstadt D the upper wing is too close to the eyes and just at eye level. A bit further off from the eye and a bit higher is comfortable for air combat."

***"Sturzflüge:** Einen senkrechten Sturzflug von über 1000 m, dabei eine Verwindungskehrtwendung muss die Maschine unbedingt aushalten; bei Albatros D III nicht immer der Fall. Maschine muss Oberdruck auf Flächen aushalten / Verwindungs-Looping/; auch verunglückte Loopings und vielfache scharfe, überrissene Kurven mit voll laufendem Motor. Bei Sturzflügen ist die Möglichkeit, grosse Fahrt zu erreichen, dringend notwending. Ein langsamer Sturzflug, wie beim Halberstädter, meistens zwecklos.*

Nosedive: The craft must tolerate a vertical dive over more than 1000 m including a half roll under all circumstances; for Albatros D III [this is] not always the case. Aircraft must bear top pressure on the wings/ aileron-looping/ even failed loopings and numerous tight turns with a full-speed engine. The potential to gain fast speed in dives is most essential. A slow dive, as with the Halberstadt, is mostly useless."

Furthermore, Ferko lists that by 28 February *Jasta* 11 had exchanged their complement of Halberstadts for Albatros D.IIIs. With the six Albatros D.IIIs that arrived in January, the unit received at least another six Albatros D.IIIs between 25–27 February. One cannot dismiss the possibility of a Halberstadt straggler, but it appears fairly certain that by March *Jasta* 11 was an Albatros *Staffel*. Even if for some reason *Le Petit Rouge's* wings weren't repaired/replaced by March, it is unlikely that Richthofen would have flown any lesser-powered/armed/regarded Halberstadt straggler with such an availability of better powered/armed/regarded Albatros machines. Still, was mindfulness of the Albatros's recent wing trouble and recollection of his personal close-call with same the reason why he could not follow 5785's dive?

4. The crew of 5785 reported that MvR's first attack was from head-on, out of the sun. It is common knowledge that attacking from within the sun's blinding glare is the paramount attack methodology. However, as will be seen, later in his career Richthofen wrote an Air Combat Operations Manual in which he stated: "I consider it to be very dangerous to attack a two-seater from the front. In the first place, one seldom encounters the opponent [this way]. One almost never makes him incapable of fighting [this way]." It seems his attack on 5785 at least in part provided the experience upon which those comments are based.

Victory No.22 Statistics:

Two-Seaters
1st of 7 March two-seater victories
5th of 33 1917 two-seater victories
12th of 39 pre-wound two-seater victories
12th of 45 total two-seater victories

BE.2 Δ
1st of 5 March BE.2 victories
4th of 15 1917 BE.2 victories
6th of 17 total BE.2 victories

No.2 Squadron
3rd of 4 No.2 Squadron BE.2 victories
1st of 2 March No.2 Squadron victories
3rd of 4 1917 No.2 Squadron victories
3rd of 5 total No.2 Squadron victories

Wounded
No.2 Squadron:
1st and only March No.2 Squadron crewmen WiA
3rd of 3 1917 No.2 Squadron crewmen WiA
3rd of 3 total No.2 Squadron crewmen WiA

Two-Seaters:
1st and only March two-seater crewman WiA
4th of 11 1917 two-seater crewmen WiA
4th of 16 pre-wound two-seater crewmen WiA
4th of 18 total two-seater crewmen WiA

All:
1st of 2 total March crewmen WiA
4th of 19 total 1917 crewmen WiA
6th of 18 total pre-wound crewmen WiA
6th of 26 total crewmen WiA

Etc.
1st of 10 total March victories
7th of 48 total 1917 victories
22nd of 57 total pre-wound victories
5th victory in an Albatros D.III
5th of 22 two-seater victories flying from La Brayelle
6th of 28 total victories flying from La Brayelle
5th of 30 two-seater victories as *Staffelführer*
6th of 39 total victories flying as *Staffelführer*

Statistics Notes
Δ All BE.2 victories are pre-wound

Above & Above Right: Manfred and his dog Moritz in lighter moments showing a side of Manfred not normally illustrated.

Victory No.23

Above: One of A/1108's souvenired serial numbers hanging in the Richthofen Museum.

Above: Large section of A/1108's fuselage fabric in the Richthofen Museum.

4 March
Sopwith 1½ Strutter A/1108
No.43 Squadron RFC

Day/Date: Sunday, 4 March
Time: 4.20PM
Weather: Fine.
Attack Location: Acheville
Crash Location: Acheville
Side of Lines: Friendly

KOFL 6. Armee Weekly Activity Report: "4.20 Nach. 1 fdl. Sopwith bei Acheville durch Lt. Frh.v.Richthofen (als 23).

4.20PM 1 enemy Sopwith [shot down] near Acheville by *Lt. Frh.*v.Richthofen (as 23rd)"

MvR Combat Report: "Sopwith Two-seater. Occupants: Lieut. W. Reid and Lieut. H. Green; both killed, buried by local command Bois Bernard.

Accompanied by 5 of my planes I attacked an enemy squadron above Acheville. The Sopwith I had singled out flew for quite a while in my fire. After the 400th shot plane lost a wing whilst making a curve. Machine dashed downwards.

It is not worth while to have plane taken back, as parts all over Acheville and surroundings. Two machine-guns were seized by my *Staffel.* (1 Lewis Gun No.20024 and 1 Maxim Gun [Vickers] L.7500).

Frhr. v. Richthofen
Was acknowledged."

RFC Combat Casualties Report: "Left aerodrome 1.40PM. When near S. of VIMY the machine was seen by Lt. Henderson of No.40 Squadron to leave the formation and attack a group of H.A., three of whom shot it down about 3.25PM."

RFC A/C:

Make/model/serial number: Sopwith 1½ Strutter A.1108
Manufacturer: Subcontracted to and built by Vickers Ltd. (Aviation Department), Imperial Court, Basil Street, Knightsbridge, London, S.W. (production at Crayford works).
Unit: No.43 Squadron RFC
Aerodrome: Treizennes
Sortie: Patrol
Colors/markings: u/c. Generally, PC10 wings/ailerons/vertical and horizontal stabilizers/elevators; engine cowls either bare engine turned metal, PC10, or identifying Flight color; b/w/r rudder; white serial number with CDL background on vertical stabilizer.
Engine Number: R1169 WD 7782
Guns: One Vickers, one Lewis. Nos. 7500, 20024.
Manner of Victory: Machine gun fire caused in-flight breakup, followed by uncontrolled descent until terrain impact.
Parts Souvenired: a. Fabric with serial numbers, from port/stbd vertical stabilizers. Photographed on display at Roucourt, Richthofen's home bedroom in Schweidnitz, and in the post-war Richthofen Museum.
b. Fabric with C4, from stbd fuselage. Photographed on display at Roucourt, Richthofen's home in Schweidnitz, and in the post-war Richthofen Museum.

Above: Green's headstone, 2011.

Above: Reid's headstone, 2011.

c. Fabric with roundel and C4, from stbd fuselage. Photographed on display at Roucourt, Richthofen's home in Schweidnitz, and in the post-war Richthofen Museum.

RFC Crew:

Pilot: 2nd Lt. Herbert John Green, 19 (DoB 30 July 1897), KiA
RFC Combat Casualties Report: "2/Lt. A.W. Reid reported killed 4/3/17 – German source and disc forwarded from Germany (3rd Echelon Part II Orders – AFO 1810 – dated 8/9/17)."
Cause of Death: u/k. Either gunshot wound(s) or trauma associated with airplane crash.
Burial: Cabaret-Rouge British Cemetery (Souchez), France, Grave XII. E. 3.

Obs: 2nd Lt. Alexander William Reid, 20 or 21 (DoB 1896), KiA
RFC Combat Casualties Report: "2/Lt. H.J. Green Death accepted by the Army Council as having occurred on 4-3-17 on the evidence of a German list (3rd Echelon part II orders A.F. 0 1810 20-10-17)."
Cause of Death: u/k. Either gunshot wound(s) or trauma associated with airplane crash.
Burial: Cabaret-Rouge British Cemetery (Souchez), France, Grave XII. E. 4.

MvR A/C:

Make/model/serial number: Albatros D.III, possibly *Le Petit Rouge*; or Albatros D.III 2006/17.
Unit: *Jasta* 11
Commander: *Ltn.* Manfred von Richthofen
Airfield: La Brayelle
Colors/markings: *Le Petit Rouge* – See Vic. #17.
Albatros D.III 2006/16 – Factory finish shellacked and varnished 'warm straw yellow' wooden fuselage with red band encircling fuselage between cockpit and national marking. Pale greenish-gray cowl/fittings/vents/hatches/struts. Red spinner. Upper surfaces of wings/ailerons and horizontal stabilizers/elevator were factory three-color camouflage of pale green, olive green and Venetian red. Wheel covers/undersurface of

Albatros D.III, *Le Petit Rouge*. Its appearance is based on descriptions, anecdotes, and photographs from the time period.

wings/ailerons and horizontal stabilizers/elevators light blue. Rudder was either clear doped fabric or one of the camouflage colors. All crosses were black with white borders.

Damage: n/k

Initial Attack Altitude: u/k

Gunnery Range: u/k

Rounds Fired: 400

Known Staffel Participants: "Five planes," identities u/k.

Notes:

1. Richthofen's combat report submits this claim as his 22nd victory.
2. High-resolution photographs published in Lance Bronnenkant's *Manfred von Richthofen*, Volume 5 of Aeronaut's *Blue Max Airmen* series, reveal Richthofen sitting in the cockpit of Albatros D.III 2006/17 and then subsequently taking off from La Brayelle in the same machine. As will be seen, Richthofen's combat reports of 1918 consistently list the serial numbers of machines he flew during victorious sorties, and they reveal he often switched between machines. Now it is obvious he also did so during March 1917—and likely had since the beginning. Thus it must be presumed that he flew *Le Petit Rouge* without exclusivity and that during any given sortie could have flown any given Albatros. Since there is photographic provenance of him flying 2006/17, which was ultimately "handed down" to his brother Lothar after he joined *Jasta* 11, 2006/17 is an obvious alternative to associate with this combat report. However, this does not mean Richthofen did not fly an Albatros other than the two listed here. Determining which machine Richthofen flew during his victories in 1916 and 1917, aside from the very few he listed in combat reports, is uncertain and continually being researched.

Victory No.23 Statistics:

Two-Seaters

2nd of 7 March two-seater victories
6th of 33 1917 two-seater victories
13th of 39 pre-wound two-seater victories
13th of 45 total two-seater victories

Sop. Strutter Δ

1st and only March Sop. Strutter victory
1st of 3 1917 Sop. Strutter victories
1st of 3 total Sop. Strutter victories

No.43 Squadron

1st of 3 No.43 Squadron Sop. Strutter victories
1st and only March No.43 Squadron victories
1st of 3 1917 No.43 Squadron victories
1st of 3 total No.43 Squadron victories

Deaths

No.43 Squadron:
1st and 2nd of 2 March No.43 Squadron crewmen KiA
1st and 2nd of 4 1917 No.43 Squadron crewmen KiA
1st and 2nd of 4 total No.43 Squadron crewmen KiA

Two-Seaters:

1st and 2nd of 12 March two-seater crewman KiA
4th and 5th of 44 1917 two-seater crewmen KiA
15th and 16th of 54 pre-wound two-seater crewmen KiA
15th and 16th of 63 total two-seater crewmen KiA

Albatros D.III 2006/16. Its appearance is based on photographs taken during March 1917. Dates and circumstances regarding fuselage repair patches unknown.

All:
1st and 2nd of 13 total March crewmen KiA
6th and 7th of 53 total 1917 crewmen KiA
23rd and 24th of 66 total pre-wound crewmen KiA
23rd and 24th of 84 total crewmen KiA

Etc.
2nd of 10 total March victories
8th of 48 total 1917 victories
23rd of 57 total pre-wound victories
6th victory in an Albatros D.III
6th of 22 two-seater victories flying from La Brayelle
7th of 28 total victories flying from La Brayelle
6th of 30 two-seater victories as *Staffelführer*
7th of 39 total victories flying as *Staffelführer*

Statistics Notes:
Δ *All Sop. Strutter victories are pre-wound*

Below: Poignant view within Cabaret-Rouge British Cemetery, 2011.

Victory No.24

Above: Royal Aircraft Factory BE.2e. There is a single pair of interplane struts on each side, as compared to two pairs on the BE.2c.

6 March
BE.2e A2785
No.16 Squadron RFC

Day/Date: Tuesday, 6 March
Time: 5PM
Weather: Fine.
Attack Location: Souchez
Crash Location: Souchez,
Side of Lines: Enemy

RFC Communique No. 78: "Sixty-five targets were dealt with by artillery with aeroplane observation. Hostile aircraft were exceptionally active during the day, and a great number of combats took place."

KOFL 6. Armee Weekly Activity Report: "*5.0 Nach. 1 fdl. F.E. Zweisitzer bei souchez durch Lt.Frhr.v.Richthofen (als 24).*

5.00PM 1 enemy F.E. two-seater [shot down] near Souchez by *Lt. Frh.*v.Richthofen (as 24th)."

MvR Combat Report: "Details unknown, as plane landed on enemy's side.

Together with *Ltn.* Allmenroeder I attacked 2 enemy artillery flyers in a low altitude over the other side. The wings of the plane I attacked came off; it dashed down and broke on the ground.

Frhr. v. Richthofen
Was acknowledged."

RFC Combat Casualties Report: "Left Aerodrome about 1.50PM. Machine brought down by H.A. Pilot and observer killed and machine wrecked."

RFC A/C:
Make/model/serial number: Royal Aircraft Factory BE.2e A2785
Manufacturer: Subcontracted to and built by the British and Colonial Aeroplane Co., Ltd., Filton, Bristol.
Unit: No.16 Squadron RFC
Aerodrome: Bruay
Sortie: Artillery Observation
Colors/markings: u/c. Generally, either overall CDL wings/ailerons/vertical and horizontal stabilizers/elevators or the same in PC10; battleship gray engine cowl; b/w/r rudder; black

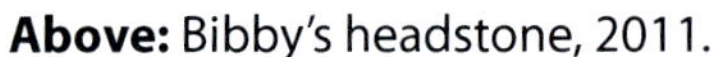

Above: Bibby's headstone, 2011.

Above: Brichta's headstone, 2011.

serial number on vertical stabilizer. No.16 Sqn markings were a triangle aft of the roundel and sometimes on the turtledeck (black on CDL machines, white on PC10).
Engine: 90 hp R.A.F. 1a
Engine Number: E1083 WD 5234
Guns: Two Lewis. Nos. 10487, 5393
Manner of Victory: Machine gun fire caused in-flight breakup, followed by uncontrolled descent until terrain impact.
Parts Souvenired: none

RFC Crew:

Pilot: 2nd Lt. Gerald Maurice Gosset-Bibby, 19 (9 April 1897), KiA
Cause of Death: u/c. Most likely trauma associated with airplane crash.
Burial: Barlin Communal Cemetery, Grave I. J. 53.
Obs: Lt. Geoffrey Joseph Oglivie Brichta, 32 (6 July 1884), KiA
Cause of Death: u/c. Most likely trauma associated with airplane crash.
Burial: Barlin Communal Cemetery, Grave I. J. 54.

MvR A/C:

Make/model/serial number: Albatros D.III. Possibly 2006/16 or 1996/16.
Unit: *Jasta* 11
Commander: *Ltn.* Manfred von Richthofen
Airfield: La Brayelle
Colors/markings: Albatros D.III 2006/16 – See Vic. #23.
Albatros D.III 1996/16 – Fuselage longitudinally quartered light blue and yellow (port upper-half yellow, lower-half light blue; reversed on starboard side). Vertical stabilizer was half-yellow (top) and half-light blue (bottom), with two light blue squares on the yellow portion and an unpainted rectangular strip in the middle that bordered the serial number. Pale greenish-gray spinner and all engine cowls. Yellow cabane struts and wheel cover, light blue interplane and landing

Above: Albatros D.III 1996/16. This airplane is a suspected candidate for Lübbert's half-blue/half-yellow machine.

gear struts (on port, likely reversed on starboard). Uppersurfaces of wings, horizontal stabilizers and elevator camouflaged pale green, olive green and Venetian red, sloped left. Undersurfaces were light blue. Salient details include square patch on fuselage near forward engine vent, square patch on fuselage just above lower port wing, ca. mid-chord, rectangular patch on fuselage *Eiserne Kreuz*, at least two patches visible on elevator, Albatros logo on rudder unpainted, heavy paint flaking on rudder control horn.

Initial Attack Altitude: u/k
Gunnery Range: u/k
Rounds Fired: u/k
Known Staffel Participants: Ltn. Allmenröder.

Notes:

1. Once again it is unclear exactly which machine Richthofen flew during this victory. Earlier that morning he had been forced to land after being shot in the fuel tanks and engine, and his autobiography suggests he had been flying an Albatros. It reveals that as Richthofen glided down he witnessed a distant airplane fall from the fight above and later wrote, "*Like mine*, it goes straight down, spinning, always spinning—then it recovers and straightens out. As it flies toward me, I see that it is *also an Albatros*." [author's emphases]). Later, when Emil Schaefer arrived by automobile to fetch Richthofen, he learned "the name of the pilot of the *other* fallen Albatros" [author's emphasis] was Ltn. Edy Lübbert, one of his pilots in *Jasta* 11. Although wounded, Lübbert's machine apparently had been either undamaged or only slightly or superficially damaged, because Richthofen stated he flew it out of the field and back to La Brayelle. Since Richthofen's Albatros D.III—presumably *Le Petit Rouge*—had extensive fuel tank and engine damage ("...I did not have a drop of fuel [left in the tanks]...the engine was likewise shot up...") that precluded it being flown from the field in a timely manner, it is possible Richthofen used Lübbert's available Albatros D.III while the wounded airman lay in a field hospital.
2. Uncertainty exists regarding whether this 1996/16 was Lübbert's. *Jasta 11* received D.1996/16 on 21 January 1917. Lübbert described his machine 3 March as 'half blue, half yellow,' and a light-and-dark quartered machine can be seen in the backgrounds of at least two photographs taken at Roucourt in April. The airplane is too distant to read the serial number but machine appears similar to a photograph of D.1996/16 when assigned to *Fl.Abt.(A)263* later that summer. Orthochromatic film used during the First World

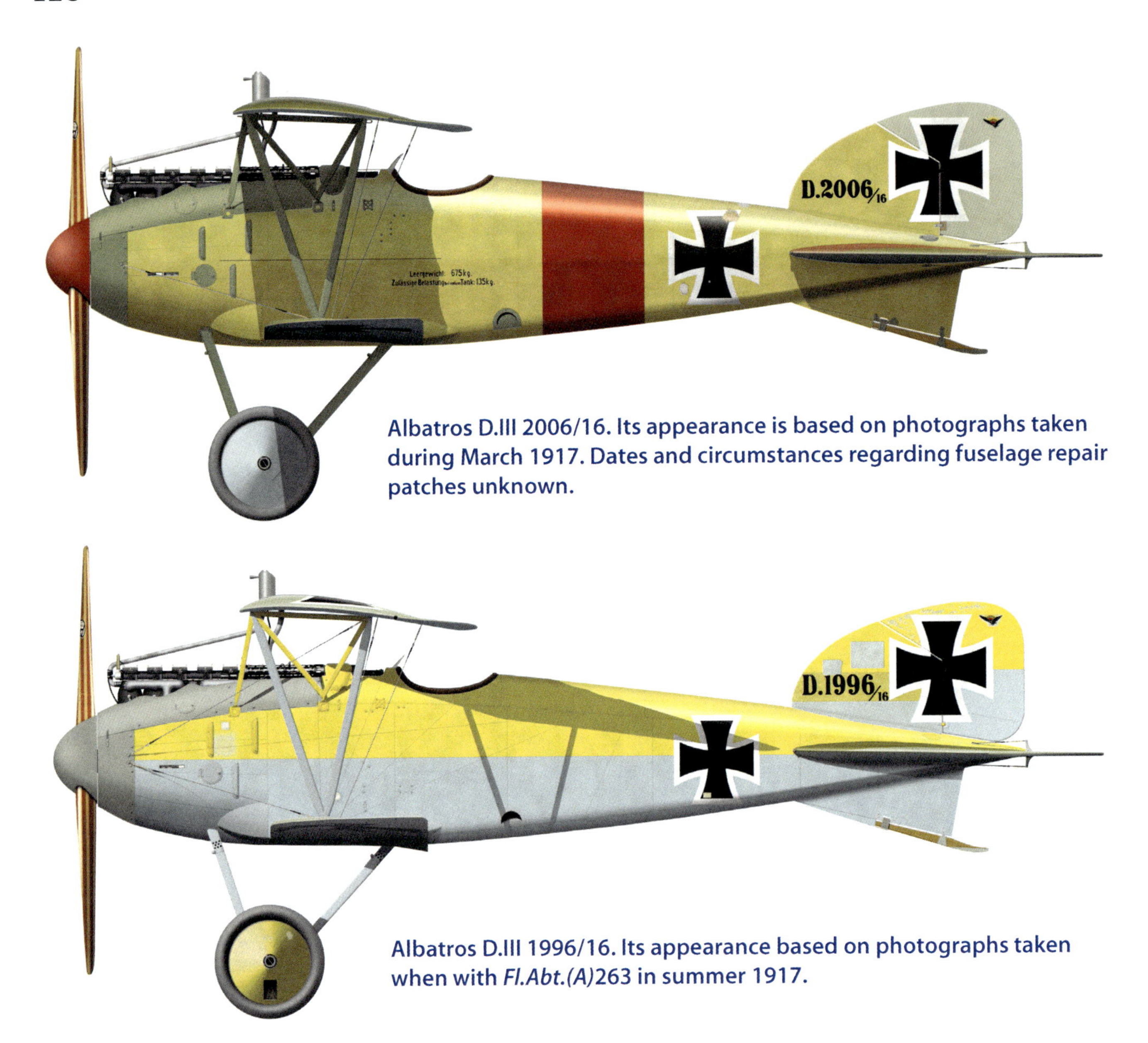

Albatros D.III 2006/16. Its appearance is based on photographs taken during March 1917. Dates and circumstances regarding fuselage repair patches unknown.

Albatros D.III 1996/16. Its appearance based on photographs taken when with *Fl.Abt.(A)*263 in summer 1917.

War generally rendered blues light gray (as the sky appears in many photographs), yellows dark gray and red darker still, so a light-blue-and-yellow airplane would appear light gray and dark gray. So it is possible 1996/16 appeared as Lübbert described, and its many patches seemingly coincide with Lübbert's nickname "the bullet catcher."

One question remains: if Lübbert was KiA before *Jasta* 11 relocated to Roucourt, why would his Albatros (presumed to be 1996/16) be there? It is not certain 1996/16 was his Albatros but if so he obviously was flying another machine when shot down.

Victory No.24 Statistics:

Two-Seaters
3rd of 7 March two-seater victories
7th of 33 1917 two-seater victories
14th of 39 pre-wound two-seater victories
14th of 45 total two-seater victories

BE2 Δ
2nd of 5 March BE.2 victories
5th of 15 1917 BE.2 victories
7th of 17 total BE.2 victories

No.16 Squadron
2nd of 6 No.16 BE.2 victories
1st of 3 March No.16 Squadron victories
1st of 6 1917 No.16 Squadron victories

1st of 6 total No.16 Squadron victories

Deaths
No.16 Squadron:
1st and 2nd of 6 March No.16 Squadron crewmen KiA
3rd and 4th of 12 1917 No.16 Squadron crewmen KiA
3rd and 4th of 12 total No.16 Squadron crewmen KiA

Two-Seaters:
3rd and 4th of 12 March two-seater crewmen KiA
6th and 7th of 44 1917 two-seater crewmen KiA
17th and 18th of 54 pre-wound two-seater crewmen KiA
17th and 18th of 63 total two-seater crewmen KiA

All:
3rd and 4th of 13 total March crewmen KiA
8th and 9th of 53 total 1917 crewmen KiA
25th and 26th of 66 total pre-wound crewmen KiA
25th and 26th of 84 total crewmen KiA

Etc.
4th of 14 victories behind enemy lines
3rd of 10 total March victories
9th of 48 total 1917 victories
24th of 57 total pre-wound victories
7th victory in an Albatros D.III
7th of 22 two-seater victories flying from La Brayelle
8th of 28 total victories flying from La Brayelle
7th of 30 two-seater victories as *Staffelführer*
8th of 39 total victories flying as *Staffelführer*

Statistics Notes
Δ *All BE.2 victories are pre-wound*

Above: Richthofen's Albatros D.III *Le Petit Rouge* after an apparent off-field forced landing. The horizontal stabilizers and elevator have been removed and both ailerons have been disconnected, perhaps in preparation for wing removal and eventual airplane towing. Any battle damage cannot be discerned in this view but it is speculative possibility this photograph shows *Le Petit Rouge* on 6 March, after having been attacked by No.40 Squadron FE.8s that morning. (Lance Bronnenkant)

Victory No.25

Above: Albatros D.II(OAW) 910/16, shortly after being downed 4 March 1916 by Lt Pearson of No.29 Squadron. Richthofen shot Pearson down five days later for his 25th victory."

9 March
DH.2 A2571
No.29 Squadron RFC

Day/Date: Friday, 9 March
Time: 11.55AM
Weather: Fine.
Attack Location: Between Roclincourt and Bailleul
Crash Location: Between Roclincourt and Bailleul, "this side of line, 500 meters behind trenches"
Side of Lines: Friendly

KOFL 6. Armee Weekly Activity Report: "12.00 Mitt. 1 fdl. Vickers-Einsitzer bei Roclincourt durch Lt.Freiherr von Richthofen, Führer von J. St. 11 (als 25.)

12.00 midday 1 enemy Vickers-single-seater [shot down] near Roclincourt by *Lt. Frh.*v.Richthofen, leader of *Jagdstaffel* 11 (as 25th)."

MvR Combat Report: "Vikkers One-Seater. Occupants: Not recognizable, as completely burnt. Plane No. On tail rudder A.M.C. 3425a.

With three of my planes I attacked several enemy planes. The machine I had singled out soon caught fire and dashed after 100 shots downwards. The plane is lying on our side, but cannot be salvaged as it nearly completely burned and too far in front.

Frhr. v. Richthofen
Was acknowledged."

RFC Combat Casualties Report: "Left aerodrome 9.20AM. Machine reported last seen diving to earth in flames."

RFC A/C:
Make/model/serial number: AMC DH.2 A2571
Manufacturer: Aircraft Manufacturing Co., Ltd., Hendon, London, N.W.
Unit: No.29 Squadron RFC

Above: Albatros D.II(OAW) 910/16, now AL 910, post-capture and in French markings. Note the overall silver-aluminum appearance, additional empennage strut, and Levasseur propeller.

Aerodrome: Le Hameau
Sortie: Escort
Colors/markings: u/k. Typically for the make/model, PC10 nacelle/wings/ailerons/horizontal stabilizers/elevator uppersurfaces; CDL undersurfaces; nacelle either CDL or PC10 with white undersurface; r/w/b rudder with black serial numbers sometimes bordered in white on the red/blue. No. 29 Squadron used colored numerals to identify flights: A Flight used red numeral with white shadow or border; B Flight used white numeral with blue shadow or border; and C Flight used blue numeral with white shadow or border.
Engine: 100 hp Gnôme Monosoupape
Engine Number: 30377 WD 4098
Gun: One Lewis, no. 6855
Manner of Victory: Uncontrolled flaming descent until terrain impact.
Items Souvenired: none

RFC Crew:
Pilot: Lt. Arthur John Pearson, 19 (DoB July 1887), KiA
Pilot Victories: 1 (forced to land)

Victory#	Date	Details
1	4 March 1917	Albatros D.II(OAW) 910/16

RFC Combat Casualties Report: Death accepted by the Army Council as having occurred on 9-3-17 on the evidence of a report transmitted by the Geneva red Cross Society in conjunction with lapse of time (3rd Echelon part II orders d/ 3-11-17).
Cause of Death: u/k Either gunshot wound(s), immolation, or trauma associated with airplane crash.
Burial: Either in-field or laid where fallen; location lost during turmoil of war.

MvR A/C:
Make/model/serial number: Albatros D.III. Possibly *Le Petit Rouge*, 2006/16, or 1996/16.
Unit: *Jasta* 11
Commander: *Ltn.* Manfred von Richthofen
Airfield: La Brayelle
Colors/markings: *Le Petit Rouge* – See Vic. #17.
Albatros D.III 2006/16 – See Vic. #23.
Albatros D.III 1996/16 – See Vic. #24.
Damage: u/k
Initial Attack Altitude: u/k
Gunnery Range: u/k
Rounds Fired: 100
Known Staffel Participants: "Three of my planes."

Notes:

1. RFC Communique No. 78 reveals that on 4 March "an Albatros Scout, which was attacked by Lieut. Pearson, No.29 Squadron, and Lieuts. Graham and Boddy, No. 11 Squadron, landed near Tilloy.

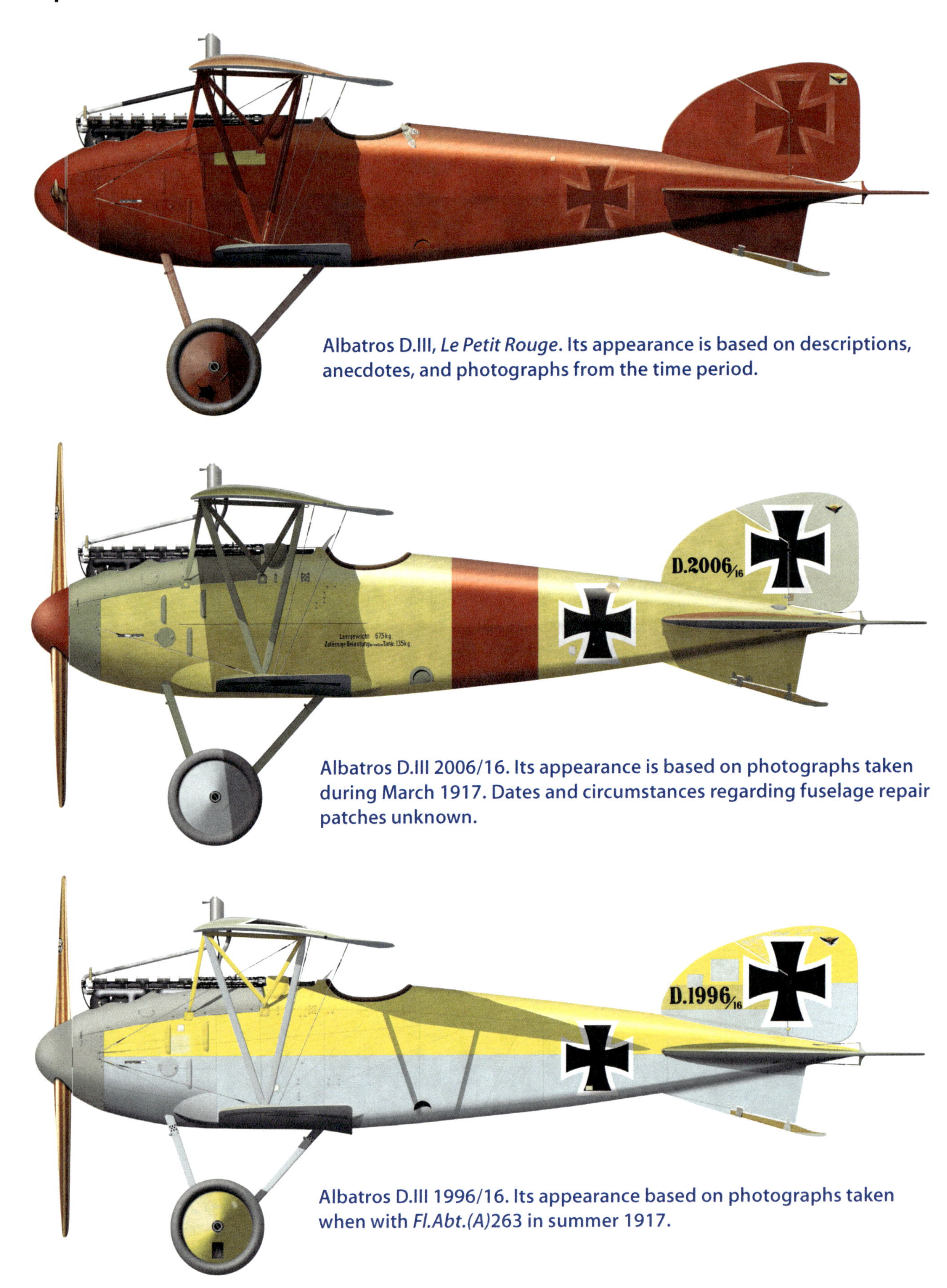

Albatros D.III, *Le Petit Rouge*. Its appearance is based on descriptions, anecdotes, and photographs from the time period.

Albatros D.III 2006/16. Its appearance is based on photographs taken during March 1917. Dates and circumstances regarding fuselage repair patches unknown.

Albatros D.III 1996/16. Its appearance based on photographs taken when with *Fl.Abt.(A)*263 in summer 1917.

The pilot was taken prisoner." This machine was Albatros D.II(OAW) 910/16, flown by *Ltn.* Max Böhme of *Jasta* 5. Seized by the British and assigned the captured aircraft number G 14, ultimately it was handed to the French and tested at Villacoublay.

2. Uncertainty exists again regarding Richthofen's Albatros. After *Le Petit Rouge* had been shot up by No.40 Squadron FE.8s on 6 March, it is unknown how long it took to prepare the machine for towing (photographs believed to be taken at that time reveal the elevator and horizontal stabilizers were removed and that the ailerons had been disconnected, likely as a precursor to removing the wings, all of which are associated with the vehicular towing of airplanes), return it to La Brayelle, and then, depending on the unknown extent of the damage, repair or even replace the fuel tanks and engine. Thus, while it is possible *Le Petit Rouge* had been repaired by the 9th, Richthofen's description of "shot up" indicates the engine may have required more than passing attention and that he needed another Albatros for this sortie. 2006/16 is a perennial possibility, as is 1996/16, since Lübbert was still convalescing from his wounds three days earlier.
3. Richthofen's combat report reference of "A.M.C. 3425a" regards A2571's batch number, not serial number. This also confirms that at least part of the rudder survived the pre/post-crash fires. Although ripe for the taking this rudder, either wholly or in part, has not been seen amongst the photographs of Richthofen's victory souvenirs.

Victory No.25 Statistics:

Single-Seaters
1st of 3 March single-seater victories
3rd of 15 1917 single-seater victories
11th of 18 pre-wound single-seater victories
11th of 35 total single-seater victories

DH.2 Δ
1st and only March DH.2 victory
1st and only 1917 DH.2 victory
5th of 5 total DH.2 victories

No.29 Squadron
3rd of 3 No.29 DH.2 victories
1st and only March No.29 Squadron victory
1st of 3 1917 No.29 Squadron victories
3rd of 5 total No.29 Squadron victories

Deaths
No.29 Squadron:
1st and only March No.29 Squadron crewman KiA
1st of 2 1917 No.29 Squadron crewmen KiA
2nd of 3 total No.29 Squadron crewmen KiA

Single-seaters:
1st and only March single-seater crewman KiA
3rd of 9 1917 single-seater crewmen KiA
9th of 12 pre-wound single-seater crewmen KiA
9th of 21 total single-seater crewmen KiA

All:
5th of 13 total March crewmen KiA
10th of 53 total 1917 crewmen KiA
27th of 66 total pre-wound crewmen KiA
27th of 84 total crewmen KiA

Etc.
4th of 10 total March victories
10th of 48 total 1917 victories
25th of 57 total pre-wound victories
8th victory in an Albatros D.III
2nd of 6 single-seater victories flying from La Brayelle
9th of 28 total victories flying from La Brayelle
2nd of 9 single-seater victories as *Staffelführer*
9th of 39 total victories flying as *Staffelführer*

Statistics Notes
Δ *All DH.2 victories are pre-wound*

Victory No.26

11 March
BE.2d 6232
No.2 Squadron RFC

Day/Date: Sunday, 11 March
Time: 12 noon
Weather: Fine in morning; cloudy in afternoon.
Attack Location: South of La Folie Wood, near Vimy
Crash Location: Front lines

RFC Communique No. 79: "Successful reconnaissances were carried out by all Brigades, and much valuable information was obtained. Hostile aircraft were extremely active."

KOFL 6. Armee Weekly Activity Report: "12.00 Mitt. 1 B.E. Zweisitzer bei Vimy diesseits unserc Linie durch Lt.Freiherr von Richthofen (als 26.)

12.00 midday 1 B.E. two-seater [shot down] near Vimy on this side of our line by *Lt.Freiherr* von Richthofen (as 26th)"

MvR Combat Report: "B.E. Two-seater. Occupants: Lieut. Byrne and Lieut. Smyth, 40th Squadron [sic]. Both killed. Plane No.6232. Details of Motor not at hand, as motor dashed into earth; cannot be dug up as locality under heaviest artillery fire.

I had lost my squad, and was flying alone, and had been observing for some time, an enemy artillery flyer. In a favorable moment I attacked the B.E. machine, and after 200 shots the body of the machine broke in half. The plane fell smoking into our lines.

The plane is lying near the forest of La Folie west of Vimy, only a few paces behind the trenches.

Frhr. v. Richthofen
Was acknowledged."

RFC Combat Casualties Report: "Left aerodrome 10.30AM. Reported to have been attacked by H.A. over GIVENCHY at 10.45AM and brought down on enemy side of lines."

RFC A/C:
Make/model/serial number: Royal Aircraft Factory BE.2d 6232
Manufacturer: Subcontracted to and built by Ruston, Proctor & Co., Ltd., Lincoln
Unit: No.2 Squadron RFC

Above: 6232's souvenired serial number hanging in the Richthofen Museum.

Aerodrome: Hesdigneul
Sortie: Photography
Colors/markings: u/c. Generally, PC10 wings/ailerons/vertical and horizontal stabilizers/elevators; battleship gray engine cowl; b/w/r rudder; black serial number on vertical stabilizer. No.2 Sqn markings were a triangle aft of the roundel and sometimes on the turtledeck (black on CDL machines, white on PC10).
Engine: 90 hp R.A.F. 1a
Engine Number: 23100 WD 2096
Guns: 17458, 17352
Manner of Victory: Machine gun bullets caused structural compromise and in-flight breakup.
Items Souvenired: Fabric with serial number. Photographed on display at Roucourt, Richthofen's home bedroom in Schweidnitz, and in the post-war Richthofen Museum.

RFC Crew:
Pilot: 2nd Lt. James Smyth, 19, KiA
Cause of Death: u/k Either gunshot wound(s), immolation, or trauma associated with airplane crash.
RFC Combat Casualties Report: Death accepted by the Army Council as having occurred on 12/3/17 on the evidence of a prisoner of war in England (3rd Echelon Part II Orders, AFO 1810, dated 21/7/17)
Burial: Cabaret-Rouge British Cemetery (Souchez), France, Grave XII. E. 9.
Obs: 2nd Lt. Edward Gordon Byrne, 35 or 36 (DoB 1881, headstone indicates age as 35), KiA
Cause of Death: u/k Either gunshot wound(s), immolation, or trauma associated with airplane crash.
RFC Combat Casualties Report: Death accepted for official purposes as having occurred in action

Above: Smyth's headstone, 2011.

on 12/3/17. (3rd Echelon Part II Orders AFO 1810 dated 29/9/17).
Burial: Cabaret-Rouge British Cemetery (Souchez), France, Grave XII. E. 8.

MvR A/C:
Make/model/serial number: Albatros D.III. Possibly *Le Petit Rouge*, 2006/16, or 1996/16.
Unit: *Jasta* 11
Commander: *Ltn.* Manfred von Richthofen
Airfield: La Brayelle
Colors/markings: *Le Petit Rouge* – See Vic. #17.
Albatros D.III 2006/16 – See Vic. #23.
Albatros D.III 1996/16 – See Vic. #24
Damage: u/k

Initial Attack Altitude: u/k
Gunnery Range: u/k
Rounds Fired: 200
Known Staffel Participants: None, solo flight

Above: Byrne's headstone, 2011.

Notes:

1. It is unknown exactly which Albatros Richthofen flew during this sortie. (See Vic. #25, Note 2.)
2. The Combat Casualties reports "accepted" death as having occurred on 12 March, which is the date listed on Smyth and Byrne's headstones, but this date was for "official purposes." Based on Richthofen's combat report that names his opponents and indicates they had been killed, their deaths most likely occurred on 11 March.

Victory No.26 Statistics:

Two-Seaters
4th of 7 March two-seater victories
8th of 33 1917 two-seater victories
15th of 39 pre-wound two-seater victories
15th of 45 total two-seater victories

BE.2 Δ
3rd of 5 March BE.2 victories
6th of 15 1917 BE.2 victories
8th of 17 total BE.2 victories

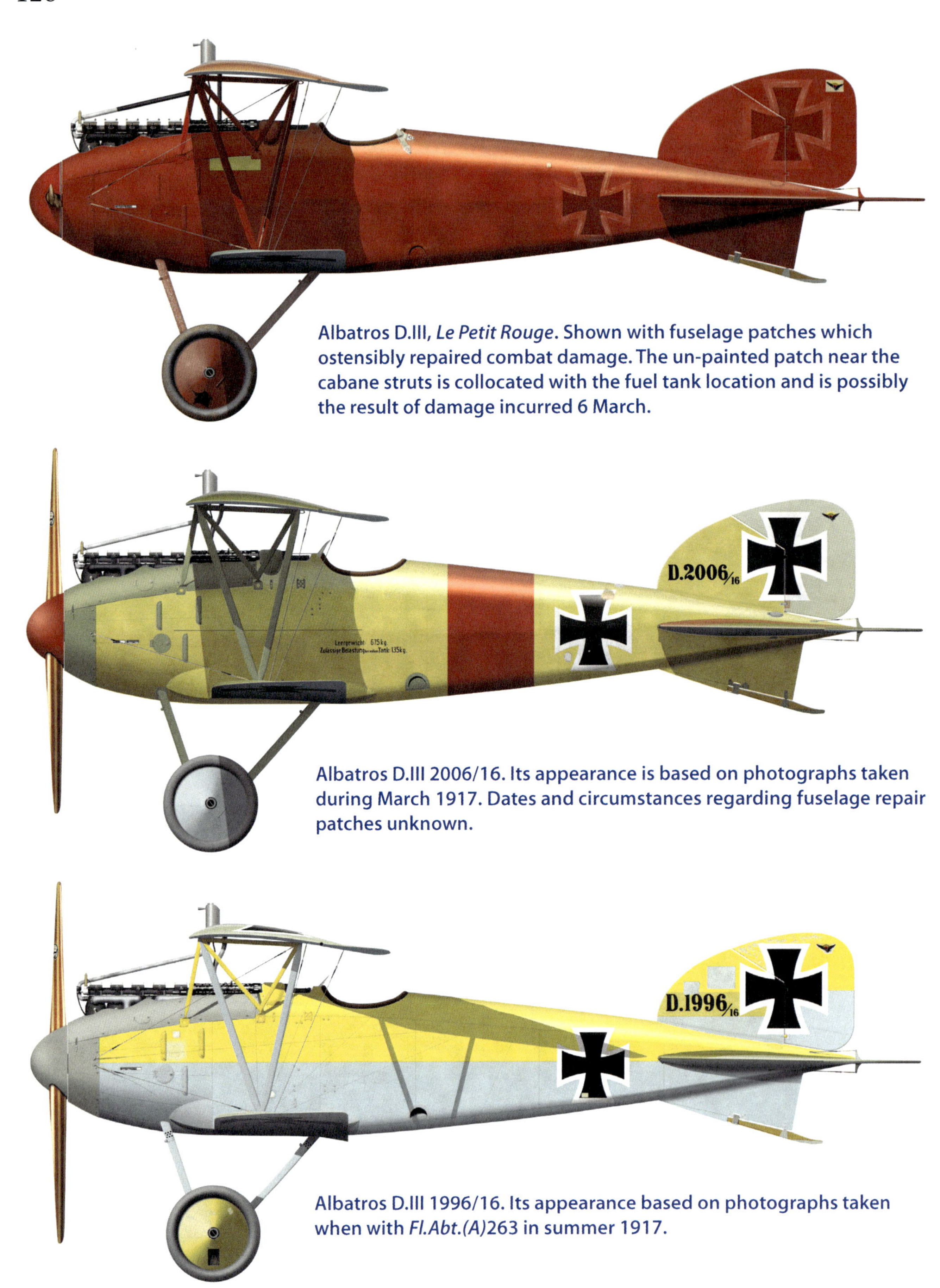

Albatros D.III, *Le Petit Rouge*. Shown with fuselage patches which ostensibly repaired combat damage. The un-painted patch near the cabane struts is collocated with the fuel tank location and is possibly the result of damage incurred 6 March.

Albatros D.III 2006/16. Its appearance is based on photographs taken during March 1917. Dates and circumstances regarding fuselage repair patches unknown.

Albatros D.III 1996/16. Its appearance based on photographs taken when with *Fl.Abt.(A)*263 in summer 1917.

No.2 Squadron
4th of 4 No.2 BE.2 victories
2nd of 2 March No.2 Squadron victories
4th of 4 1917 No.2 Squadron victories
4th of 5 total No.2 Squadron victories

Deaths
No. 2 Squadron:
1st and 2nd of 2 March No. 2 Squadron crewmen KiA
2nd and 3rd of 3 1917 No. 2 Squadron crewmen KiA
2nd and 3rd of 5 total No. 2 Squadron crewmen KiA

Two-Seaters:
5th and 6th of 12 March two-seater crewmen KiA
8th and 9th of 44 1917 two-seater crewmen KiA
19th and 20th of 54 pre-wound two-seater crewmen KiA
19th and 20th of 63 total two-seater crewmen KiA

All:
6th and 7th of 13 total March crewmen KiA
11th and 12th of 53 total 1917 crewmen KiA
28th and 29th 66 total pre-wound crewmen KiA
28th and 29th of 84 total crewmen KiA

Etc.
5th of 10 total March victories
11th of 48 total 1917 victories
26th of 57 total pre-wound victories
9th victory in an Albatros D.III
8th of 22 two-seater victories flying from La Brayelle
10th of 28 total victories flying from La Brayelle
8th of 30 two-seater victories as *Staffelführer*
10th of 39 total victories flying as *Staffelführer*

Statistics Notes
Δ *All BE.2 victories are pre-wound*

Above: Richthofen's office. Although often identified as an Albatros D.Va, this fine cockpit view is actually that of an Albatros D.III (to be clear, not Richthofen's), as corroborated by the rounded shape of the fuselage near cockpit (as opposed to the more ovoid D.Va), asymmetrical machine gun aperture, location of the starboard Maxim ammunition feeder chute, and slab-side internal vertical cockpit framing. The engine tachometer is at top-center of the cockpit, with fuel valves and water pump greaser at right, and field-mod altimeter at left. Forward are metal bins that store ammunition and collect spent ammunition belts, one of which can be seen feeding into the bin vertically from behind the tachometer.

Victory No.27

17 March
FE.2b A5439
No.25 Squadron RFC

Day/Date: Saturday, 17 March
Time: 11.30AM
Weather: Fine all day; ground mist early morning.
Attack Location: Oppy
Crash Location: Oppy
Side of Lines: Friendly

RFC Communique No. 79: "Reconnaissances were carried out by all Brigades and 9th Wing. 18 aeroplanes of the 1st Brigade taking photographs in the rear of the enemy's lines encountered 19 hostile machines."

KOFL 6. Armee Weekly Activity Report: "bei Bailleul-Oppy (disseits) durch Lt. Frhr. v.Richthofen (als 27)

Near Bailleul-Oppy (this side [of lines]) by *Lt.Freiherr* von Richthofen (as 27th)"

MvR Combat Report: "Vickers Two-seater. Occupants: Both killed, no identity discs, names found on maps were Smith and Heanly. Plane A.3439 [sic]. Motor Aero Engine 854. Machine-gun [sic]: 19633 and 19901.

About 11.30 I attacked with nine of my machines an enemy squad of 16 units. During the fight I managed to force a Vikkers Two-seater aside which I then, after 800 shots, brought down. In my machine-gun fire the plane lost its open work fuselage.

The occupants were killed and were taken for burial by local Commanders at Oppy.

Frhr. v. Richthofen.
Was acknowledged."

RFC Combat Casualties Report: "Left aerodrome at 9AM. An entry in the German papers states that F.E.2.b. A.5439 was brought down behind their lines in March and that both occupants were dead."

RFC A/C:
Make/model/serial number: Royal Aircraft Factory FE.2b A5439
Manufacturer: Boulton & Paul, Ltd., Norwich
Unit: No.25 Squadron RFC
Aerodrome: Lozinghem

Above: FE.2b A5439's souvenired engine placard.

Sortie: Escort to photography. ANNOELLIN-VITRY.
Colors/markings: Presentation a/c *Zanzibar No. 10*. Generally, PC10 nacelle with one, two, or three horizontal white-bordered black bands across nose (corresponding with A, B, or C Flight); likely white "ZANZIBAR 10" on sides of nacelle; wing uppersurfaces PC10 with CDL undersurfaces; likely upperwing roundels had white surrounds.
Engine: 160 hp Beardmore
Engine Number: 854 WD 7464
Guns: Two Lewis. Nos. 19901,19633
Manner of Victory: Machine gun bullets caused structural compromise and in-flight breakup.
Items Souvenired: Engine placard. Photographed on display at Roucourt, Richthofen's home bedroom in Schweidnitz, and in the post-war Richthofen Museum.

RFC Crew:
Pilot: Lt. Arthur Elsdale Boultbee, 19 (DoB 1897), KiA
Cause of Death: u/k. Either gunshot wound(s) or trauma associated with airplane crash.
RFC Combat Casualties Report: "Death of Lt. A.E. Boultbee accepted by the Army Council as having occurred on 17/3/17 on the evidence of a report published in the German Press together with the lapse of time (3rd Echelon Part II Orders AFO 1810 d-15/12/17)."
Burial: Canadian cemetery No.2 (Neuville St-Vaast), France, Grave 11. A. 2.
Obs: 2nd Class Air Mechanic Frederick King, 22 (DoB 1894), KiA

Above: Boultbee's headstone, 2011.

Above: King's headstone, 2011.

Cause of Death: u/k. Either gunshot wound(s) or trauma associated with airplane crash.
Burial: Canadian cemetery No.2 (Neuville St-Vaast), France, Grave 11. A. 1.

MvR A/C:
Make/model/serial number: Albatros D.III. Possibly *Le Petit Rouge*, 2006/16, or 1996/16.
Unit: *Jasta* 11
Commander: *Ltn.* Manfred von Richthofen
Airfield: La Brayelle
Colors/markings: *Le Petit Rouge* – See Vic. #17.
Albatros D.III 2006/16 – See Vic. #23.
Albatros D.III 1996/16 – See Vic. #24
Damage: u/k

Initial Attack Altitude: u/k
Gunnery Range: u/k
Rounds Fired: 800
Known Staffel Participants: "Attacked with nine of my machines"

Notes:

1. Crew names on the maps referenced in Richthofen's combat reports indicate the maps were borrowed.
2. Albatros identification for this sortie again is uncertain. It could have been any of these machines, or even others. In Ferko's book *Richthofen*, he states Richthofen flew Lübbert's Albatros (i.e., presumed to be 1996/16) during this sortie. There is no provenance for this association other than a theory that the source is Lübbert's diary (which the author did not find during several trips to examine Ferko's records at the University of Texas Dallas).

Victory No.27 Statistics:

Pushers*
1st of 2 March pusher victories
1st and only March two-seater pusher victory
2nd of 7 1917 two-seater pusher victories
7th of 12 total two-seater pusher victories

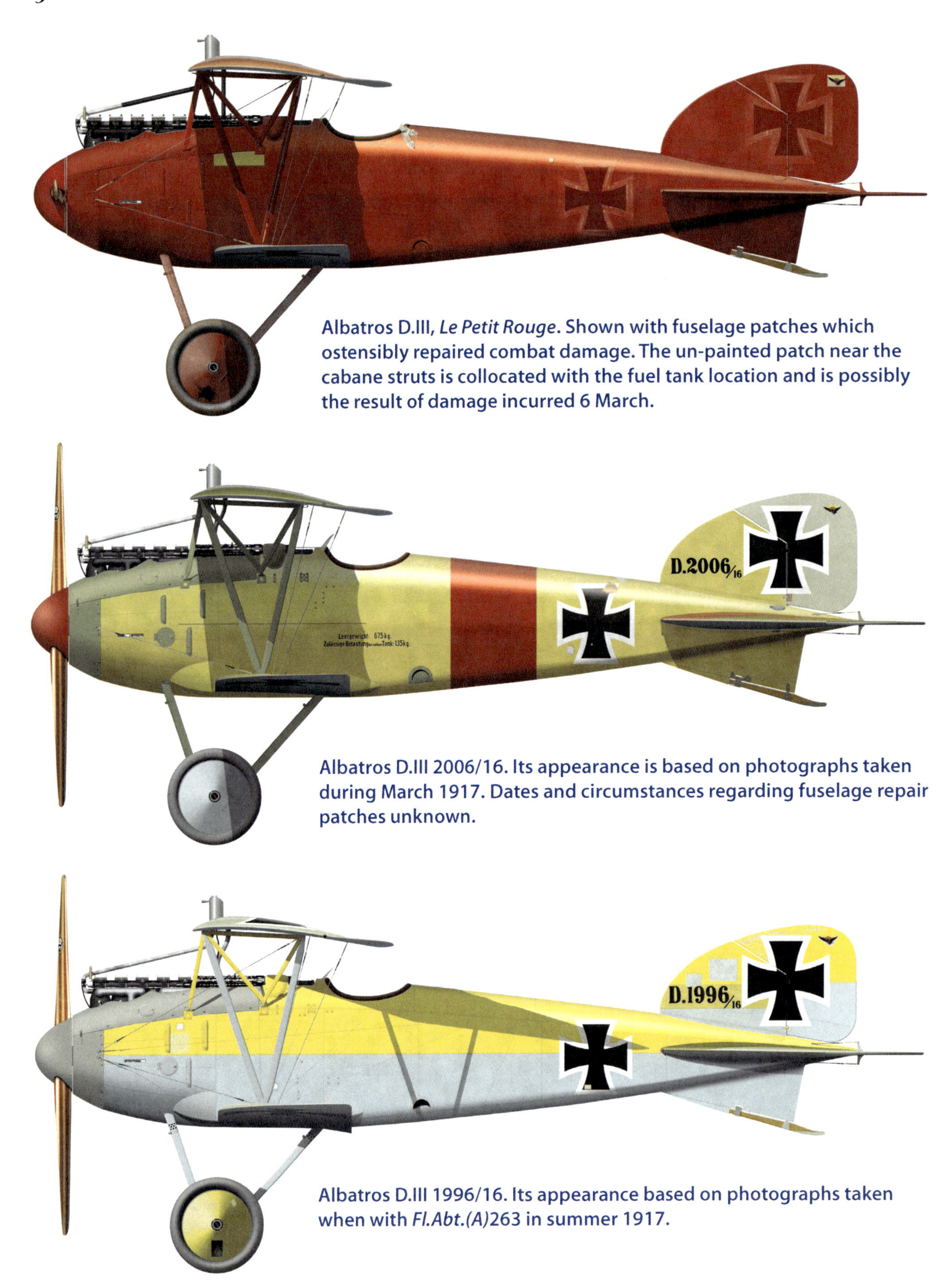

Albatros D.III, *Le Petit Rouge*. Shown with fuselage patches which ostensibly repaired combat damage. The un-painted patch near the cabane struts is collocated with the fuel tank location and is possibly the result of damage incurred 6 March.

Albatros D.III 2006/16. Its appearance is based on photographs taken during March 1917. Dates and circumstances regarding fuselage repair patches unknown.

Albatros D.III 1996/16. Its appearance based on photographs taken when with *Fl.Abt.(A)*263 in summer 1917.

13th of 18 total pusher victories

Two-Seaters
5th of 7 March two-seater victories
9th of 33 1917 two-seater victories
16th of 39 pre-wound two-seater victories
16th of 45 total two-seater victories

FE.2 Δ
1st and only March FE.2 victory
2nd of 7 1917 FE.2 victories
7th of 12 total FE.2 victories

No.25 Squadron ±
1st of 4 No.25 FE.2 victories
1st and only March No.25 Squadron victory
2nd of 4 1917 No.25 Squadron victories
2nd of 4 total No.25 Squadron victories

Deaths †
No. 25 Squadron:
1st and 2nd of 2 March No. 25 Squadron crewmen KiA
1st and 2nd of 5 1917 No. 25 Squadron crewmen KiA
1st and 2nd of 5 total No. 25 Squadron crewmen KiA

Pushers:
1st and 2nd of 2 March two-seater pusher crewmen KiA
1st and 2nd of 7 1917 two-seater pusher crewmen KiA
11th and 12th of 17 total two-seater pusher crewmen KiA

Two-Seaters:
7th and 8th of 12 March two-seater crewmen KiA
10th and 11th of 44 1917 two-seater crewmen KiA
21st and 22nd of 54 pre-wound two-seater crewmen KiA
21st and 22nd of 63 total two-seater crewmen KiA

All:
8th and 9th of 13 total March crewmen KiA
13th and 14th of 53 total 1917 crewmen KiA
30th and 31st of 66 total pre-wound crewmen KiA
30th and 31st of 84 total crewmen KiA

Etc.
6th of 10 total March victories
12th of 48 total 1917 victories
27th of 57 total pre-wound victories
10th victory in an Albatros D.III
9th of 22 two-seater victories flying from La Brayelle
11th of 28 total victories flying from La Brayelle
9th of 30 two-seater victories as *Staffelführer*
11th of 39 total victories flying as *Staffelführer*

Statistics Notes
* *All pusher victories are pre-wound*
Δ *All FE.2 victories are pre-wound*
± *All No.25 Squadron victories are pre-wound*
† *All No.25 Squadron deaths are pre-wound*

Below: Albatros D.III 1996/16 apparently being run-up sans chocks. Note the absence of the forward engine cowl panels, perhaps in an effort to increase cooling.

Victory No.28

Above: 2nd Lt. George Watt's headstone, 2011.

Above: Sgt. Ernest Howlett's headstone, 2011.

17 March
BE.2g 2814
No.16 Squadron RFC

Day/Date: Saturday, 17 March
Time: 5PM
Weather: Fine all day; ground mist early morning.
Attack Location: Above trenches west of Vimy
Crash Location: No Man's Land

KOFL 6. Armee Weekly Activity Report: "5,00 nach bei Souchez (jenseits) durch Lt. Frhr v. Richthofen (als 28.)

5.00PM near Souchez (beyond [the lines]) by *Lt. Freiherr* von Richthofen (as 28th)"

MvR Combat Report: "B.E. Two-seater. No details, as plane landed between lines.

I had spotted an enemy infantry flyer. Several attacks directed from above produced no results, especially as my adversary did not accept fight and was protected from above by other machines. Therefore I went down to 700 meters and attacked my adversary who was flying in 900 meters from below. After a short fight my opponent's plane lost both wings and fell. The machine crashed into no man's land and was fired at by our infantry.

Frhr. v. Richthofen.
Was acknowledged."

RFC Combat Casualties Report: "Left aerodrome at 3.25PM. Machine was attacked and brought down by H.A. at S.27a.5.9 It collapsed in the air at a height of about 1000 ft. Pilot and observer killed. Machine wrecked."

RFC A/C:

Make/model/serial number: Royal Aircraft Factory BE.2g 2814
Unit: No.16 Squadron RFC
Aerodrome: Bruay
Sortie: Artillery Observation. FARBUS.
Colors/markings: u/c. Generally, either overall CDL wings/ailerons/vertical and horizontal stabilizers/elevators or the same in PC10; battleship gray engine cowl; b/w/r rudder; black serial number on vertical stabilizer. No.16 Sqn markings were two vertical bands that encircled the fuselage, one on either side of the roundel (black on CDL machines, white on PC10).
Engine: 90 H.P. R.A.F. 1a
Engine Number: E640 WD 2989
Guns: Two Lewis. Nos. 17440, 15754
Manner of Victory: Machine gun bullets caused structural compromise and in-flight breakup.
Parts Souvenired: none

RFC Crew:

Pilot: 2nd Lt. George MacDonald Watt, 27 (DoB 8 January 1890), KiA
Cause of Death: u/k. Either gunshot wound(s) or trauma associated with airplane crash.
Burial: Bruay Communal Cemetery Extension, Neuville St-Vaast, France, Grave D. 7.
Obs: Sgt. Ernest Adam Howlett, 26, KiA
Cause of Death: u/k. Either gunshot wound(s) or trauma associated with airplane crash.
Burial: Bruay Communal Cemetery Extension, Neuville St-Vaast, France, Grave D. 8.

MvR A/C:

Make/model/serial number: Albatros D.III. Possibly *Le Petit Rouge*, 2006/16, or 1996/16.
Unit: *Jasta* 11
Commander: *Ltn.* Manfred von Richthofen
Airfield: La Brayelle
Colors/markings: *Le Petit Rouge* – See Vic. #17.
Albatros D.III 2006/16 – See Vic. #23.
Albatros D.III 1996/16 – See Vic. #24
Damage: u/k

Initial Attack Altitude: From 700 to 900 meters
Gunnery Range: u/k
Rounds Fired: u/k
Known Staffel Participants: u/k

Victory No.28 Statistics:

Two-Seaters
6th of 7 March two-seater victories
10th of 33 1917 two-seater victories
17th of 39 pre-wound two-seater victories
17th of 45 total two-seater victories

BE.2 ∆
4th of 5 March BE.2 victories
7th of 15 1917 BE.2 victories
9th of 17 total BE.2 victories

No.16 Squadron ±
3rd of 6 No. 16 BE.2 victories
2nd of 3 March No.16 Squadron victories
3rd of 6 1917 No.16 Squadron victories
3rd of 6 total No.16 Squadron victories

Deaths †
No.16 Squadron:
3rd and 4th of 6 March No.16 Squadron crewmen KiA
5th and 6th of 12 1917 No.16 Squadron crewmen KiA
5th and 6th of 12 total No.16 Squadron crewmen KiA

Two-Seaters:
9th and 10th of 12 March two-seater crewman KiA
12th and 13th of 44 1917 two-seater crewmen KiA
23rd and 24th of 54 pre-wound two-seater crewmen KiA
23rd and 24th of 63 total two-seater crewmen KiA

All:
10th and 11th of 13 total March crewmen KiA
15th and 16th of 53 total 1917 crewmen KiA
32nd and 33rd of 66 total pre-wound crewmen KiA
32nd and 33rd of 84 total crewmen KiA

Etc.
7th of 10 total March victories
13th of 48 total 1917 victories
28th of 57 total pre-wound victories
11th victory in an Albatros D.III
10th of 22 two-seater victories flying from La Brayelle
12th of 28 total victories flying from La Brayelle
10th of 30 two-seater victories as *Staffelführer*
12th of 39 total victories flying as *Staffelführer*

Statistics Notes
∆ *All BE.2 victories are pre-wound*
± *All No.16 Squadron victories are pre-wound*
† *All No.16 Squadron deaths are pre-wound*

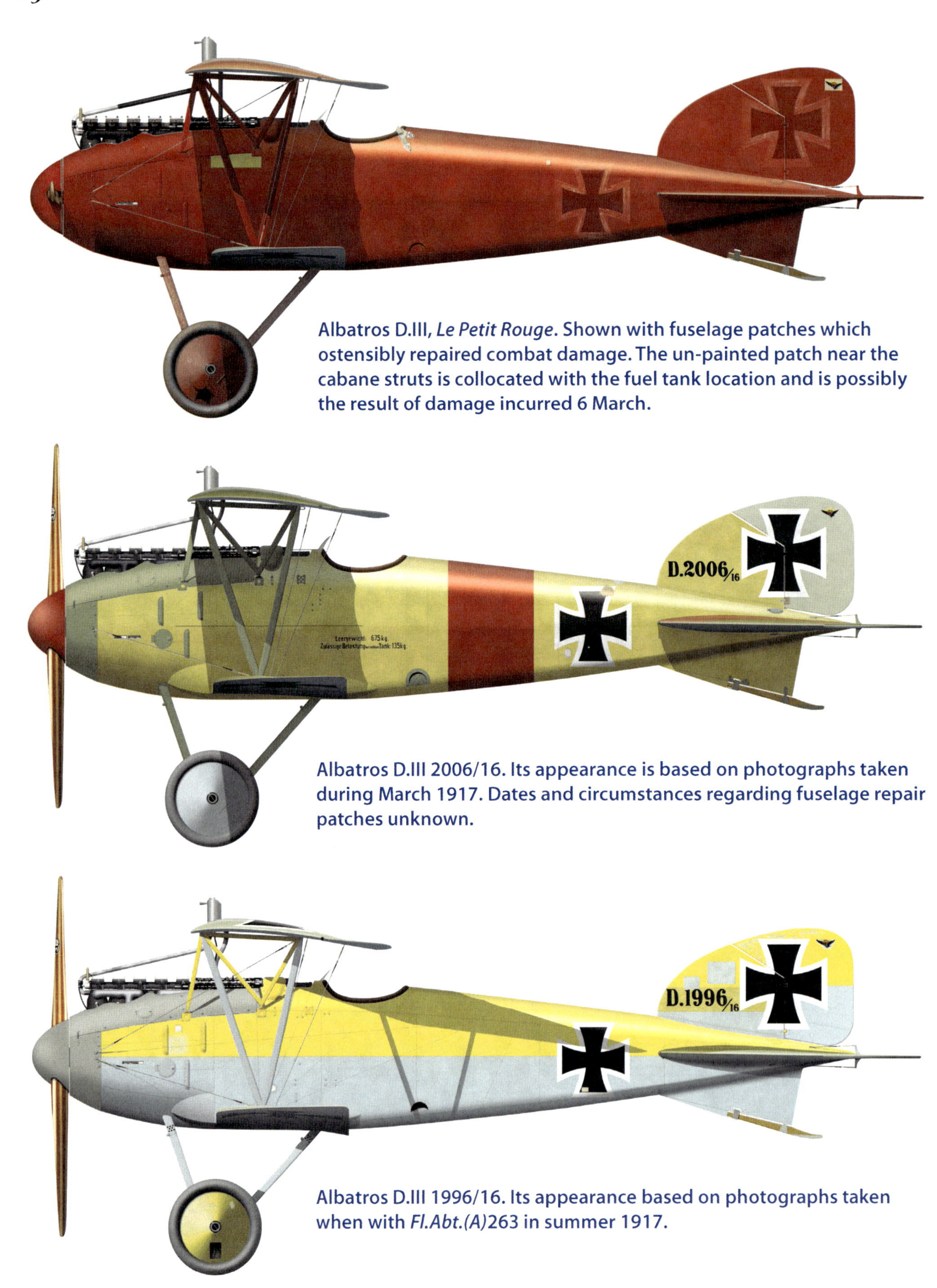

Albatros D.III, *Le Petit Rouge*. Shown with fuselage patches which ostensibly repaired combat damage. The un-painted patch near the cabane struts is collocated with the fuel tank location and is possibly the result of damage incurred 6 March.

Albatros D.III 2006/16. Its appearance is based on photographs taken during March 1917. Dates and circumstances regarding fuselage repair patches unknown.

Albatros D.III 1996/16. Its appearance based on photographs taken when with *Fl.Abt.(A)*263 in summer 1917.

Victory No.29

Above: Flt. Sgt. Sidney Quicke's headstone, 2011. Leaning on the flowers is a laminated sheet of paper upon which was printed two photos: one with Quicke sitting with a woman, and another with four multi-generational women holding a newborn baby. The physical resemblance between Quicke and the women in both photos suggests they are related. Furthermore, the resemblance between the oldest woman in the group shot and the woman next to Quicke suggests they are the same person. Considering the passage of time it seems likely the group photograph was taken several years prior to 2011, but regardless of specifics and relationships, 94 years after Quicke's death his grave seems to still be tended by loved ones.

21 March
BE.2f A3154
No.16 Squadron RFC

Day/Date: Wednesday, 21 March
Time: 5.30PM
Weather: Low clouds and rain during morning; clearing in places in the afternoon.
Attack Location: "One kilometer beyond our lines."
Crash Location: Hill 123, north of Neuville
Side of Lines: Enemy

RFC Communique No. 80: "Nineteen targets were dealt with by artillery with aeroplane observation. Very few hostile machines seen."

KOFL 6. Armee Weekly Activity Report: "5,25 nach bei Neuville (jenseits) durch Lt. Frhr.v.Richtofen (als 29.)

5.25PM near Neuville (beyond [the lines]) by Lt. Frhr. von Richthofen (as 29th)."

MvR Combat Report: "B.E. Two-seater. Plane details unknown, as plane came down on enemy's territory.

Messages came through that enemy planes had been seen in 1000 meters altitude in spite of bad weather and strong east wind.

I went up by myself intending to bring down an infantry or artillery flyer.

After one hour I spotted in 800 meters a large number of enemy artillery flyers beyond the lines. They sometimes approached our front, but never passed it. After several vain attempts I managed, half hidden by clouds, to take one of these B.E.'s by surprise and to attack him in 600 meters 1 kilometer beyond our lines. The adversary made the mistake of flying in a straight line when he tried to evade me, and thus he was just a wink too long in my fire. (500 shots). Suddenly he made two uncontrolled curves and dashed, smoking, to the ground. The plane was completely ruined. It fell in section F.3.

Frhr. v. Richthofen.
Was acknowledged."

RFC Combat Casualties Report: "Left aerodrome 3PM. At 4.30 message was received from the batteries that the machine had been shot down by H.A. Pilot killed; observer died of wounds 22/3/17."

RFC A/C:
Make/model/serial number: Royal Aircraft Factory BE.2f A3154
Unit: No.16 Squadron RFC
Aerodrome: Bruay
Sortie: Artillery Registration – A.10.d. (51.b.)

Albatros D.III, *Le Petit Rouge*. Shown with fuselage patches which ostensibly repaired combat damage. The un-painted patch near the cabane struts is collocated with the fuel tank location and is possibly the result of damage incurred 6 March.

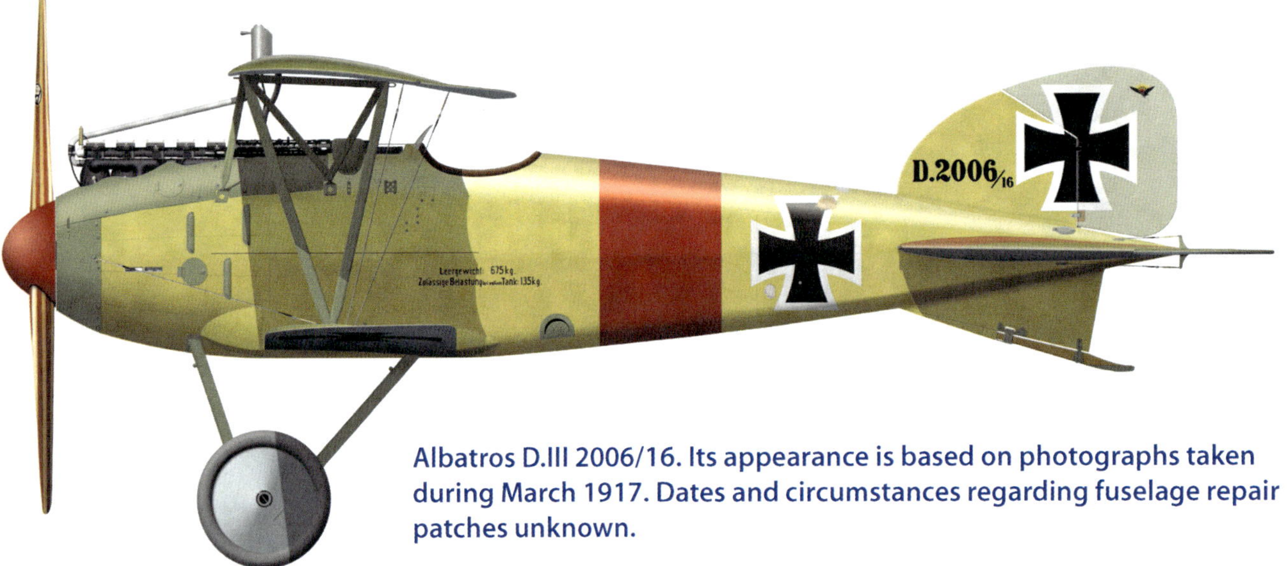

Albatros D.III 2006/16. Its appearance is based on photographs taken during March 1917. Dates and circumstances regarding fuselage repair patches unknown.

Colors/markings: u/c. Generally, either overall CDL wings/ailerons/vertical and horizontal stabilizers/elevators or the same in PC10; battleship gray engine cowl; b/w/r rudder; black serial number on vertical stabilizer. No.16 Sqn markings were two vertical bands that encircled the fuselage, one on either side of the roundel (black on CDL machines, white on PC10).
Engine: 90 H.P. R.A.F. 1a
Engine Number: 2021 WD 1704
Guns: E10689; E2173
Manner of Victory: Smoking spiral descent until terrain impact.
Items Souvenired: none

RFC Crew:

Pilot: Flt Sgt Sidney Herbert Quicke, KiA
Cause of Death: u/k. Either gunshot wound(s) or trauma associated with airplane crash.
Burial: Bruay Communal Cemetery (Pas de Calais), France, Grave D. 12.

Obs: 2nd Lt. William John Lidsey, 21 (DoB June 1895), DoW 22.3.17
Cause of Death: Gunshot wounds to head and legs, with possible trauma associated with airplane crash.
Burial: Aubigny Communal Cemetery Extension (Pas de Calais), France, Grave V. A. 38.

MvR A/C:

Make/model/serial number: Albatros D.III. Possibly *Le Petit Rouge*, or 2006/16
Unit: *Jasta* 11
Commander: *Ltn*. Manfred von Richthofen

Airfield: La Brayelle
Colors/markings: *Le Petit Rouge* – See Vic. #17.
Albatros D.III 2006/16 – See Vic. #23.
Damage: u/k
Initial Attack Altitude: 600 meters
Gunnery Range: u/k
Rounds Fired: 500
Known Staffel Participants: n/a, solo flight

Notes:
1. Entire stalk and attack occurred across enemy lines in British territory. This is an ostentatious example of Richthofen's willingness to fly beyond the lines well before being wounded and over a year before his final flight.
2. The exact Albatros flown during this sortie is unknown. 2006/16 is shown to illustrate that again it could be machines other than *Le Petit Rouge*. However, if 1996/16 was Lübbert's machine it is presumed he would have been flying it since his return to *Jasta* 11 flying status on 19 March.

Victory No.29 Statistics:

Two-Seaters
7th of 7 March two-seater victories
11th of 33 1917 two-seater victories
18th of 39 pre-wound two-seater victories
18th of 45 total two-seater victories

BE.2 Δ
5th of 5 March BE.2 victories
8th of 15 1917 BE.2 victories
10th of 17 total BE.2 victories

No.16 Squadron ±
4th of 6 No.16 BE.2 victories
3rd of 3 March No.16 Squadron victories
4th of 6 1917 No.16 Squadron victories
4th of 6 total No.16 Squadron victories

Deaths †
No.16 Squadron:
5th and 6th of 6 March No.16 Squadron crewmen KiA
7th and 8th of 12 1917 No.16 Squadron crewmen KiA
7th and 8th of 12 total No.16 Squadron crewmen KiA

Two-Seaters:
11th and 12th of 12 March two-seater crewmen KiA
14th and 15th of 44 1917 two-seater crewmen KiA
25th and 26th of 54 pre-wound two-seater crewmen KiA
25th and 26th of 63 total two-seater crewmen KiA

All:
12th and 13th of 13 total March crewmen KiA
17th and 18th of 53 total 1917 crewmen KiA
34th and 35th of 66 total pre-wound crewmen KiA
34th and 35th of 84 total crewmen KiA

Etc.
5th of 14 victories behind enemy lines
8th of 10 total March victories
14th of 48 total 1917 victories
29th of 57 total pre-wound victories
12th victory in an Albatros D.III
11th of 22 two-seater victories flying from La Brayelle
13th of 28 total victories flying from La Brayelle
11th of 30 two-seater victories as *Staffelführer*
13th of 39 total victories flying as *Staffelführer*

Statistics Notes
Δ *All BE.2 victories are pre-wound*
± *All No.16 Squadron victories are pre-wound*
† *All No.16 Squadron deaths are pre-wound; KiA includes DoW*

Below: *Le Petit Rouge* during the takeoff roll.

Victory No.30

Above: SPAD VII A6706 after its encounter with Richthofen.

24 March
SPAD VII C.1 A6706
No.19 Squadron RFC

Day: Saturday, 24 March
Time: 11.55AM
Weather: Fine all day.
Attack Location: Near Givenchy
Crash Location: Near Givenchy
Side of Lines: Friendly

RFC Communique No. 80: "Hostile aircraft were active in the neighbourhood of Lens and Arras."

KOFL 6. Armee Weekly Activity Report: "11.55 vorm. 1 Nieuport-Spad Einsitzer mit Hispano-Suizia-Motor bei Vimy (diesseits) dch. Oblt. Frhr.v.Richthofen, J.St. 11 (als 30) Insasse gefangen, Apparat zertrümmert.

11.55AM 1 Nieuport-SPAD single-seater with Hispano-Suiza engine near Vimy (this side [of lines]) by *Oblt. Frhr.* von Richthofen, *Jagdstaffel* 11 (as 30th). Inmate [pilot] captured, apparatus [airplane] smashed."

MvR Combat Report: "Occupant: Lieut. Baker. Plane: A Spad with Hispano Motor. The first here encountered. Plane No:6607 [*sic*].
Motor: Hispano-Suiza 140 H.P. Machine-gun No. Maxim 4810 [sic].

I was flying with several of my gentlemen when I observed an enemy squad passing our front. Aside from this squad two new one-seaters which I did not know were flying. They were extremely fast and handy. I attacked one of them and ascertained that my machine was the better one. After a long fight I managed to hit the adversary's tank. The propeller stopped running. The plane had to go down. As the fight had taken place above the trenches, adversary tried to escape, but I managed to force him to land behind our lines near Givenchy. The plane turned, in a shell-hole, upside down. The plane was taken by our troops.

Frhr. v. Richthofen.
Was acknowledged."

RFC Combat Casualties Report: "Machine was last seen by 2/Lt.Harding about 5 miles S.E. of LENS."

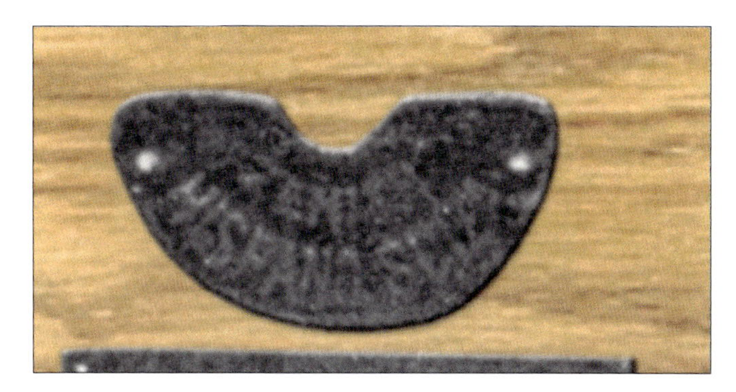

Above: Hispano-Suiza placard souvenired from A6706.

Above: Richthofen taking off in Albatros D.III 2006/16, La Brayelle, early April 1917.

RFC A/C:

Make/model/serial number: SPAD (*Société Pour L'Aviation et ses Dérivés*) VII C.1 A6706
Unit: No.19 Squadron RFC
Aerodrome: Fienvillers
Sortie: Offensive Patrol LENS-HARNIES-RENIN-LIETARD-QUIERY-BAILLEUL.
Colors/markings: Overall French beige/clear doped linen. White-bordered blue/white/red British roundel on fuselage. Black "dumbbell" squadron marking on fuselage, aft of roundel. Black serial number on vertical stabilizer. Blue/white/red rudder.
Engine: 140 hp Hispano-Suiza 8A
Engine Number: 5687
Gun: L4810
Manner of Victory: Gliding descent until deadstick forced landing.
Items Souvenired: Hispano-Suiza engine placard. Photographed on display at Roucourt, Richthofen's home bedroom in Schweidnitz, and in the post-war Richthofen Museum.

RFC Crew:

Pilot: Lt. Richard Plunkett Baker, 28, WiA/PoW
Manner of Injury: Gunshot wound to the right knee.
RFC Combat Casualties Report: "A letter from 2/Lt.R.P.Baker received by his brother states that the former is a prisoner (wounded) in Germany. He was forced to land in enemy lines owing to engine failure."
Repatriation: 12 December 1918

MvR A/C:

Make/model/serial number: Albatros D.III. Possibly *Le Petit Rouge*, or 2006/16.
Unit: *Jasta* 11
Commander: *ObLt*. Manfred von Richthofen
Airfield: La Brayelle
Colors/markings: *Le Petit Rouge* – See Vic. #17. Albatros D.III 2006/16 – See Vic. #23.
Damage: u/k

Initial Attack Altitude: u/k
Gunnery Range: u/k
Rounds Fired: u/k
Known Staffel Participants: u/k

Notes:

1. This was Richthofen's first victory after his 23 March promotion to *Oberleutnant*: *Besonderes: Am 23.3.17 wurde der Führer von J.St. 11 Frhr.v.Richthofen durch A.K. O ausser der Reihe zum Oberleutnant beföndert.*
   Special: On 23 March 17 the leader of *Jagdstaffel* 11 Frhr.v.Richthofen was promoted to *Oberleutnant* out of sequence by A.K. O ." ["Out of sequence" meaning outside of the normal sequence of promotion.]
2. Post-landing photograph of A6706 reveals damage consistent with an overturned landing (crumpled upper wing, broken propeller with mud on the hub, broken landing gear struts, bent and shorn off wheels), as Richthofen described.

Victory No.30 Statistics:

Single-Seaters
2nd of 3 March single-seater victories
4th of 15 1917 single-seater victories
12th of 18 pre-wound single-seater victories
12th of 35 total single-seater victories

SPAD
1st and only March SPAD victory
1st of 4 1917 SPAD victories
1st of 4 total SPAD victories

No.19 Squadron
1st of 3 No.19 SPAD victories
1st and only March No.19 Squadron victory
1st of 3 1917 No.19 Squadron victories
2nd of 4 total No.19 Squadron victories

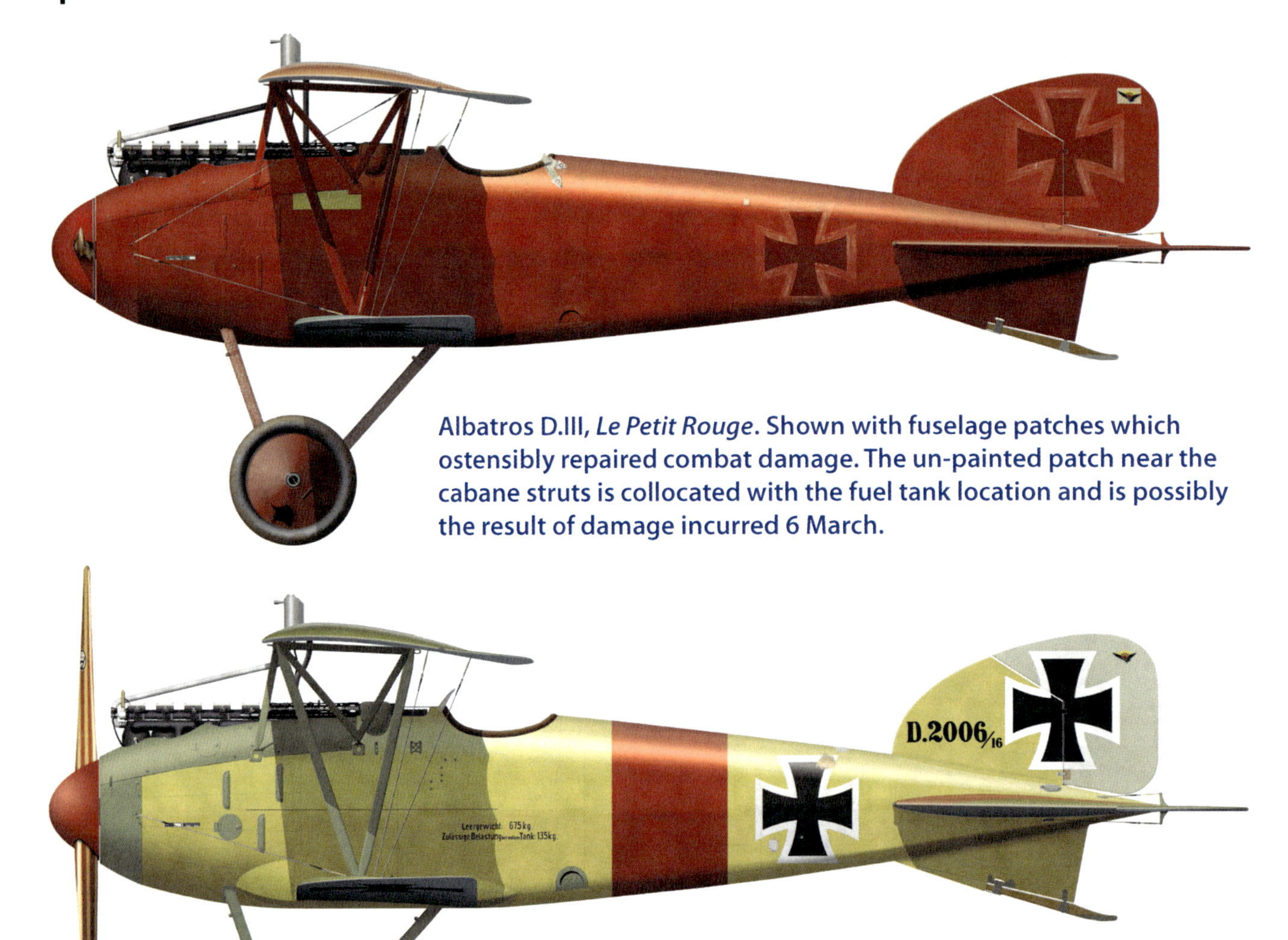

Albatros D.III, *Le Petit Rouge*. Shown with fuselage patches which ostensibly repaired combat damage. The un-painted patch near the cabane struts is collocated with the fuel tank location and is possibly the result of damage incurred 6 March.

Albatros D.III 2006/16. Its appearance is based on photographs taken during March 1917. Dates and circumstances regarding fuselage repair patches unknown.

Wounded

No.19 Squadron:
1st and only March No.19 Squadron crewman WiA
1st and only 1917 No.19 Squadron crewman WiA
1st and only total No.19 Squadron crewmen WiA

Single-Seaters:
1st and only March single-seater crewman WiA
1st of 3 1917 single-seater crewmen WiA
1st and only pre-wound single-seater crewman WiA
2nd of 8 total single-seater crewmen WiA

All:
2nd of 2 total March crewmen WiA
5th of 19 1917 crewmen WiA
7th of 18 total pre-wound crewmen WiA
7th of 26 total crewmen WiA

Etc.
9th of 10 total March victories
15th of 48 total 1917 victories
30th of 57 total pre-wound victories
13th victory in an Albatros D.III
3rd of 6 single-seater victories flying from La Brayelle
14th of 28 total victories flying from La Brayelle
3rd of 9 single-seater victories as *Staffelführer*
14th of 39 total victories flying as *Staffelführer*

Right: Albatros D.III 2006/16.

Victory No.31

Above: A Nieuport 23.

25 March
Nieuport 23 A6689
No.29 Squadron RFC

Day/Date: Sunday, 25 March
Time: 8.20AM
Weather: Clear in the morning with occasional clouds.
Attack Location: Over Douai
Crash Location: Near Tilley
Side of Lines: Friendly

RFC Communique No. 81: "A reconnaissance of No.11 Squadron encountered seven hostile aircraft in the Scarpe Valley."

KOFL 6. Armee Weekly Activity Report: "8.20 vorm. 1 Nieuport-Einsitzer bei Tilloy (diesseits) durch Oblt.Frhr.v.Richthofen J.St.11, (als 31.) 1 engl. Offizier gefangen, Apparat verbrannt.

8.20AM. 1 Nieuport single-seater near Tilloy (this side [of lines]) by *Oblt. Frhr.* von Richthofen, *Jagdstaffel* 11 (as 31st). 1 English officer captured, apparatus [airplane] burned."

MvR Combat Report: "Nieuport One-seater. Tilley. Inmate: Lieut. Grivert [sic]—English. Plane: Burnt.

An enemy squad had passed our lines. I went up, overtaking their last machine. After only a very few shots, the enemy's propeller stopped running. The adversary landed near Tilley upsetting his plane thereby. I observed that some moments later the plane began to burn.

Frhr. v. Richthofen.
Was acknowledged."

RFC Combat Casualties Report: "Left aerodrome 7-5 am. Officially reported prisoner of war (3rd Echelon Part II Orders, AFO.1810, dated 30/6/17)."

RFC A/C:
Make/model/serial number: Nieuport 23 A6689 (SFA N2874)
Manufacturer: Société Anonyme des Établissements Nieuport
Unit: No.29 Squadron RFC
Aerodrome: Le Hameau
Sortie: Escort.
Colors/markings: Generally, overall silver aluminum with b/w/r roundels on wings and fuselage; white-bordered black serial number on r/w/b rudder. An unofficial squadron marking of a red band squadron often was found aft of the fuselage roundel, and individual markings consisted of numbers ahead of the roundel.
Engine: Le Rhône 9J rotary
Engine: 110 hp Le Rhône 9J
Engine Number: T7401 J
Gun: One Lewis, no. 6632

Manner of Victory: Punctured fuel tank precipitated gliding descent until deadstick forced landing.
Items Souvenired: Wing fabric with SFA number. Photographed on display at Roucourt, Richthofen's home bedroom in Schweidnitz, and in the post-war Richthofen Museum.

RFC Crew:
Pilot: Lt. Christopher Guy Gilbert, PoW
Incarceration: Crefeld, Schwarmstedt, Holzminden, Graudenz
Repatriation: 2 December 1918

MvR A/C:
Make/model/serial number: Albatros D.III. Possibly *Le Petit Rouge*, or 2006/16.
Unit: *Jasta* 11
Commander: *ObLt.* Manfred von Richthofen
Airfield: La Brayelle
Colors/markings: *Le Petit Rouge* – See Vic. #17. Albatros D.III 2006/16 – See Vic. #23.
Damage: u/k
Initial Attack Altitude: u/k
Gunnery Range: u/k
Rounds Fired: "Very few shots"
Known Staffel Participants: u/k

Notes:
1. A discrepancy exists regarding the exact model of this Nieuport, whether it was a 17 or 23. The two were nearly identical, with the major difference being that the latter employed a 120 hp engine, rather than the former's 110 hp—yet it seems early examples of the Nieuport 23 still employed the 110 hp engine. In the Cross and Cockade International book *Nieuports in RNAS, RFC and RAF Service*, A6689's serial and SFA numbers are listed as those of a Nieuport 23. However, the same publication also says the RFC did not receive their first Nieuport 23 until May 1917—after Gilbert was shot down. The strongest association with a Nieuport 17 is found on the swatch of fabric souvenired from this machine and displayed in the post-war Richthofen Museum, which reads "2874 TYPE 17." Yet other identified N23s have been photographed with N17 stenciling, and records often list individual machines as both types, so a bit of uncertainty remains.

Above: A6689's souvenired SFA number, as seen in the Richthofen Museum.

Victory No.31 Statistics:

Single-Seaters
3rd of 3 March single-seater victories
5th of 15 1917 single-seater victories
13th of 18 pre-wound single-seater victories
13th of 35 total single-seater victories

Nieuport*
1st and only March Nieuport victory
1st of 4 1917 Nieuport victories
1st of 4 total Nieuport victories

No.29 Squadron
1st of 2 No.29 Nieuport victories*
2nd of 2 March No.29 Squadron victories
2nd of 3 1917 No.29 Squadron victories
3rd of 5 total No.29 Squadron victories

Etc.
10th of 10 total March victories
16th of 48 total 1917 victories
31st of 57 total pre-wound victories
14th victory in an Albatros D.III
4th of 6 single-seater victories flying from La Brayelle
15th of 28 total victories flying from La Brayelle
4th of 9 single-seater victories as *Staffelführer*
15th of 39 total victories flying as *Staffelführer*

Statistics Notes
**Includes N17 and N23 models.*

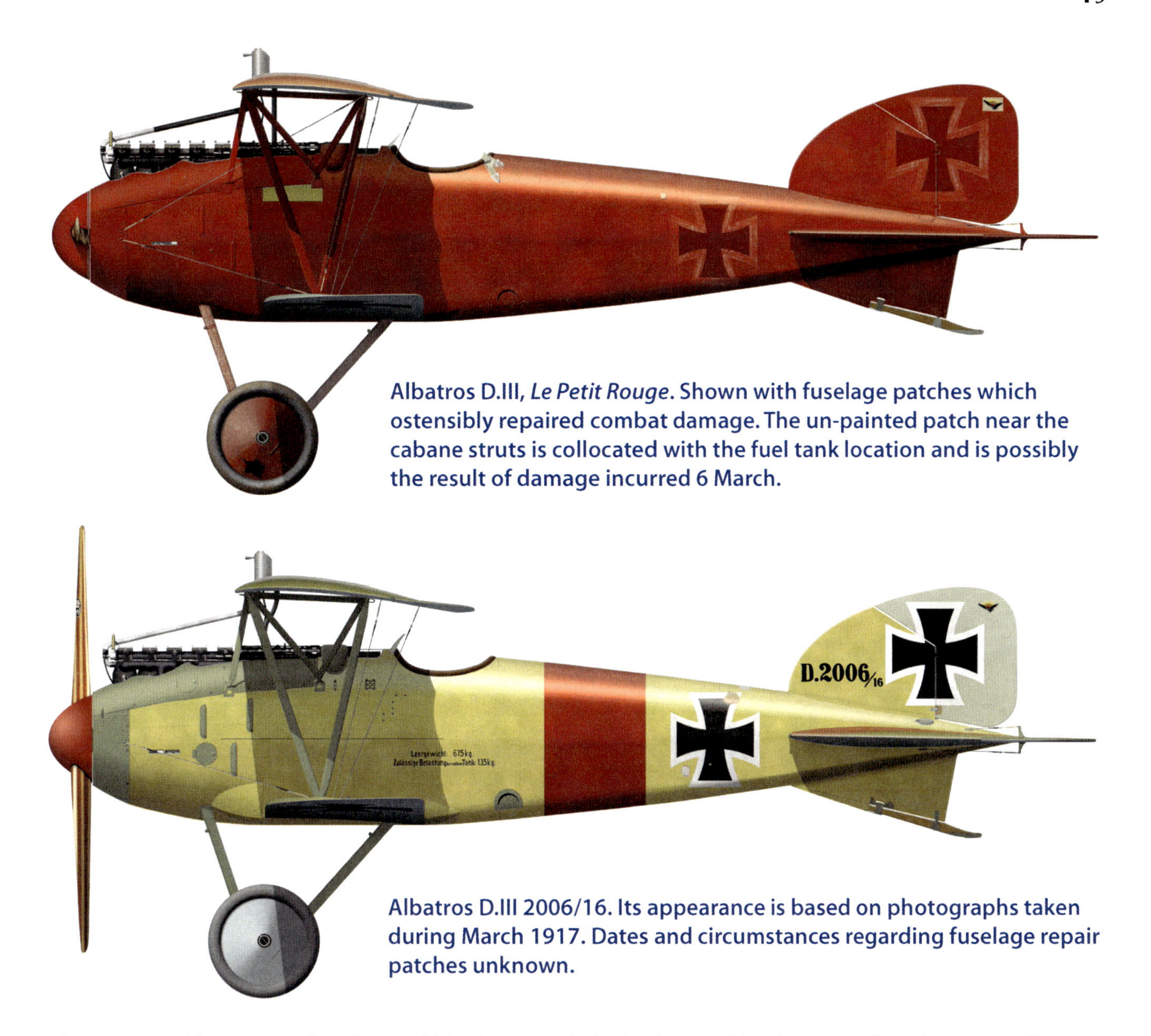

Albatros D.III, *Le Petit Rouge*. Shown with fuselage patches which ostensibly repaired combat damage. The un-painted patch near the cabane struts is collocated with the fuel tank location and is possibly the result of damage incurred 6 March.

Albatros D.III 2006/16. Its appearance is based on photographs taken during March 1917. Dates and circumstances regarding fuselage repair patches unknown.

Below: *Jasta* 11 Albatros D.IIIs face the wind (the direction of which is betrayed by blowing uniforms) at La Brayelle. The foreground machine preparing to depart is D.2006/16, and a German-language indication on the reverse of this photograph identifies Richthofen as the pilot. The wide-open expanse of La Brayelle is evident. (Lance Bronnenkant)

Victory No.32

2 April
BE.2d 5841
No.13 Squadron RFC

Day/Date: Monday, 2 April
Time: 8.35AM
Weather: Wind, rain, and low clouds.
Attack Location: Farbus
Crash Location: Farbus
Side of Lines: Friendly

KOFL 6. Armee Weekly Activity Report: "8.35 vorm. 1 F.E. Zweisitzer bei Farbus (diesseits) durch Oblt. Frhr.von Richthofen, J.St.11, (als 32.) 2 Insassen tot.

8.35AM. 1 F.E. two-seater near Farbus (this side [of lines]) by *Oblt. Frhr.* von Richthofen, *Jagdstaffel* 11, (as 32nd). 2 occupants dead."

MvR Combat Report: "Occupants: Both killed. Name of one – Lieut.Powell. The second inmate had no documents of identification. Plane NO. 5841. Motor P.D.1345/80.

I attacked an enemy artillery flyer. After a long fight I managed to force adversary nearly on to the ground, but without putting him out of action. The strong and gusty wind had driven the enemy plane over our lines. My adversary tried by jumping over trees and other objects to escape. Then I forced him to land in the village of Farbus where the machine was smashed against a house. The observer kept on shooting until the machine touched ground.

Frhr. v. Richthofen.
Was acknowledged."

RFC Combat Casualties Report: "Left aerodrome 7-47 am. An unconfirmed report in a German publication states that pilot and observer were both killed. German message states: Lt. P.T.G. [sic] Powell, Pilot, killed; A.M.P. Bonner, Obs. killed."

RFC A/C:
Make/model/serial number: Royal Aircraft Factory BE.2d 5841, built by the British and Colonial Aeroplane Co., Ltd., Filton, Bristol
Unit: No.13 Squadron RFC
Aerodrome: Savy
Sortie: Photography
Colors/markings: u/c. Generally, PC10 wings/ailerons/vertical and horizontal stabilizers/elevators; battleship gray engine cowl; b/w/r rudder; black serial number on vertical stabilizer. No.13 Sqn markings were a narrow horizontal bar running down the fuselage from observer's cockpit to the tail (black on CDL machines, white on PC10).
Engine: 90 H.P. R.A.F. 1a
Engine Number: 21962 WD 278
Gun: 19594, 2966
Manner of Victory: Descending fight led to low-altitude pursuit until airplane fatally impacted a house.
Items Souvenired: Port and starboard white serial number on PC10 fabric. Photographed on display at Roucourt, Richthofen's home bedroom in Schweidnitz, and in the post-war Richthofen Museum.

Above: Fabric swatch cut from 5841, as seen on Richthofen's wall at Roucourt.

RFC Crew:
Pilot: Lt. Patrick John Gordon Powell, 21 (DoB 20 September 1896), KiA
Cause of Death: Either gunshot wound(s) or trauma associated with airplane crash.
Burial: Either in-field or laid where fallen; location lost during turmoil of war.
Obs: 1st Class Air Mechanic Percy Bonner, 23, KiA
Cause of Death: Trauma associated with airplane crash.
Burial: Either in-field or laid where fallen; location lost during turmoil of war.

MvR A/C:
Make/model/serial number: Albatros D.III, *Le*

Albatros D.III, *Le Petit Rouge*. Shown with fuselage patches which ostensibly repaired combat damage. The un-painted patch near the cabane struts is collocated with the fuel tank location and is possibly the result of damage incurred 6 March.

Petit Rouge
Unit: *Jasta* 11
Commander: *ObLt*. Manfred von Richthofen
Airfield: La Brayelle
Colors/markings: See Vic. #17.
Damage: Richthofen reported his plane was struck by return fire but did not specify where and to what extent. Lothar von Richthofen described his brother's Albatros as having "a few hits."

Initial Attack Altitude: u/k. Final stages were below tree-height.
Gunnery Range: u/k
Rounds Fired: u/k
Known Staffel Participants: None. Solo flight.

Notes:

1. It is nearly certain Richthofen flew *Le Petit Rouge* during this sortie, based on his account of this victory in *Der Rote Kampfflieger*:

"I was still in bed when the orderly rushed in crying: "Herr *Leutnant*, the English are here!" Still somewhat sleepy, I looked out the window and there circling over the field were my dear 'friends.' I got out of bed, quickly put my things on and got ready. My *red bird* [author's emphasis] was all set to begin the morning's work. My mechanics knew that I would not let this favorable moment go by without taking advantage of it. Everything was ready. I quickly donned my flight suit and was off.

"Even so, I was last to start. My comrades got much closer to the enemy. Then, suddenly, one of the impudent fellows fell on me, attempting to force me down. I calmly let him come on, and then we began a merry dance. At one point my opponent flew on his back, then he did this, then that. It was a two-seater. I was superior to him, and he soon realized that he could not escape me. During a pause in the fighting I looked around and saw that we were alone. Therefore, whoever shot better, whoever had the greatest calm and the best position in the moment of danger, would win.

"It did not take long. I squeezed under him and fired, but without causing serious damage. We were at least two kilometers from the Front and I thought he would land, but I miscalculated my opponent. When only a few meters above the ground, he suddenly leveled off and flew straight ahead, seeking to escape me. That was too bad for him. I attacked him again and went so low that I feared I would touch the houses in the village beneath me. The Englishman kept fighting back. Almost at the end I felt a hit on my machine. I must not let up now; he must fall. He crashed at full speed into a block of houses, and there was not much left. It was again a case of splendid daring. He defended himself right up to the end.

"Very pleased with the results of my red 'bicycle' in the morning's work, I turned back. My comrades were still in the air and were very surprised when, as we later sat down to breakfast, I told them of my number thirty-two."

Although undated, Lothar has an account in *Ein Heldenleben* that apparently describes the same day:

"We sat in the ready room between four and

five o'clock [am] all set to jump up, when the telephone rang: 'Six Bristols from Arras to Douai.' We dashed out to our planes and took off.

"At three thousand meters there was a break in the clouds. We saw the Englishman right under the cloud cover in the vicinity of our airfield. My brother's *red bird* [author's emphasis] ["der rote Vogel"] stood on the field ready to go, his mechanics nearby. There was nothing to be seen of my brother...We landed at our field after about an hour. My brother's red bird still stood on the field, but even from far away the work of the mechanics and the position of the machine showed that he had been in action. Back in the ready room, we were told: Yes the *Rittmeister* [*sic*] had started about five minutes after the squadron. He had been in bed when the reports came. He had quickly put his flight suit on over his pajamas and was off. He returned after twenty minutes, and in that time had shot down an Englishman on this side of the lines. When we came back again, he was in bed, sleeping again, as if nothing had happened. Only some hits in his machine and reports which poured in about the airplane shot down gave evidence of his flight."

Lothar's post-war account is not free of discrepancy. During those events he described participating in his first aerial combat [as fighter pilot], attacking an English airplane, and causing a fuel leak in same. In a letter to his mother, Manfred wrote, "Yesterday I shot down my thirty-first, and my thirtieth the day before... Lothar had his first aerial combat yesterday..." This dates Lothar's account as taking place on 25 March, the date of Manfred's 31st victory, not 2 April. Also, Lothar recounts that Manfred's victory had "plunged down burning on our side," although neither Manfred's 31st nor 32nd victories crashed in flames.

It is likely these discrepancies can be attributed to the passage of time and memory coalescence. Overall, Lothar's account largely matches Manfred's, and both men described Manfred's plane as a "red bird." Although *Jasta* 11 would soon adopt red as the base unit identifying color, evidence suggests that in very early April Manfred's Albatros was still the only one so painted.

Victory No.32 Statistics:

Two-Seaters
1st of 17 April two-seater victories
12th of 33 1917 two-seater victories
19th of 39 pre-wound two-seater victories
19th of 45 total two-seater victories

BE.2 Δ
1st of 7 April BE.2 victories
9th of 15 1917 BE.2 victories
11th of 17 total BE.2 victories

No.13 Squadron ±
1st of 4 No.13 BE.2 victories
1st of 4 April No.13 Squadron victories
1st of 4 1917 No.13 Squadron victories
1st of 4 total No.13 Squadron victories

Deaths †
No.13 Squadron:
1st and 2nd of 4 April No.13 Squadron crewmen KiA
1st and 2nd of 4 1917 No.13 Squadron crewmen KiA
1st and 2nd of 4 total No.13 Squadron crewmen KiA

Two-Seaters:
1st and 2nd of 20 April two-seater crewmen KiA
16th and 17th of 44 1917 two-seater crewmen KiA
27th and 28th of 54 pre-wound two-seater crewmen KiA
27th and 28th of 63 total two-seater crewmen KiA

All:
1st and 2nd of 23 total April crewmen KiA
19th and 20th of 53 total 1917 crewmen KiA
36th and 37th of 66 total pre-wound crewmen KiA
36th and 37th of 84 total crewmen KiA

Etc.
1st of 21 total April victories
17th of 48 total 1917 victories
32nd of 57 total pre-wound victories
15th victory in an Albatros D.III
12th of 22 two-seater victories flying from La Brayelle
16th of 28 total victories flying from La Brayelle
12th of 30 two-seater victories as *Staffelführer*
16th of 39 total victories flying as *Staffelführer*

Statistics Notes
Δ *All BE.2 victories are pre-wound*
± *All No.13 Squadron victories are pre-wound*
† *All No.13 Squadron deaths are pre-wound; KiA includes DoW*

Victory No.33

Above: A Sopwith 1½ Strutter.

2 April
Sopwith 1½ Strutter A2401
No.43 Squadron RFC

Day/Date: Monday, 2 April
Time: 11.15AM
Weather: Wind, rain, snow flurries and low clouds
Attack Location: East of Vimy, "above closed cloud cover on the enemy's side."
Forced Landing Location: 300 metres east of Givenchy (MvR) near Avion (Warren)
Side of Lines: Friendly

KOFL 6. Armee Weekly Activity Report: "11.20 vorm. 1 Sopwith-Zweisitzer bei Givenchy (diesseits) durch Oblt.Frhr.von Richthofen, J.St.11 (als 33.) 1 Insasse tot, der andere gefangen."

11.20AM. 1 Sopwith two-seater near Givenchy (this side [of lines]) by *Oblt. Frhr.* von Richthofen, *Jagdstaffel* 11 (as 33rd). 1 occupant dead, the other captured.

MvR Combat Report: "Sopwith Two-Seater. Givenchy. Plane A. 2401. Motor Clerget Blin without number, Type 2. Occupants: Sergt. Dunn and Lieut. Warrens.

Together with *Lts.*Voss and Lothar *Frhr.* v. Richthofen I attacked an enemy squad of 8 Sopwiths above the closed cover of clouds on the enemy's side. The plane I had singled out was driven away from its squad and gradually came over to our side. The enemy plane tried to escape and hide in the clouds after I had holed its benzine tank. Below the clouds I immediately attacked him again, thereby forcing him to land 300 meters east of Givenchy. But as yet my adversary would not surrender and even as his machine was on the ground, he kept shooting at me, thereby hitting my machine very severely in an altitude of 5 meters. I once more attacked him, already on the ground, and killed one of the occupants."

RFC Combat Casualties Report: "Left aerodrome 10.30AM. Machine engaged H.A. about 11.10AM and was last seen by other pilots in a spiral but under control E. of VIMY and disappeared into the cloud still under control. German message states Lt. P. Warren is an unwounded prisoner and Sergt.Dunn killed."

RFC A/C:

Make/model/serial number: Sopwith 1½ Strutter A2401

Manufacturer: Subcontracted to and built by Ruston, Proctor & Co., Ltd., Lincoln.
Unit: No.43 Squadron RFC
Aerodrome: Treizennes
Sortie: Photography Patrol E. of VIMY
Colors/markings: u/c. Generally, PC10 wings/ailerons/vertical and horizontal stabilizers/elevators; engine cowls either bare "engine turned" metal, PC10, or identifying Flight color; b/w/r rudder; black serial number with white border.
Engine: Clerget
Engine Number: R1072 WD 6283
Guns: One Vickers, one Lewis. Nos. 21341, L7765.
Manner of Victory: Gliding descent and forced dead-stick landing.
Items Souvenired: Black serial number with white border on PC10 background. Photographed on display at Roucourt, Richthofen's home bedroom in Schweidnitz, and in the post-war Richthofen Museum.

RFC Crew:
Pilot: 2nd Lt. Algernon Peter Warren, 19, PoW RFC
Combat Casualties Report: "Letter d-15/5/17 from the father of Lt.L.Dodson states that 2/Lt. A.P. Warren is a prisoner of war."
Incarceration: Karlsruhe, Schwarmstedt
Repatriation: 17 December 1918
Obs: Sgt Reuel Dunn, 24, PoW/DoW
Obs. Victories: 1 (destroyed)

Victory#	Date	Details
1	4 March 1917	Type u/k

RFC Combat Casualties Report: "An unconfirmed report in a German publication states that Sergt. Dunn was killed."
Cause of Death: Gunshot wound(s) to abdomen.
Burial: Cabaret Rouge British Cemetery, Souchez, France, Grave XV M 24

MvR A/C:
Make/model/serial number: Albatros D.III, *Le Petit Rouge*
Unit: *Jasta* 11
Commander: *ObLt*. Manfred von Richthofen
Airfield: La Brayelle
Colors/markings: See Vic. #17.
Damage: Richthofen reported his plane was hit "very severely" but did not specify where and to what extent.

Initial Attack Altitude: ca. 12,000 (Warren).
Gunnery Range: u/k
Rounds Fired: u/k
Known Staffel Participants: Ltn. Lothar von Richthofen, *Ltn*. Werner Voss (*Jasta* Boelcke)

Above: Dunn's headstone, 2011.

Notes:

1. Richthofen discusses this victory in *Der Rote Kampfflieger* but the account edits the strafing, likely because it would "not look good" to civilians in general and the British in particular, who it was known would read the book eventually. And indeed they did; in 1918 it was translated into English and published in London as *The Red Air Fighter*. In it, the account of this victory reads, in part:

"When he had come to the ground [landed] I flew over him at an altitude of about thirty-feet in order to ascertain whether I had killed him or not. What did the rascal do? He took his machine-gun and shot holes into my machine. Afterwards, Voss told me if that had happened to him he would have shot the aviator on the ground. As a matter of fact I ought to have done so, for he had not surrendered. He was one of the few fortunate fellows who escaped with their lives."

Albatros D.III, *Le Petit Rouge*. Shown with fuselage patches which ostensibly repaired combat damage. The un-painted patch near the cabane struts is collocated with the fuel tank location and is possibly the result of damage incurred 6 March.

Strafing a downed enemy is considered by many to be "unfair" and on occasion Richthofen is regarded negatively for having done so. Curiously, similar sentiments normally are not extended to others, such as Werner Voss and Albert Ball, who also strafed downed machines.

2. Although strafed, neither Warren nor Dunn were wounded in this attack. In fact, during a post-war interview with Floyd Gibbons, Warren revealed he did not realize they had been strafed. "I managed to flatten out somehow in the landing and piled up with an awful crash. As I hit the ground, the red machine swooped over me, but I don't remember him firing on me when I was on the ground." He also said that immediately afterward he was struggling to remove the wounded Dunn from the observer's cockpit and that neither of them had fired on Richthofen's low-flying Albatros.

Warren also recalled Richthofen's initial attack was from out of the sun and caught the British crew by surprise. "The first notice I had of the attack was when I heard Dunn from his seat behind me shout something at me, and at the same time a spray of bullets went over my shoulder from behind and splintered the dashboard almost in front of my face." Warren took immediate evasive action and upon stealing a look over his shoulder saw that "Dunn was not in sight. I did not know whether he had been thrown out of the plane in my quick dive or was lying dead at the bottom of his cockpit." Employing his usual no-quarter *modus operandi* Richthofen attacked the Sopwith all the way down, long after its engine had stopped and its trailing fuel vapor revealing the tanks had been hit and a forced landing was inevitable: "I was busy with the useless controls all the time [the elevator control cables had been severed by bullets], and going down at a frightful speed, but the red machine seemed to be able to keep itself poised just above and behind me all the time, and its machine guns were working every minute. I found later that bullets had gone through both of my sleeves and both of my boot legs, but in all of the firing, not one of them touched me, although they came uncomfortably close."

3. Post-war Warren recalled Dunn may have had "three Hun machines to his credit." At least one is mentioned in RFC Communique No.78: "March 4. Hostile aircraft were destroyed by the following, during the day: One by 2nd-Lieut. C.P. Thornton and Sergt. R. Dunn, No.43 Squadron."

Victory No.33 Statistics:

Two-Seaters
2nd of 17 April two-seater victories
13th of 33 1917 two-seater victories
20th of 39 pre-wound two-seater victories
20th of 45 total two-seater victories

Sop. Strutter Δ
1st of 2 April Sop. Strutter victories
2nd of 3 1917 Sop. Strutter victories
2nd of 3 total Sop. Strutter victories

No.43 Squadron ±
2nd of 3 No.43 Sop. Strutter victories
1st of 2 April No.43 Squadron victories
2nd of 3 1917 No.43 Squadron victories
2nd of 3 total No.43 Squadron victories

Deaths †
No.43 Squadron:
1st of 2 April No.43 Squadron crewmen KiA
3rd of 4 1917 No.43 Squadron crewmen KiA
3rd of 4 total No.43 Squadron crewmen KiA

Two-Seaters:
3rd of 20 April two-seater crewmen KiA
18th of 44 1917 two-seater crewmen KiA
29th of 54 pre-wound two-seater crewmen KiA
29th of 63 total two-seater crewmen KiA

All:
3rd of 23 total April crewmen KiA
21st of 53 total 1917 crewmen KiA
38th of 66 total pre-wound crewmen KiA
38th of 84 total crewmen KiA

Etc.
2nd of 21 total April victories
18th of 48 total 1917 victories
33rd of 57 total pre-wound victories
16th victory in an Albatros D.III
13th of 22 two-seater victories flying from La Brayelle
17th of 28 total victories flying from La Brayelle
13th of 30 two-seater victories as *Staffelführer*
17th of 39 total victories flying as *Staffelführer*

Statistics Notes
∆ All Sop. Strutter victories are pre-wound
± All No.43 Squadron victories are pre-wound
† All No.43 Squadron deaths are pre-wound; KiA includes DoW

Above: Richthofen (center, walking toward camera) in front of Albatros D.III 2051/16. At left is Rumpler C.I 4633/15, and at right is Prinz Karl's Albatros D.I.

Victory No.34

Above: A6382's souvenired serial number from the port side of the rudder, hanging in Richthofen's room at Roucourt.

Above: A6382's starboard side rudder serial number, as seen at the Richthofen Museum.

3 April
FE.2d A6382
No.25 Squadron RFC

Day/Date: Tuesday, 3 April
Time: 4.15PM
Weather: Storms and low clouds
Attack Location: Between Lens and Lieven
Forced Landing Location: "Near Lieven" (MvR) "Just outside Lens on the southeast side." (McDonald)
Side of Lines: Friendly

KOFL 6. Armee Weekly Activity Report: "4.15 nachm. 1 F.E. Zweisitzer bei Lens (diesseits) durch Oblt. Frhr.von Richthofen, J.St.11 (als 34.)

4.15PM. 1 F.E. two-seater near Lens (this side [of lines]) by *Oblt. Frhr.* von Richthofen, *Jagdstaffel* 11 (as 34th)."

MvR Combat Report: "Occupants: Pilot: Lieut. O'Bierne, killed. Observer: McDonald. Plane A.6382. Motor unrecognisable.

Together with Lieut. Schaefer and Lieut.v. Richthofen Lothar, I attacked three enemy planes. The plane I myself attacked was forced to land near Lieven. After a short fight the motor began to smoke and the observer ceased shooting. I followed adversary to the ground.

Frhr. v. Richthofen
Was acknowledged."

RFC Combat Casualties Report: "Left aerodrome 3.12PM. Machine was driven down by H.A. in the vicinity of LENS in enemy territory. Letter d.15/5/17 from the father of Lt.L.Dodson states that 2/Lt.D.P.McDonald is a prisoner of war and that 2/Lt.J.I.M.O'Beirne was killed."

RFC A/C:
Make/model/serial number: Royal Aircraft Factory FE.2d A6382
Manufacturer: Subcontracted to and built by Boulton & Paul Ltd and Howes & Sons Ltd, Norwich.
Unit: No.25 Squadron RFC
Aerodrome: Lozinghem
Sortie: Photography MERICOURT - GRAVELLE
Colors/markings: Generally, PC10 nacelle with one, two, or three horizontal white-bordered black bands across nose (corresponding with A, B, or C Flight); nacelle; wing/ailerons/horizontal stabilizers/elevators uppersurfaces PC10 with CDL undersurfaces; likely upperwing roundels had white surrounds.
Engine: Rolls Royce
Engine Number: 1/275/73 WD13896
Guns: Two Lewis. Nos. 26751, 24934.
Manner of Victory: Gliding descent and forced dead-stick landing, during which machine overturned.
Items Souvenired: a. A6382 port side of rudder. Black numbers/white borders on r/w/b background. Photographed on display at Roucourt, Richthofen's home bedroom in Schweidnitz, and in the post-war Richthofen Museum.
b. Boulton & Paul Ltd/Howes & Sons Ltd manufacture placard. Photographed on display at Roucourt and the post-war Richthofen Museum.
c. Rolls Royce engine placard. Photographed on display at Roucourt and the post-war Richthofen Museum.

Above: A6382's souvenired manufacture placard.

RFC Crew:

Pilot: 2nd Lt. Donald Peter McDonald, 21 or 22 (DoB 1895), PoW
Incarceration: Saarbrücken
Repatriation: December 1918
Obs: 2nd Lt. John Ingram Mullaniffe O'Beirne, 23 (DoB 24 April 1893), KiA
Cause of Death: Gunshot wound to head.
Burial: Either in-field or laid where fallen; location lost during turmoil of war.

MvR A/C:

Make/model/serial number: Possibly *Le Petit Rouge*, or 2006/16.
Unit: *Jasta* 11
Commander: *ObLt*. Manfred von Richthofen
Airfield: La Brayelle
Colors/markings: *Le Petit Rouge* – See Vic. #17. Albatros D.III 2006/16 – See Vic. #23.
Damage: n/a

Initial Attack Altitude: u/k
Gunnery Range: u/k
Rounds Fired: u/k
Known Staffel Participants: Ltn. Lothar von Richthofen, *Ltn*. Emil Schaefer

Notes:

1. In a letter written to O'Beirne's mother, published in Gibbons's *The Red Knight of Germany*, McDonald indicated Richthofen "...followed me right down to the ground, firing all the time, till he almost shot away every control I had." This is consistent with Richthofen's usual no-quarter *modus operandi*.
2. Regarding O'Beirne's fate, McDonald wrote, "Jack was firing over my top plane [wing] when suddenly a close burst from behind hit him right in the head and he dropped down in his seat... When...I...looked over to see how he was, I saw he was hit right through the side of the head, and death must have been instantaneous."

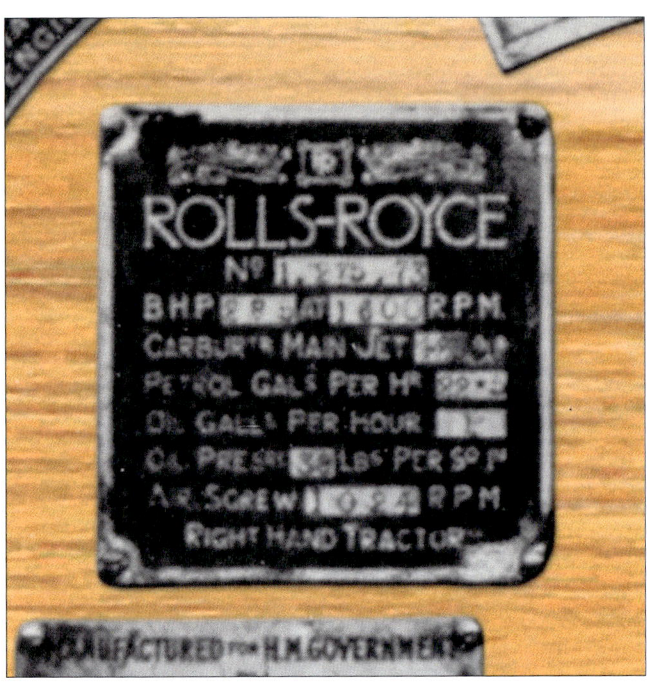

Above: A6382's souvenired Rolls Royce engine placard.

3. About his forced landing, McDonald wrote, "I made for a good-looking field, but, as luck would have it, beneath the long grass were barbed-wire entanglements, and with all my instruments shot away, I landed at a faster pace than I should have done and caught in the wire, going head over heels... Fortunately, the machine went right over the top of me when I was thrown out." He then describes last seeing O'Beirne "laid out on the grass beside the wreckage of my machine..."
4. Richthofen's combat report described A6382's engine as "unrecognizable." It is not known if he was unfamiliar with the Rolls Royce or if this indicated the engine was badly damaged during the tumbling crash landing.
5. The identity of Richthofen's Albatros for this victory is elusive. Although *Le Petit Rouge* and 2006/16 are suggested, it could have been yet another machine.
6. In his combat report, Richthofen misidentified the crew positions.

Victory No.34 Statistics:

Pushers*
1st of 5 April pusher victories
1st of 5 April two-seater pusher victory
3rd of 7 1917 two-seater pusher victories
8th of 12 total two-seater pusher victories
14th of 18 total pusher victories

Albatros D.III, *Le Petit Rouge*. Shown with fuselage patches which ostensibly repaired combat damage. The un-painted patch near the cabane struts is collocated with the fuel tank location and is possibly the result of damage incurred 6 March.

Albatros D.III 2006/16. Its appearance is based on photographs taken during March 1917. Dates and circumstances regarding fuselage repair patches unknown.

Two-Seaters
3rd of 17 April two-seater victories
14th of 33 1917 two-seater victories
21st of 39 pre-wound two-seater victories
21st of 45 total two-seater victories

FE.2 Δ
1st of 5 April FE.2 victories
3rd of 7 1917 FE.2 victories
8th of 12 total FE.2 victories

No.25 Squadron ±
3rd of 4 No.25 FE.2 victories
1st of 2 April No.25 Squadron victories
3rd of 4 1917 No.25 Squadron victories
3rd of 4 total No.25 Squadron victories

Deaths †
No.25 Squadron:
1st of 3 April No.25 Squadron crewmen KiA
3rd of 5 1917 No.25 Squadron crewmen KiA
3rd of 5 total No.25 Squadron crewmen KiA

Pushers:
1st of 3 April two-seater pusher crewmen KiA
3rd of 7 1917 two-seater pusher crewmen KiA
13th of 17 total two-seater pusher crewmen KiA

Two-Seaters:
4th of 20 April two-seater crewmen KiA
19th of 44 1917 two-seater crewmen KiA
30th of 54 pre-wound two-seater crewmen KiA
30th of 63 total two-seater crewmen KiA

All:
4th of 23 total April crewmen KiA
22nd of 53 total 1917 crewmen KiA
39th of 66 total pre-wound crewmen KiA
39th of 84 total crewmen KiA

Etc.
3rd of 21 total April victories
19th of 48 total 1917 victories
34th of 57 total pre-wound victories
17th victory in an Albatros D.III
14th of 22 two-seater victories flying from La Brayelle
18th of 28 total victories flying from La Brayelle
14th of 30 two-seater victories as *Staffelführer*
18th of 39 total victories flying as *Staffelführer*

Statistics Notes
* *All pusher victories are pre-wound*
Δ *All FE.2 victories are pre-wound*
± *All No.25 Squadron victories are pre-wound*
† *All No.25 Squadron deaths are pre-wound*

Above: The more powerful FE.2d was visually distinguished from the earlier FE.2b by the cut down pilot nacelle and exposed radiator jutting into the relative wind behind the pilot. This example is FE.2d A5, shown here unarmed at Johannisthal after being captured 1 June 1916. (DEHLA Collection)

Below: The Rolls Royce engine in FE.2d A6516 enabled it to carry three machine guns, including a fixed gun for the pilot. This presentation aeroplane, *The Colony of Mauritius No.13*, was flown by 20 Squadron in the summer of 1916.

Victory No.35

Above: Bristol F.2A

5 April
Bristol F.2A A3340
No.48 Squadron RFC

Day/Date: Thursday, 5 April
Time: 11.15AM
Weather: Misty and cloudy
Attack Location: Lewarde, southeast of Douai
Forced Landing Location: Near Lewarde.
Side of Lines: Friendly

KOFL 6. Armee Weekly Activity Report: "11.00 vorm. 1 fdl.Flz. bei Quincy (diesseits) durch Oblt. Frhr. von Richthofen. J.St.11 (als 35.)

11.00AM. 1 enemy airplane [*feindlichen Flugzeug*, or "*fdl Flz*"] near Quincy (this side [of lines]) by *Oblt. Frhr.* von Richthofen, *Jagdstaffel* 11 (as 35th)."

MvR Combat Report: "Bristol Two-Seater. Occupants: Lieut. McLickler [*sic*] and Lieut. George; both seriously wounded. Plane No.3340. Motor No.10443.

It was foggy and altogether very bad weather when I attacked an enemy squad on its flying between Douai – Valenciennes. Up to this point it had managed to advance without being fired upon. I attacked with four planes of my *Staffel*. I personally singled out the last machine which I forced to land after a short fight near Lewarde. The occupants burnt their machine. It was a new type of plane which we had not known as yet; it appears to be quick and rather handy. A powerful motor, V-shaped, 12 cyl; its name could not be recognisable.

The [Albatros] D.III was both in speed and in ability to rise, undoubtedly superior.

Of the enemy squad which consisted of 6 planes, four were forced to land on our side by my *Staffel*.

Frhr. v. Richthofen.
Was acknowledged."

RFC Combat Casualties Report: "2/Lt.H.B.Griffith, one of the patrol, reports that about 10.30AM he saw one machine diving emitting smoke over DOUAI. When the formation was last seen about 11AM three of the machines were heavily engaged with a superior number of H.A., this during his second combat over DOUAI."

Above: A-3340's souvenired serial number at Roucourt.

Above: A-3340's souvenired serial number now mounted on a backing and hanging at the Richthofen Museum.

RFC A/C:

Make/model/serial number: Bristol F2.A A3340
Manufacturer: British & Colonial Aeroplane Co., Ltd., Filton and Brislington, Bristol
Unit: No.48 Squadron RFC
Aerodrome: Bertangles
Sortie: Offensive Patrol - DOUAI
Colors/markings: Generally PC10 fuselage and uppersurfaces of wings/ailerons/horizontal stabilizers/elevators, with CDL undersurfaces. PC10 vertical stabilizer with white serial number. Battleship gray metal engine cowls. b/w/r rudder
Engine: Rolls Royce Falcon I
Engine Number: 1/190/30 WD 10443
Guns: One Vickers, one Lewis. Nos. 22089, A1194.
Manner of Victory: Gliding descent and forced dead-stick landing, after which crew burned the machine.
Items Souvenired: Serial number "A3340" from the vertical stabilizer (side u/k). White numbers on PC10 background. Photographed on display at Roucourt, Richthofen's home bedroom in Schweidnitz, and in the post-war Richthofen Museum.

RFC Crew:

Pilot: 2nd Lt. Arthur Norman Lechler, 27 (DoB 13 February 1890), WiA/PoW
RFC Combat Casualties Report: "Wire received by relatives through Geneva Red Cross states that 2/Lt. Leckler and Lt. George are wounded prisoners of war."
Manner of Injury: Gunshot wounds to head (presumably glancing) and leg
Internment: 9 April 1918 into neutral Holland for treatment of wounds
Repatriation: 7 September 1918
Obs: 2nd Lt. Herbert Duncan King George, 19 (DoB 23 July 1897), PoW/DoW 6 April 1917.
RFC Combat Casualties Report: "German message states that Lt.N.A.Leckler is wounded and Lt.George was killed."
Cause of Death: Gunshot wounds to leg and back.
Burial: Douai Cemetery, Douai, France, Grave D.7.

MvR A/C:

Make/model/serial number: Albatros D.III, possibly *Le Petit Rouge*, or 2006/16
Unit: *Jasta* 11
Commander: *ObLt.* Manfred von Richthofen
Airfield: La Brayelle
Colors/markings: *Le Petit Rouge* – See Vic. #17. Albatros D.III 2006/16 – See Vic. #23.
Damage: n/a

Initial Attack Altitude: u/k
Gunnery Range: u/k
Rounds Fired: u/k
Known Staffel Participants: Vzfw. Sebastian Festner, *Ltn.* Georg Simon

Victory No.35 Statistics:

Two-Seaters
4th of 17 April two-seater victories
15th of 33 1917 two-seater victories
22nd of 39 pre-wound two-seater victories
22nd of 45 total two-seater victories

BF.2 Δ
1st of 2 April BF.2 victories
1st of 2 1917 BF.2 victories
1st of 3 total BF.2 victories

No.48 Squadron ±
1st of 2 No.48 BF.2 victories
1st of 2 April No.48 Squadron victories
1st of 2 1917 No.48 Squadron victories
1st of 2 total No.48 Squadron victories

Above: George's headstone, 2011.

Deaths †
No.48 Squadron:
1st and only April No.48 Squadron crewman KiA
1st and only 1917 No.48 Squadron crewman KiA
1st and only total No.48 Squadron crewmen KiA

Two-Seaters:
5th of 20 April two-seater crewmen KiA
20th of 44 1917 two-seater crewmen KiA
31st of 54 pre-wound two-seater crewmen KiA
31st of 63 total two-seater crewmen KiA

All:
5th of 23 total April crewmen KiA
23rd of 53 total 1917 crewmen KiA
40th of 66 total pre-wound crewmen KiA
40th of 84 total crewmen KiA

Wounded
No. 48 Squadron:
1st of 2 April No. 48 Squadron crewmen WiA
1st of 2 1917 No. 48 Squadron crewman WiA
1st of 2 total No. 48 Squadron crewmen WiA

Two-Seaters:
1st of 11 April two-seater crewmen WiA
5th of 16 1917 two-seater crewmen WiA
6th of 16 pre-wound two-seater crewmen WiA
6th of 18 total two-seater crewmen WiA

All:
1st of 11 total April crewmen WiA
6th of 19 total 1917 crewmen WiA
8th of 18 total pre-wound crewmen WiA
8th of 26 total crewmen WiA

Etc.
4th of 21 total April victories
20th of 48 total 1917 victories
35th of 57 total pre-wound victories
18th victory in an Albatros D.III
15th of 22 two-seater victories flying from La Brayelle
19th of 28 total victories flying from La Brayelle
15th of 30 two-seater victories as *Staffelführer*
19th of 39 total victories flying as *Staffelführer*

Statistics Notes
Δ *All BF.2 victories are pre-wound*
± *All No.48 Squadron victories are pre-wound*
† *All No.48 Squadron deaths are pre-wound*

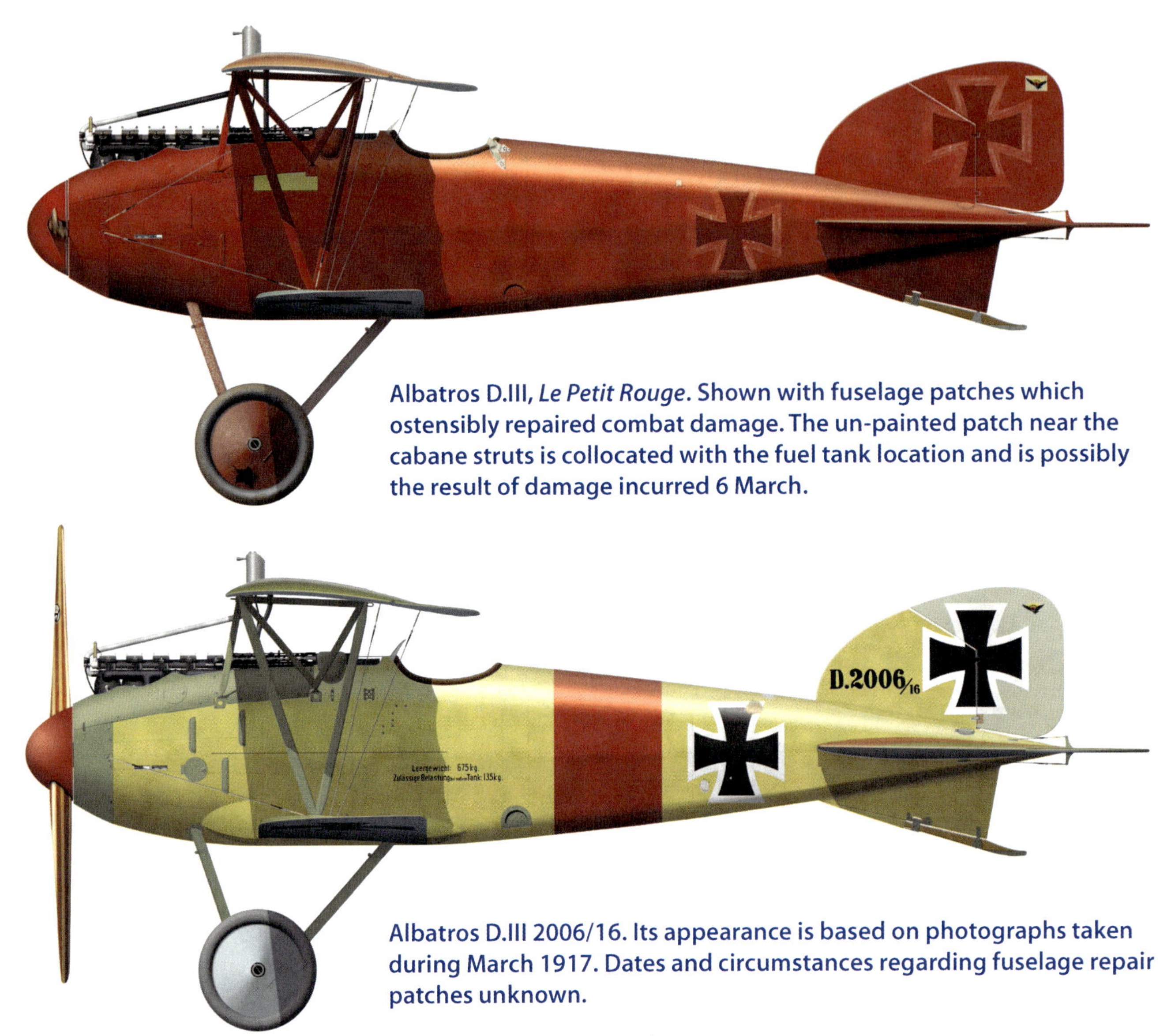

Albatros D.III, *Le Petit Rouge*. Shown with fuselage patches which ostensibly repaired combat damage. The un-painted patch near the cabane struts is collocated with the fuel tank location and is possibly the result of damage incurred 6 March.

Albatros D.III 2006/16. Its appearance is based on photographs taken during March 1917. Dates and circumstances regarding fuselage repair patches unknown.

Above: Closeup of *Le Petit Rouge* at Roucourt.

Victory No.36

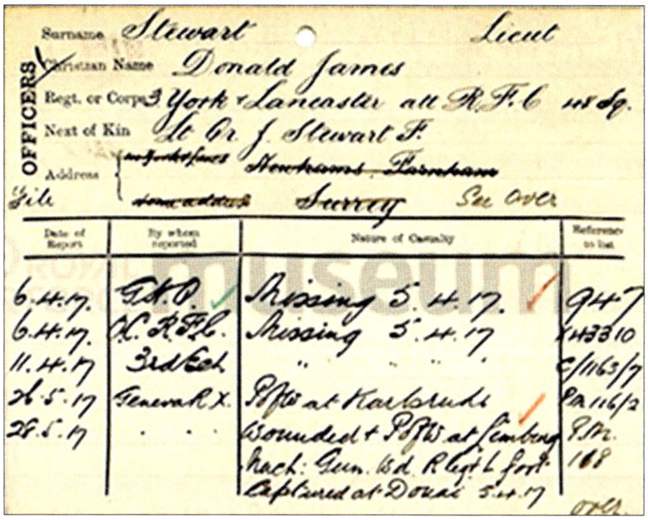
Surname Stewart Lieut
Christian Name Donald James
Regt. or Corps 3 York & Lancaster att R.F.C 48 Sq.
Next of Kin Lt Col. J. Stewart
Address Surrey See over

Date of Report	By whom reported	Nature of Casualty	Reference to list
6.4.17.	G.H.Q.	Missing 5.4.17.	947
6.4.17.	H.RFC.	Missing 5.4.17	H3310
11.4.17	3rd Ech	" " "	C/1163/7
26.5.17	Geneva R.X.	PoW at Karlsruhe	PA 116/2
26.5.17	" "	Wounded + PoW at [illegible] Captured at Douai 5.4.17	P.M. 168

Above: Lt. Donald James Stewart's Casualty Card, indicating he was "wounded and PoW." (RAF Museum)

5 April
Bristol F2.A A3343
No.48 Squadron RFC

Day/Date: Thursday, 5 April
Time: 11.30AM
Weather: Misty and cloudy
Attack Location: Cuincy
Forced Landing Location: Near Cuincy
Side of Lines: Friendly

KOFL 6. Armee Weekly Activity Report: "11.00 vorm. 1 fdl.Flz. bei Lewarde (diesseits) durch Oblt. Frhr.von Richthofen. J.St.11 (als 36.)

11.00AM. 1 enemy airplane near Leward (this side [of lines]) by *Oblt. Frhr.* von Richthofen. *Jagdstaffel* 11 (as 36th)."

MvR Combat Report: "Bristol Two-Seater. Occupants: Pilot Lieut. Adams, Oberser Lieut. Steward [*sic*] (unwounded). Plane details not at hand as machine was burned.

After having put the first adversary near Leward [*sic*] out of action, I pursued the remaining part of the enemy squad and overtook the last plane above Douai. After a rather long fight the adversary surrendered. I forced him to land near Quincy [*sic*]. The occupants burnt their machine to ashes.

Frhr. v. Richthofen.
Was acknowledged."

RFC Combat Casualties Report: "2/Lt.H.B.Griffith, one of the patrol, reports that about 10.50AM he saw one of our machines diving emitting smoke under control over DOUAI. When the formation was last seen about 11AM three of the machines were heavily engaged with a superior number of H.A., this during his second combat over DOUAI."

RFC A/C:
Make/model/serial number: Bristol F.2A A3343
Manufacturer: British & Colonial Aeroplane Co., Ltd., Filton and Brislington, Bristol
Unit: No.48 Squadron RFC
Aerodrome: Bertangles
Sortie: Offensive Patrol - DOUAI
Colors/markings: Generally PC10 fuselage and uppersurfaces of wings/ailerons/horizontal stabilizers/elevators, with CDL undersurfaces. PC10 vertical stabilizer with white serial number. Battleship gray metal engine cowls. b/w/r rudder
Engine: Rolls Royce Falcon I
Engine Number: 1/190/22/ WD 1046
Guns: One Vickers, one Lewis. Nos. 21294, A1194
Manner of Victory: Spiral descent and forced dead-stick landing, after which crew burned the machine.
Items Souvenired: n/a

RFC Crew:
Pilot: Lt. Alfred Terence Adams, 20 (DoB 1 June 1896), PoW
RFC Combat Casualties Report: "Wire received by relatives through Geneva Red Cross states Lts. Adams and Stewart are prisoners at Karlsruhe."
Incarceration: Karlsruhe
Repatriation: 14 December 1918
Obs: Lt. Donald James Stewart, WiA/PoW
RFC Combat Casualties Report: "German message states that Lt.A.T. Adams, pilot, is unwounded and Lt. D.J.Stewart, observer, is wounded."
Manner of Injury: u/k
Incarceration: Karlsruhe
Repatriation: 31 December 1918

MvR A/C:
Make/model/serial number: Albatros D.III, possibly *Le Petit Rouge*, or 2006/16.
Unit: *Jasta* 11
Commander: *ObLt.* Manfred von Richthofen
Airfield: La Brayelle
Colors/markings: *Le Petit Rouge* – See Vic. #17.

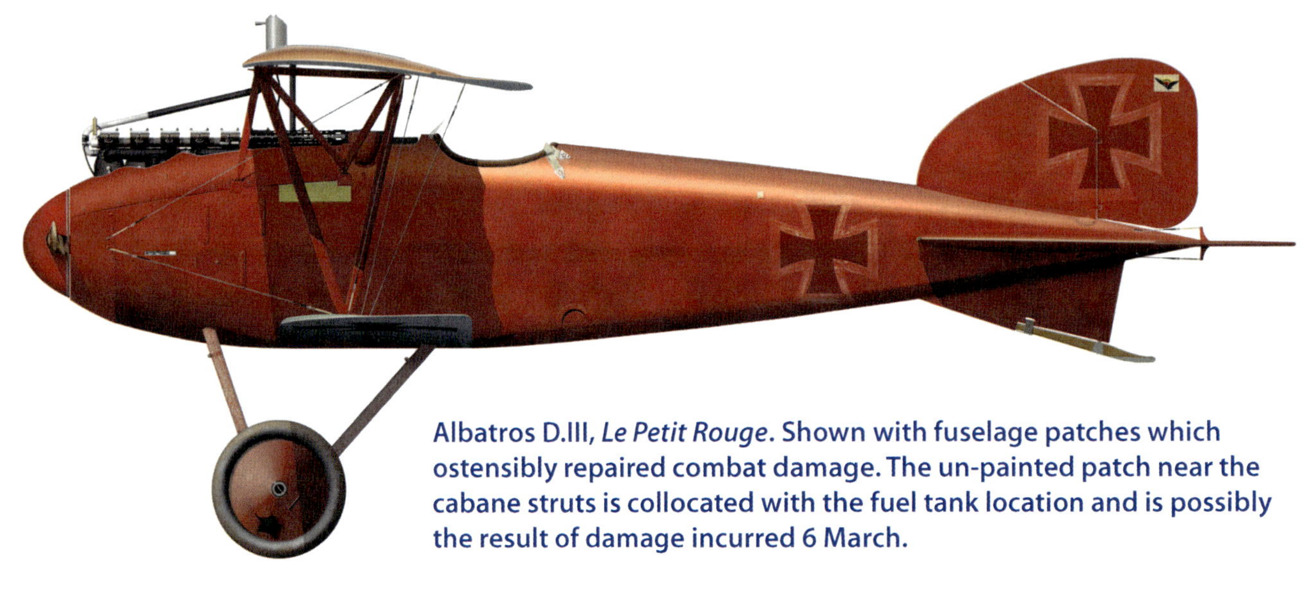

Albatros D.III, *Le Petit Rouge*. Shown with fuselage patches which ostensibly repaired combat damage. The un-painted patch near the cabane struts is collocated with the fuel tank location and is possibly the result of damage incurred 6 March.

Albatros D.III 2006/16. Its appearance is based on photographs taken during March 1917. Dates and circumstances regarding fuselage repair patches unknown.

Damage: n/a
Initial Attack Altitude: u/k
Gunnery Range: u/k
Rounds Fired: u/k
Known Staffel Participants: Vzfw. Sebastian Festner, *Ltn.* Georg Simon

Notes:

1. The RFC Casualty Report state Stewart was wounded, as does his Combat Casualty Card. However, details are u/k.
2. This was Richthofen's last victory as *Oberleutnant*.
3. Gun no. A1194 is also listed as belonging to Vic. #35 A.3340.

Victory No.36 Statistics:

Two-Seaters
5th of 17 April two-seater victories
16th of 33 1917 two-seater victories
23rd of 39 pre-wound two-seater victories
23rd of 45 total two-seater victories

F.2A Δ
2nd of 2 April F.2A victories
2nd of 2 1917 F.2A victories
2nd of 3 total F.2A victories

No.48 Squadron ±
2nd of 2 No.48 F.2A victories
2nd of 2 April No.48 Squadron victories
2nd of 2 1917 No.48 Squadron victories
2nd of 2 total No.48 Squadron victories

Wounded †
No.48 Squadron:
2nd of 2 April No.48 Squadron crewmen WiA
2nd of 2 1917 No.48 Squadron crewman WiA
2nd of 2 total No.48 Squadron crewmen WiA

Two-Seaters:
2nd of 11 April two-seater crewmen WiA
6th of 16 1917 two-seater crewmen WiA
7th of 16 pre-wound two-seater crewmen WiA
7th of 18 total two-seater crewmen WiA

All:
2nd of 11 total April crewmen WiA
7th of 19 total 1917 crewmen WiA
9th of 18 total pre-wound crewmen WiA
9th of 26 total crewmen WiA

Etc.
5th of 21 total April victories
21st of 48 total 1917 victories
36th of 57 total pre-wound victories
19th victory in an Albatros D.III
16th of 22 two-seater victories flying from La Brayelle
20th of 28 total victories flying from La Brayelle
16th of 30 two-seater victories as *Staffelführer*
20th of 39 total victories flying as *Staffelführer*

Statistics Notes
Δ *All BF.2 victories are pre-wound*
± *All No.48 Squadron victories are pre-wound*
† *All No.48 Squadron WiA are pre-wound*

Above: Manfred von Richthofen converses with brother Lothar prior to (or just after) a flight. They are standing next to an Albatros D.III but there is not enough of the machine visible to identify which one.

Victory No.37

Above: Although not taken during the 7/8 April raids that Richthofen described (see Notes), this photograph shows an earlier daylight bombing raid on LaBrayelle by No.23 Squadron FE.2bs on 13 August 1916. Bomb explosions (casting long shadows) can be seen amongst the landing area and structures in the middle of the photograph. (H. Kilmer)

7 April
Nieuport 17 A6645
No.60 Squadron RFC

Day/Date: Saturday, 7 April
Time: 5.45PM
Weather: Low clouds and rain
Attack Location: "Mercatel, other side of our lines."
Crash Location: Near Mercatel
Side of Lines: Enemy

KOFL 6. Armee Weekly Activity Report: "5.45 nachm. 1 Nieuport-Einsitzer bei Mercatel (jenseits) durch Rittmeister Frhr.von Richthofen, J.St.11 (als 37.)

5.45PM. 1 Nieuport one-seater near Mercatel (other side [of lines]) by *Rittmeister Frhr.* von Richthofen, *Jagdstaffel* 11 (as 37th)."

MvR Combat Report: "Nieuport One-seater. Nercatel [*sic*], other side of our lines. Details not at hand.

I attacked, together with 4 of my gentleman [*sic*], an enemy squad of six Nieuport machines south of Arras and behind the enemy lines. The plane I had singled out tried to escape six times by various manoevres. When he was doing this for the 7th time, I managed to hit him, whereupon the engine began to smoke and the plane itself went down head first twisting and twisting. At first I thought it was another manoeuvre, but then I saw that the plane dashed without catching itself to the ground near Mercatel.

Frhr. v. Richthofen.
Was acknowledged."

RFC Combat Casualties Report: "Left aerodrome 4.40PM. 3rd Bde tel:SC.643 d-18/4/17 states that the machine of 2/Lt.Smart has been found at N.7.d.53 burnt."

RFC A/C:
Make/model/serial number: Nieuport 17 A6645 (SFA N2483)
Manufacturer: Société Anonyme des Établissements Nieuport
Unit: No.60 Squadron RFC
Aerodrome: Filescamp Farm
Sortie: Offensive Patrol
Colors/markings: Generally, overall silver aluminum with b/w/r roundels on wings and fuselage; white-bordered black serial number on r/w/b rudder; likely an identifying number (u/k) aft of the fuselage roundel.
Engine: Le Rhône 9J rotary
Engine Number: T3557J
Gun: One Lewis, No.5868
Manner of Victory: Uncontrolled spiral descent until terrain impact.
Items Souvenired: n/a

RFC Crew:
Pilot: 2nd Lt. George Orme Smart, 30 (DoB 17 August 1886), KiA
RFC Combat Casualties Report: "The pilot had been burnt to death. He has been buried in a shell hole near his machine and a cross is being erected by his squadron."
Cause of Death: Immolation. Gunshot wound(s) and trauma associated with airplane crash are possible contributing factors.
Burial: In a shell hole near the crash site, the location of which was eventually lost during turmoil of war.

MvR A/C:
Make/model/serial number: Albatros D.III, possibly *Le Petit Rouge*, or 2006/16.
Unit: *Jasta* 11

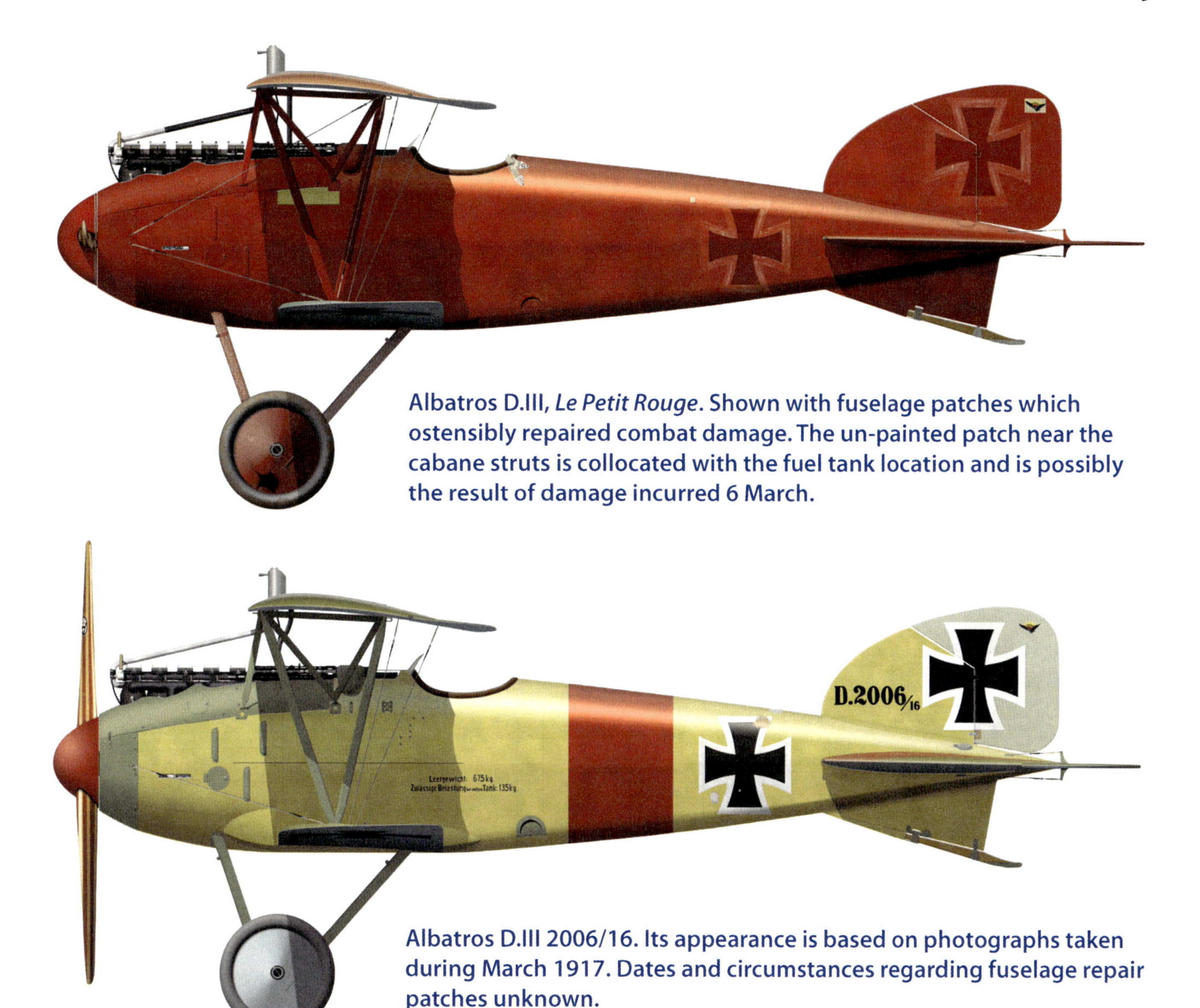

Albatros D.III, *Le Petit Rouge*. Shown with fuselage patches which ostensibly repaired combat damage. The un-painted patch near the cabane struts is collocated with the fuel tank location and is possibly the result of damage incurred 6 March.

Albatros D.III 2006/16. Its appearance is based on photographs taken during March 1917. Dates and circumstances regarding fuselage repair patches unknown.

Commander: *Rittm.* Manfred von Richthofen
Airfield: La Brayelle
Colors/markings: *Le Petit Rouge* – See Vic. #17. Albatros D.III 2006/16 – See Vic. #23.
Damage: n/a
Initial Attack Altitude: u/k
Gunnery Range: u/k
Rounds Fired: u/k
Known Staffel Participants: Ltn. Kurt Wolff, *Ltn.* Karl Schaefer

Notes:
1. Richthofen's first victory as *Rittmeister*.
2. This victory is another example of Richthofen operating beyond the lines, months before being wounded and over a year prior to his final flight.
3. La Brayelle was attacked on the night of 7/8 April by No. 100 Sqn FE.2s. From *Royal Flying Corps Communique No. 83*:

"On the first raid, one phosphorus and forty-five 20-lb. bombs were dropped. Three hangers [sic] were destroyed and buildings near the aerodrome were hit… On the second raid, which took place about 2.40AM, a fourth hangar was destroyed. The bombs were dropped from an average height of 600 feet."

In *Der Rote Kampfflieger* Richthofen described a two-phase bombing attack "during the full-moon nights of April 1917" that appears to match that just described: "One [bomb] landed a few meters from *'Le petit rouge,'* but did not harm it. Later that night this fun [bombing] was repeated. I was already in bed and fast asleep when, as if in a dream, I heard antiaircraft fire. I woke up only to learn that the dream was really happening. One fellow flew so low over my room that I pulled the covers over my head in fright. The next minute there was a terrific noise near my window, and the panes of glass fell victim to the bomb. Quickly I jumped into my shirt in order to get a few shots at him [via a carbine]. Outside he was the target of vigorous shooting. Unfortunately, I had outslept my comrades."

Victory No.37 Statistics:

Single-Seaters
1st of 4 April single-seater victories
6th of 15 1917 single-seater victories
14th of 18 pre-wound single-seater victories
14th of 35 total single-seater victories

Nieuport *
1st of 2 April Nieuport victories
2nd of 4 1917 Nieuport victories
2nd of 4 total Nieuport victories

No.60 Squadron ±
1st of 2 No.60 Nieuport victories
1st of 2 April No.60 Squadron victories
1st of 2 1917 No.60 Squadron victories
1st of 2 total No.60 Squadron victories

Deaths †
No.60 Squadron:
1st and only April No.60 Squadron crewman KiA
1st and only 1917 No.60 Squadron crewman KiA
1st and only total No.60 Squadron crewmen KiA

Single-Seaters:
1st of 3 April single-seater crewmen KiA
4th of 9 1917 single-seater crewmen KiA
10th of 12 pre-wound single-seater crewmen KiA
10th of 21 total single-seater crewmen KiA

All:
6th of 23 total April crewmen KiA
24th of 53 total 1917 crewmen KiA
41st of 66 total pre-wound crewmen KiA
41st of 84 total crewmen KiA

Etc.
6th of 14 victories behind enemy lines
6th of 21 total April victories
22nd of 48 total 1917 victories
37th of 57 total pre-wound victories
20th victory in an Albatros D.III
5th of 6 single-seater victories flying from La Brayelle
21st of 28 total victories flying from La Brayelle
5th of 9 single-seater victories as *Staffelführer*
21st of 39 total victories flying as *Staffelführer*

Statistics Notes
* Includes N17 and N23 models
± *All No.60 Squadron victories are pre-wound*
† *All No.60 Squadron deaths are pre-wound*

Above: With a touch of forward elevator pressure to lift the tail, Richthofen guns Albatros D.III 2006/16 into its takeoff roll at La Brayelle. Note the downslope of the terrain, as well as its convex surface that obscures all but the tops of the distant hangars. (Lance Bronnenkant)

Victory No.38

Albatros D.III, *Le Petit Rouge*. Shown with fuselage patches which ostensibly repaired combat damage. The un-painted patch near the cabane struts is collocated with the fuel tank location and is possibly the result of damage incurred 6 March.

8 April
Sopwith Strutter A2406
No.43 Squadron RFC

Day/Date: Sunday, 8 April
Time: 11.40AM
Weather: Fine but cloudy
Attack Location: "Above Farbus."
Crash Location: "Near Farbus."
Side of Lines: Friendly

KOFL 6. Armee Weekly Activity Report: "11.40 vorm. 1 Sopwith-Zweisitzer bei Farbus (diesseits) durch Rittm.Frhr.von Richthofen, J.St.11 (als 38.)

11.40AM. 1 Sopwith two-seater near Farbus (this side [of lines]) by *Rittm Frhr.* von Richthofen, *Jagdstaffel* 11 (as 38th)."

MvR Combat Report: "Sopwith Two-seater. Occupants: Lieut.Heagerty, wounded; Lieut.Health-Cantle, killed. Details of plane not at hand, as plane is lying in shellfire and is also dashed to pieces.

With three of my planes I attacked 3 Sopwith above Farbus. The plane I singled out, soon made a right-hand curve downwards. The observer ceased shooting. I followed the adversary to the ground where he dashed to pieces.

Frhr. v. Richthofen.
Was acknowleged."

RFC Combat Casualties Report: "Machine was last seen about 11.35 am gliding down through clouds S.E.of LENS after fight with H.A."

RFC A/C:
Make/model/serial number: Sopwith Strutter A2406
Manufacturer: Subcontracted to and built by Ruston, Proctor & Co., Ltd., Lincoln.
Unit: No.43 Squadron RFC
Aerodrome: Treizennes
Sortie: Line Patrol
Colors/markings: u/c. Generally, PC10 wings/ailerons/vertical and horizontal stabilizers/elevators; engine cowls either bare engine turned metal, PC10, or identifying Flight color; b/w/r rudder; black serial number with white border.
Engine: Le Rhône 9J rotary
Engine Number: 1154 WD 7767
Guns: One Vickers, one Lewis. Nos. 28517, L690
Manner of Victory: Uncontrolled descent until terrain impact.
Items Souvenired: n/a

RFC Crew:
Pilot: 2nd Lt. John Seymour Heagerty, 20, WiA/PoW
Manner of Injury: Trauma associated with airplane crash, including fractured jaw, displaced teeth, lacerated eyelid, u/k foot injury, and general body aches.
RFC Combat Casualties Report: "German message dropped in our lines states that Lt. Heagerty is wounded."
Repatriation: 17 December 1918
Obs: Lt. Leonard Heath Cantle, 21 (10 August 1895), KiA
RFC Combat Casualties Report: "13th Corps A.749 d-16/4/17 states that the body of an officer was found beside aeroplane A.2406 at B.13.d.7.4 and

buried at the Commandant's House. Subsequent correspondence failed to elicit the identification of the officer.
A postcard dated 11/6/17 from the father of Lt. L.H. Cantle states he has been informed through the International Red Cross, Geneva, that his son was killed on 8th April near LENS.
German message dropped in our lines states that…Lt. H.Cantle was killed."
Cause of Death: Either gunshot wound(s) or trauma associated with airplane crash.
Burial: In-field; location lost during turmoil of war.

MvR A/C:
Make/model/serial number: Albatros D.III, likely *Le Petit Rouge*
Unit: *Jasta* 11
Commander: *Rittm*. Manfred von Richthofen
Airfield: La Brayelle
Colors/markings: *Le Petit Rouge* – See Vic. #17. Heagerty described his attacker as "red."
Initial Attack Altitude: "Somewhere around ten thousand feet." (Heagerty)
Gunnery Range: u/k
Rounds Fired: u/k
Known Staffel Participants: u/k

Notes:
1. Heagerty stated in a post-war interview with Gibbons that after Richthofen's first attack the combatants entered a spiraling dogfight until the Strutter's flight controls were shot away, apparently in the same burst that incapacitated or killed Cantle. Heagerty noted: "From a glide, we went into a dive. All the way down the red machine, or some machine, kept right in back of me, ripping bursts after burst of machine-gun bullets into the plane from the rear." Most likely another example of Richthofen's no-quarter *Modus-Operandi* to continuously attack descending and/or disabled aeroplanes if they were still airborne.

Victory No.38 Statistics:
Two-Seaters
6th of 17 April two-seater victories
17th of 33 1917 two-seater victories
24th of 39 pre-wound two-seater victories
24th of 45 total two-seater victories

Sop. Strutter ∆
2nd of 2 April Sop. Strutter victories
3rd of 3 1917 Sop. Strutter victories
3rd of 3 total Sop. Strutter victories

No.43 Squadron ±
3rd of 3 No.43 Sop. Strutter victories
2nd of 2 April No.43 Squadron victories
3rd of 3 1917 No.43 Squadron victories
3rd of 3 total No.43 Squadron victories

Deaths †
No.43 Squadron:
2nd of 2 April No.43 Squadron crewmen KiA
4th of 4 1917 No.43 Squadron crewmen KiA
4th of 4 total No.43 Squadron crewmen KiA

Two-Seaters:
6th of 20 April two-seater crewmen KiA
21st of 44 1917 two-seater crewmen KiA
32nd of 54 pre-wound two-seater crewmen KiA
32nd of 63 total two-seater crewmen KiA

All:
7th of 23 total April crewmen KiA
25th of 53 total 1917 crewmen KiA
42nd of 66 total pre-wound crewmen KiA
42nd of 84 total crewmen KiA

Wounded
No. 43 Squadron:
1st and only April No. 43 Squadron crewman WiA
1st and only 1917 No. 43 Squadron crewman WiA
1st and only total No. 43 Squadron crewmen WiA

Two-Seaters:
3rd of 11 April two-seater crewmen WiA
7th of 16 1917 two-seater crewmen WiA
8th of 16 pre-wound two-seater crewmen WiA
8th of 18 total to-seater crewmen WiA

All:
3rd of 11 total April crewmen WiA
8th of 19 total 1917 crewmen WiA
10th of 18 total pre-wound crewmen WiA
10th of 26 total crewmen WiA

Etc.
7th of 21 total April victories
23rd of 48 total 1917 victories
38th of 57 total pre-wound victories
21st victory in an Albatros D.III
17th of 22 two-seater victories flying from La Brayelle
22nd of 28 total victories flying from La Brayelle
17th of 30 two-seater victories as *Staffelführer*
22nd of 39 total victories flying as *Staffelführer*

Statistics Notes
∆ *All Sop. Strutter victories are pre-wound*
± *All No.43 Squadron victories are pre-wound*
† *All No.43 Squadron deaths are pre-wound; KiA includes DoW*

Victory No.39

Albatros D.III, *Le Petit Rouge*. Shown with fuselage patches which ostensibly repaired combat damage. The un-painted patch near the cabane struts is collocated with the fuel tank location and is possibly the result of damage incurred 6 March.

8 April
BE.2e A2815
No.16 Squadron RFC

Day/Date: Sunday, 8 April
Time: 4.40PM
Weather: Fine but cloudy
Attack Location: Vimy
Crash Location: 1000 yards west of Vimy
Side of Lines: Friendly

KOFL 6. Armee Weekly Activity Report: "4.40 nachm. 1 Bristol-D.D. bei Vimy (diesseits) durch Rittm.Frhr.von Richthofen, J.St.11 (als 39.)

4.40PM. 1 Bristol biplane near Vimy (this side [of lines]) by Rittm. Frhr. von Richthofen, *Jagdstaffel* 11 (as 39th)."

MvR Combat Report: "BE 2. Occupants: Both killed, name of one – Davidson [*sic*]. Plane No. A.2815. Remnants distributed on more than one kilometer.

I was flying and surprised an English artillery flyer. After a very few shots the plane broke to pieces and fell near Vimy on this side of the lines.

Frhr. v. Richthofen.
Acknowledged."

RFC Combat Casualties Report: "Left aerodrome 3 pm. A B.E. is reported to have been shot down by H.A. at 4.40PM, falling and crashing about 1000 yards W. of VIMY."

RFC A/C:
Make/model/serial number: Royal Aircraft Factory BE.2e A2815
Manufacturer: Subcontracted to and built by the British & Colonial Aeroplane Co., Ltd., Filton and Brislington, Bristol.
Unit: No.16 Squadron RFC
Aerodrome: Bruay
Sortie: Photography FARBUS
Colors/markings: u/c. Generally, PC10 wings/ailerons/vertical and horizontal stabilizers/elevators; battleship gray engine cowl; b/w/r rudder; black serial number on vertical stabilizer. No. 16 Sqn markings were vertical bands encircling the fuselage on either side of the roundel (black on CDL machines, white on PC10).
Engine: 90 H.P. R.A.F. 1a
Engine Number: 1707 WD 2644
Guns: Two Lewis. Nos. 17031, 17446
Manner of Victory: Gunfire-induced in-flight breakup
Items Souvenired: n/a

RFC Crew:
Pilot: 2nd Lt Keith Ingleby MacKenzie, 18 (DoB 26 June 1898), KiA
Cause of Death: u/k. Either gunshot wound(s), trauma associated with airplane crash, or terrain impact after free-fall from altitude.
Burial: Bois-Carre British Cemetery, Thelus, France, Grave III. B.12.
Obs: 2nd Lt Guy Everingham, 22 (DoB 28 June 1894), KiA
RFC Combat Casualties Report: "The bodies of 2/

Albatros D.III 2006/16. Its appearance is based on photographs taken during March 1917. Dates and circumstances regarding fuselage repair patches unknown.

Lts. Mackenzie and Everingham have been found near BOIS-DE-BONVAL (map 36.c.S.30.c.) (1st Bde tel:A.67 dated 15/4/17)."
Cause of Death: u/k. Either gunshot wound(s), trauma associated with airplane crash, or terrain impact after free-fall from altitude.
Burial: Bois-Carre British Cemetery, Thalus, France, Grave III. B.13.

MvR A/C:
Make/model/serial number: Albatros D.III, possibly *Le Petit Rouge*, or 2006/16.
Unit: *Jasta* 11
Commander: *Rittm*. Manfred von Richthofen
Airfield: La Brayelle
Colors/markings: *Le Petit Rouge* – See Vic. #17.
Albatros D.III 2006/16 – See Vic. #23.
Damage: n/a
Initial Attack Altitude: u/k
Gunnery Range: u/k
Rounds Fired: "Very few shots."
Known Staffel Participants: u/k

Notes:

1. Lothar von Richthofen described either this or a similar inflight breakup of one of Manfred's victories appearing "as if someone had shaken out a sack of large and small bits of paper."
2. A discrepancy exists regarding the exact model of BE.2 shot down. Although identified as a BE.2 "g" in the combat casualty report, A.2815 falls within Contract No. 87/A/571 for the BE.2.*e*. The BE.2e had a redesigned tail and shorter span lower wings than previous models, and thus had one less set of interplane struts. To meet the demand for the BE.2e, earlier BE.2 c and d models were modified to BE.2e standards. This begat a supply and maintenance problem that resulted in a directive issued October 1916 stating modified BE.2c airplanes would be re-designated BE.2f, while rebuilt BE.2d airplanes would be re-designated as BE.2g.

   Thus, if A2815 was built as an e, how/why would it be modified to a g—a machine upgraded to e standards—if it were an e already? That would mean upgrading an e to e standards, which makes no sense. The reason for the combat casualty listing as a B.E.2g is unknown and likely nothing more than simple error.

Victory No.39 Statistics:

Two-Seaters
7th of 17 April two-seater victories
18th of 33 1917 two-seater victories
25th of 39 pre-wound two-seater victories
25th of 45 total two-seater victories

BE.2 ∆
2nd of 7 April BE.2 victories
10th of 15 1917 BE.2 victories
12th of 17 total BE.2 victories

No.16 Squadron ±
5th of 6 No.16 BE.2 victories
1st of 2 April No.16 Squadron victories
5th of 6 1917 No.16 Squadron victories
5th of 6 total No.16 Squadron victories

Deaths †
No.16 Squadron:
1st and 2nd of 4 April No.16 Squadron crewmen KiA

9th and 10th of 12 1917 No.16 Squadron crewmen KiA
9th and 10th of 12 total No.16 Squadron crewmen KiA

Two-Seaters:
7th and 8th of 20 April two-seater crewmen KiA
22nd and 23rd of 44 1917 two-seater crewmen KiA
33rd and 34th of 54 pre-wound two-seater crewmen KiA
33rd and 34th of 63 total two-seater crewmen KiA

All:
8th and 9th of 23 total April crewmen KiA
26th and 27th of 53 total 1917 crewmen KiA
43rd and 44th of 66 total pre-wound crewmen KiA
43rd and 44th of 84 total crewmen KiA

Etc.
8th of 21 total April victories
24th of 48 total 1917 victories
39th of 57 total pre-wound victories
22nd victory in an Albatros D.III
18th of 22 two-seater victories flying from La Brayelle
23rd of 28 total victories flying from La Brayelle
18th of 30 two-seater victories as *Staffelführer*
23rd of 39 total victories flying as *Staffelführer*

Statistics Notes
Δ *All BE.2 victories are pre-wound*
± *All No.16 Squadron victories are pre-wound*
† *All No.16 Squadron deaths are pre-wound; KiA includes DoW*

Left: Aviation artist Michael Backus' depiction of one of Richthofen's more famous poses. (Michael Backus)

Victory No.40

Albatros D.III, *Le Petit Rouge*. Shown with fuselage patches which ostensibly repaired combat damage. The un-painted patch near the cabane struts is collocated with the fuel tank location and is possibly the result of damage incurred 6 March.

11 April
BE.2.c 2501
No.13 Squadron RFC

Day/Date: Wednesday, 11 April
Time: 9.25AM
Weather: High wind, low clouds, and snow
Attack Location: Willerval
Crash Location: Front lines
Side of Lines: No Man's Land

RFC Communique No. 83: "Twenty-five targets were dealt with by artillery with aeroplane observation, and three explosions of ammunition were caused."

KOFL 6. Armee Weekly Activity Report: "9.25 vorm. 1 B.E. D.D. bei Willerval (diesseits) durch Rittm. Frhr. v.Richthofen, J.St.11 (als 40.)

9.25AM. 1 B.E. biplane near Willerval (this side [of lines]) by Rittm. Frhr. v. Richthofen, *Jagdstaffel* 11 (as 40th)."

MvR Combat Report: "BE Two-seater. Willerval, this side of the lines. Concerning machine and inmates, details cannot be given, as English attacked this part of front, thus making communication with front lines impossible.

Flying with Lt. Wolff I attacked English Infantry Flyer in low height. After a short fight the enemy plane fell into a shell-hole. When dashing to ground the wings of the plane broke off.

Frhr. v. Richthofen
Acknowledged."

RFC Combat Casualties Report: n/a

RFC A/C:
Make/model/serial number: Royal Aircraft Factory BE.2.c 2501
Manufacturer: Subcontracted to and built by Wolseley Motors, Ltd., Adderley Park, Birmingham.
Unit: No.13 Squadron RFC
Aerodrome: Savy
Sortie: Artillery Observation
Colors/markings: u/c. Generally, PC10 wings/ailerons/vertical and horizontal stabilizers/elevators; battleship gray engine cowl; b/w/r rudder; black serial number on vertical stabilizer. No. 13 Sqn markings were a narrow horizontal bar running down the fuselage from observer's cockpit to the tail (black on CDL machines, white on PC10).
Engine: 90 H.P. R.A.F. 1a
Engine Number: 22930 WD853
Guns: Two Lewis. Nos. 2170, 17353
Manner of Victory: Uncertain, but apparently low-altitude in-flight breakup.
Items Souvenired: n/a

RFC Crew:
Pilot: Lt. Edward Claude England Derwin, 22 or 23 (DoB 1894), WiA
Manner of Injury: u/c. Either gunshot wound(s) or trauma associated with airplane crash.

Albatros D.III 2006/16. Its appearance is based on photographs taken during March 1917. Dates and circumstances regarding fuselage repair patches unknown.

Obs: Air Mechanic H. Pierson, age u/k, WiA
Manner of Injury: u/c. Either gunshot wound(s) or trauma associated with airplane crash.

MvR A/C:
Make/model/serial number: Albatros D.III, possibly *Le Petit Rouge*, or 2006/16.
Unit: *Jasta* 11
Commander: *Rittm.* Manfred von Richthofen
Airfield: La Brayelle
Colors/markings: *Le Petit Rouge* – See Vic. #17.
Albatros D.III 2006/16 – See Vic. #23.
Damage: n/a
Initial Attack Altitude: "Low height"
Gunnery Range: u/k
Rounds Fired: u/k
Known Staffel Participants: Ltn. Kurt Wolff

Notes:
1. With this victory Richthofen tied Oswald Boelcke's total victory score.

Victory No.40 Statistics:

Two-Seaters
8th of 17 April two-seater victories
19th of 33 1917 two-seater victories
26th of 39 pre-wound two-seater victories
26th of 45 total two-seater victories

BE.2 Δ
3rd of 7 April BE.2 victories
11th of 15 1917 BE.2 victories
13th of 17 total BE.2 victories

No.13 Squadron ±
2nd of 4 No.13 BE.2 victories
2nd of 4 April No.13 Squadron victories
2nd of 4 1917 No.13 Squadron victories
2nd of 4 total No.13 Squadron victories

Wounded
No.13 Squadron:
1st and 2nd of 4 April No.13 Squadron crewmen WiA
1st and 2nd of 4 1917 No.13 Squadron crewmen WiA
1st and 2nd of 4 total No.13 Squadron crewmen WiA

Two-Seaters:
4th and 5th of 11 April two-seater crewmen WiA
8th and 9th of 16 1917 two-seater crewmen WiA
9th and 10th of 16 pre-wound two-seater crewmen WiA
9th and 10th of 18 total two-seater crewmen WiA

All:
4th and 5th of 11 total April crewmen WiA
9th and 10th of 19 total 1917 crewmen WiA
11th and 12th of 18 total pre-wound crewmen WiA
11th and 12th of 26 total crewmen WiA

Etc.
9th of 21 total April victories
25th of 48 total 1917 victories
40th of 57 total pre-wound victories
23rd victory in an Albatros D.III
19th of 22 two-seater victories flying from La Brayelle
24th of 28 total victories flying from La Brayelle
19th of 30 two-seater victories as *Staffelführer*
24th of 39 total victories flying as *Staffelführer*

Statistics Notes
Δ *All BE.2 victories are pre-wound*
± *All No.13 Squadron victories are pre-wound*
† *All No.13 Squadron deaths are pre-wound; KiA includes DoW*

Victory No.41

Above: Royal Aircraft Factory RE.8.

13 April
RE.8 A3190
No.59 Squadron RFC

Day/Date: Friday, 13 April
Time: 8.58AM
Weather: Fine but cloudy
Attack Location: Between Vitry and Brebières
Crash Location: Between Vitry and Brebières
Side of Lines: Friendly

KOFL 6. Armee Weekly Activity Report: "8.56 [sic] *vorm. 1 F.E.* [sic] *D.D. bei Vitry (diesseits) durch Rittm. Frhr.von Richthofen, J. St. 11 (als 41.)*

8.56AM. 1 F.E. biplane near Vitry (this side [of lines]) by *Rittm. Frhr.* von Richthofen, *Jagdstaffel* 11, (as 41st)."

MvR Combat Report: "New Body DD. Occupants: Lieut. M.A.Woat [*sic*] and Steward (Thomas) [*sic*], both killed. Plane burnt. Motor No.3759 Fixed Motor V-shaped, 12 cyl.

With six planes of my *Staffel* I attacked an enemy squadron of the same force. The plane I had singled out fell after a short fight, to the ground between Vitry and Brebieres. On touching the ground both occupants and machine burnt to ashes.

Frhr. v. Richthofen.
Acknowledged."

RFC Combat Casualties Report: n/a

RFC A/C:
Make/model/serial number: Royal Aircraft Factory RE.8 A3190
Manufacturer: Subcontracted to and built by The Austin Motor Co. (1914), Ltd., Northfield, Birmingham.
Unit: No.59 Squadron RFC
Aerodrome: La Bellevue
Sortie: Photography ETAING
Colors/markings: u/c. Generally, PC10 wings/ailerons/vertical and horizontal stabilizers/elevators; battleship gray engine cowl; b/w/r rudder; black serial number on vertical stabilizer. No. 59 Sqn markings were two narrow black or red vertical bands encircling the fuselage behind the roundel.
Engine: 150 hp RAF 4a
Engine Number: 643 WD3759
Guns: One Vickers, One Lewis. Nos 23593, A111
Manner of Victory: u/c. Apparently an uncontrolled descent until terrain impact.
Items Souvenired: n/a

RFC Crew:
Pilot: Capt. James Maitland Stuart, 20 (DoB 16 September 1896), KiA
Cause of Death: u/c. Either gunshot wound(s), immolation, or trauma associated with airplane crash. Additionally, remote possibility of

Albatros D.III, *Le Petit Rouge*. Shown with fuselage patches which ostensibly repaired combat damage. The un-painted patch near the cabane struts is collocated with the fuel tank location and is possibly the result of damage incurred 6 March.

electrocution (see notes).
Burial: Either in-field or laid where fallen; location lost during turmoil of war.
Obs: Lt Maurice Herbert Wood, 23 (25 June 1893), KiA
Cause of Death: u/c. Either gunshot wound(s), immolation, or trauma associated with airplane crash. Additionally, remote possibility of electrocution.
Burial: Either in-field or laid where fallen; location lost during turmoil of war.

MvR A/C:
Make/model/serial number: Albatros D.III, possibly *Le Petit Rouge*.
Unit: *Jasta* 11
Commander: *Rittm.* Manfred von Richthofen
Airfield: La Brayelle
Colors/markings: *Le Petit Rouge* – See Vic. #17.
Damage: n/a
Initial Attack Altitude: u/k
Gunnery Range: u/k
Rounds Fired: u/k
Known Staffel Participants: Ltn. Kurt Wolff, *Vzfw.* Sebastian Festner, *Ltn.* Lothar von Richthofen.

Notes:
1. With this victory Richthofen exceeded Oswald Boelcke's total victory score.
2. Gibbons wrote that when he visited the Richthofen home in Schweidnitz he came across a photo of two burned corpses. On the back was written, apparently by Richthofen: "My 41st. D.D. two-seater, 12 cyl. Beadmore [*sic*] motor. Two occupants dead. Shot down and burned on the electric wires on April 13, 1917, about ten o'clock in the morning at the canal between Brebières and Vitry." These comments suggest the possibility, albeit remote and unusual, of electrocution as the cause of death.
3. Because of Lothar's association flying 2006/16 it is presumed he did so during this sortie, while Manfred flew *Le Petit Rouge*. This is strictly conjecture and by no means absolute.

Victory No.41 Statistics:
Two-Seaters
9th of 17 April two-seater victories
20th of 33 1917 two-seater victories
27th of 39 pre-wound two-seater victories
27th of 45 total two-seater victories

RE.8
1st and only April RE.8 victory
1st of 5 1917 RE.8 victories
1st of 7 total RE.8 victories

No.59 Squadron ±
1st and only No.59 RE.8 victory
1st and only April No.59 Squadron victory
1st and only 1917 No.59 Squadron victory
1st and only total No.59 Squadron victories

Deaths †
No.59 Squadron:
1st and 2nd of 2 April No.59 Squadron crewmen KiA
1st and 2nd of 2 1917 No.59 Squadron crewmen KiA
1st and 2nd of 2 total No.59 Squadron crewmen KiA

Two-Seaters:
9th and 10th of 20 April two-seater crewmen KiA
24th and 25th of 44 1917 two-seater crewmen KiA
35th and 36th of 54 pre-wound two-seater crewmen KiA

35th and 36th of 63 total two-seater crewmen KiA

All:
10th and 11th of 23 total April crewmen KiA
28th and 29th of 53 total 1917 crewmen KiA
45th and 46th of 66 total pre-wound crewmen KiA
45th and 46th of 84 total crewmen KiA

Etc.
10th of 21 total April victories
26th of 48 total 1917 victories
41st of 57 total pre-wound victories
24th victory in an Albatros D.III
20th of 22 two-seater victories flying from La Brayelle
25th of 28 total victories flying from La Brayelle
20th of 30 two-seater victories as *Staffelführer*
25th of 39 total victories flying as *Staffelführer*

Statistics Notes
± *All No.59 Squadron victories are pre-wound*
† *All No.59 Squadron deaths are pre-wound*

Left: Closeup view of Albatros D.III 2006/16. Lothar von Richthofen sits in the cockpit, conversing with *Ltn*. Emil Schaefer on the ladder. Note the non-standard windshield with central frame.

Victory No.42

Albatros D.III, *Le Petit Rouge*. Shown with fuselage patches which ostensibly repaired combat damage. The un-painted patch near the cabane struts is collocated with the fuel tank location and is possibly the result of damage incurred 6 March.

13 April
FE.2b A.831
No.11 Squadron RFC

Day/Date: Friday, 13 April
Time: 12.45PM
Weather: Fine but cloudy.
Attack Location: Between Monchy and Feuchy
Crash Location: Between Monchy and Feuchy
Side of Lines: Enemy

KOFL 6. Armee Weekly Activity Report: "12.45 nachm. 1 F.E. D.D. westl. Monchy (jenseits) durch Rittm. Frhr. von Richthofen, J.St.11 (als 42.)

12.45PM. 1 F.E. biplane west of Monchy (other side [of lines]) by *Rittm. Frhr.* von Richthofen, *Jagdstaffel* 11 (as 42nd)."

MvR Combat Report: "Vikkers Two-Seater. Occupants and plane unknown, as plane went down above enemy lines.

Together with Lieut. Simon I attacked a Vikkers Two-Seater coming back from German territory. After rather a long fight, during which the adversary could discharge not one shot, the enemy plane dashed down to the ground between Monchy and Feuchy.

Frhr. v. Richthofen.
Acknowledged."

RFC Combat Casualties Report: "Left aerodrome 11-25AM. Pilot and observer now reported wounded."

RFC A/C:
Make/model/serial number: Royal Aircraft Factory FE.2b A831
Manufacturer: Subcontracted to and built by G. & J. Weir, Ltd., Cathcart, Glasgow.
Unit: No.11 Squadron RFC
Aerodrome: Izel-le-Hameau
Sortie: Offensive Patrol
Colors/markings: Generally, PC10 nacelle, commonly with open triangle unit markings on nose; wing/ailerons/horizontal stabilizers/ elevators uppersurfaces PC10 with CDL undersurfaces; likely upperwing roundels had white surrounds.
Engine: 160 hp Beardmore
Engine Number: 985/WD7595
Guns: Two Lewis. Nos. 19776, 19554
Manner of Victory: Forced landing just beyond lines in British territory. Subsequently shelled by artillery and destroyed.
Items Souvenired: n/a

RFC Crew:
Pilot: Sgt. James Allen Cunniffe, 21 or 22 (DoB 1895), WiA
Manner of injury: u/k. Either gunshot wound(s) and/or trauma associated with airplane crash.
Obs: Air Mechanic 2nd Class W. J. Batten, age u/k, WiA
Manner of injury: u/k. Either gunshot wound(s) and/or trauma associated with airplane crash.

MvR A/C:
Make/model/serial number: Albatros D.III,

Albatros D.III 2006/16. Its appearance is based on photographs taken during March 1917. Dates and circumstances regarding fuselage repair patches unknown.

possibly *Le Petit Rouge*, or 2006/16.
Unit: *Jasta* 11
Commander: *Rittm.* Manfred von Richthofen
Airfield: La Brayelle
Colors/markings: *Le Petit Rouge* – See Vic. #17.
Albatros D.III 2006/16 – See Vic. #23.
Damage: n/a

Initial Attack Altitude: u/k
Gunnery Range: u/k
Rounds Fired: u/k
Known Staffel Participants: Ltn. Georg Simon

Victory No.42 Statistics:

Pushers *
2nd of 5 April pusher victories
2nd of 5 April two-seater pusher victories
4th of 7 1917 two-seater pusher victories
9th of 12 total two-seater pusher victories
15th of 18 total pusher victories

Two-Seaters
10th of 17 April two-seater victories
21st of 33 1917 two-seater victories
28th of 39 pre-wound two-seater victories
28th of 45 total two-seater victories

FE.2 Δ
2nd of 5 April FE.2 victories
4th of 7 1917 FE.2 victories
9th of 12 total FE.2 victories

No.11 Squadron ±
3rd of 4 No.11 FE.2 victories
1st of 2 April No.11 Squadron victories
1st of 2 1917 No.11 Squadron victories
3rd of 4 total No.11 Squadron victories

Wounded
No. 11 Squadron:
1st and 2nd of 4 April No. 11 Squadron crewmen WiA
1st and 2nd of 4 1917 No. 11 Squadron crewmen WiA
1st and 2nd of 4 total No. 11 Squadron crewmen WiA

Pushers:
1st of 4 April two-seater pusher crewmen WiA
2nd of 5 1917 two-seater pusher crewmen WiA
2nd of 5 total two-seater pusher crewmen WiA

Two-Seaters:
6th and 7th of 11 April two-seater crewmen WiA
10th and 11th of 16 1917 two-seater crewmen WiA
11th and 12th of 16 pre-wound two-seater crewmen WiA
11th and 12th of 18 total two-seater crewmen WiA

All:
6th and 7th of 11 total April crewmen WiA
11th and 12th of 19 total 1917 crewmen WiA
13th and 14th of 18 total pre-wound crewmen WiA
13th and 14th of 26 total crewmen WiA

Etc.
7th of 14 victories behind enemy lines
11th of 21 total April victories
27th of 48 total 1917 victories
42nd of 57 total pre-wound victories
25th victory in an Albatros D.III
21st of 22 two-seater victories flying from La Brayelle
26th of 28 total victories flying from La Brayelle
21st of 30 two-seater victories as *Staffelführer*
26th of 39 total victories flying as *Staffelführer*

Statistics Notes
* *All pusher victories are pre-wound*
Δ *All FE.2 victories are pre-wound*
± *All No.11 Squadron victories are pre-wound*
† *All No.11 Squadron deaths are pre-wound*

Victory No.43

Above: Swatch of souvenired fabric from 4997's port rudder, mounted and hanging in the Richthofen Museum.

13 April
FE.2b 4997
No.25 Squadron RFC

Day/Date: Friday, 13 April
Time: 7.35PM
Weather: Fine but cloudy.
Attack Location: Noyelles-Godault, near Henin Liétard
Crash Location: Near Noyelles-Godault
Side of Lines: Friendly

KOFL 6. Armee Weekly Activity Report: "7.30 N. 1 F.E. –D.D. bei Henin-Lietard (diess.) durch Rittm. Frhr.v.Richthofen, J.St.11 (als 43).

7.30PM. 1 F.E. biplane near Henin-Lietard (this side [of lines]) by *Rittm. Frhr.* von Richthofen, *Jagdstaffel* 11 (as 43rd)."

MvR Combat Report: "Vikkers Two-Seater. Occupants: Leiuts. Bates and Barnes, both killed. Plane No.4997. Motor No.917, 8 cyl. stand. motor

With three planes of my *Staffel* I attacked an enemy bombing squadron consisting of Vikkers (old type) above Henin-Lietard. After a short fight my adversary began to glide down and finally dashed into a house near Noyelle-Godault. The occupants were both killed and the machine destroyed.

Frhr. v. Richthofen.
Acknowledged."

RFC Combat Casualties Report: "On returning from bomb raid machine is believed to have been shot down by H.A. in enemy territory."

RFC A/C:
Make/model/serial number: Royal Aircraft Factory FE.2b 4997.
Manufacturer: Subcontracted to and built by G. & J. Weir, Ltd., Cathcart, Glasgow.
Unit: No.25 Squadron RFC
Aerodrome: Lozinghem
Sortie: Bomb Raid – HEININ LIETARD
Colors/markings: Presentation a/c *Baroda No 17*. Generally, PC10 nacelle with one, two, or three horizontal white-bordered black bands across nose (corresponding with A, B, or C Flight); likely white "Baroda 17" on sides of nacelle; wing/ailerons/horizontal stabilizers/elevators uppersurfaces PC10 with CDL undersurfaces; likely upperwing roundels had white surrounds.
Engine: 160 hp Beardmore
Engine Number: 917 WD 7527
Guns: Two Lewis. Nos. 23118, 26632
Manner of Victory: Gliding descent until fatal collision with house.
Items Souvenired: Swatches of port/stbd rudder fabric with serial number. Photographed on display at Roucourt, Richthofen's home bedroom in Schweidnitz, and in the post-war Richthofen Museum.

RFC Crew:
Pilot: 2nd Lt. Allan Harold Bates, 20 (DoB May 1896), KiA
RFC Combat Casualties Report: "An unconfirmed report in a German publication states that pilot and observer were killed."
Cause of Death: u/k. Likely trauma associated with airplane crash.
Burial: Noyelles-Godault Communal Cemetery, Noyelles-Godault, France. South end, Grave 1.
Obs: Sgt William Alfred Barnes, 32 (DoB 1895), KiA
RFC Combat Casualties Report: "German message dropped in our lines states that A.H. Bates and W.Barnes were both killed."
Cause of Death: u/k. Likely trauma associated with airplane crash.
Burial: Noyelles-Godault Communal Cemetery, Noyelles-Godault, France. South end, Grave 2.

MvR A/C:
Make/model/serial number: Albatros D.III, possibly *Le Petit Rouge*, or 2006/16.
Unit: *Jasta* 11
Commander: *Rittm.* Manfred von Richthofen

Above: Bates's headstone, 2011.

Above: Barnes's headstone, 2011.

Airfield: La Brayelle
Colors/markings: *Le Petit Rouge* – See Vic. #17.
Albatros D.III 2006/16 – See Vic. #23.
Damage: n/a
Initial Attack Altitude: u/k
Gunnery Range: u/k
Rounds Fired: u/k
Known Staffel Participants: "Three planes of my *staffel*."

Notes:

1. This victory completed Richthofen's first triple (three victories in a single day).

Victory No.43 Statistics:

Pushers *

3rd of 5 April pusher victories
3rd of 5 April two-seater pusher victories
5th of 7 1917 two-seater pusher victories
10th of 12 total two-seater pusher victories
16th of 18 total pusher victories

Two-Seaters

11th of 17 April two-seater victories
22nd of 33 1917 two-seater victories
29th of 39 pre-wound two-seater victories
29th of 45 total two-seater victories

FE.2 Δ

3rd of 5 April FE.2 victories
5th of 7 1917 FE.2 victories
10th of 12 total FE.2 victories

No.25 Squadron ±

4th of 4 No.25 FE2 victories
2nd of 2 April No.25 Squadron victories
2nd of 2 1917 No.25 Squadron victories
4th of 4 total No.25 Squadron victories

Deaths

No.25 Squadron:
2nd and 3rd of 3 April No. 25 Squadron crewmen KiA
4th and 5th of 5 1917 No. 25 Squadron crewmen KiA
4th and 5th of 5 total No. 25 Squadron crewmen KiA

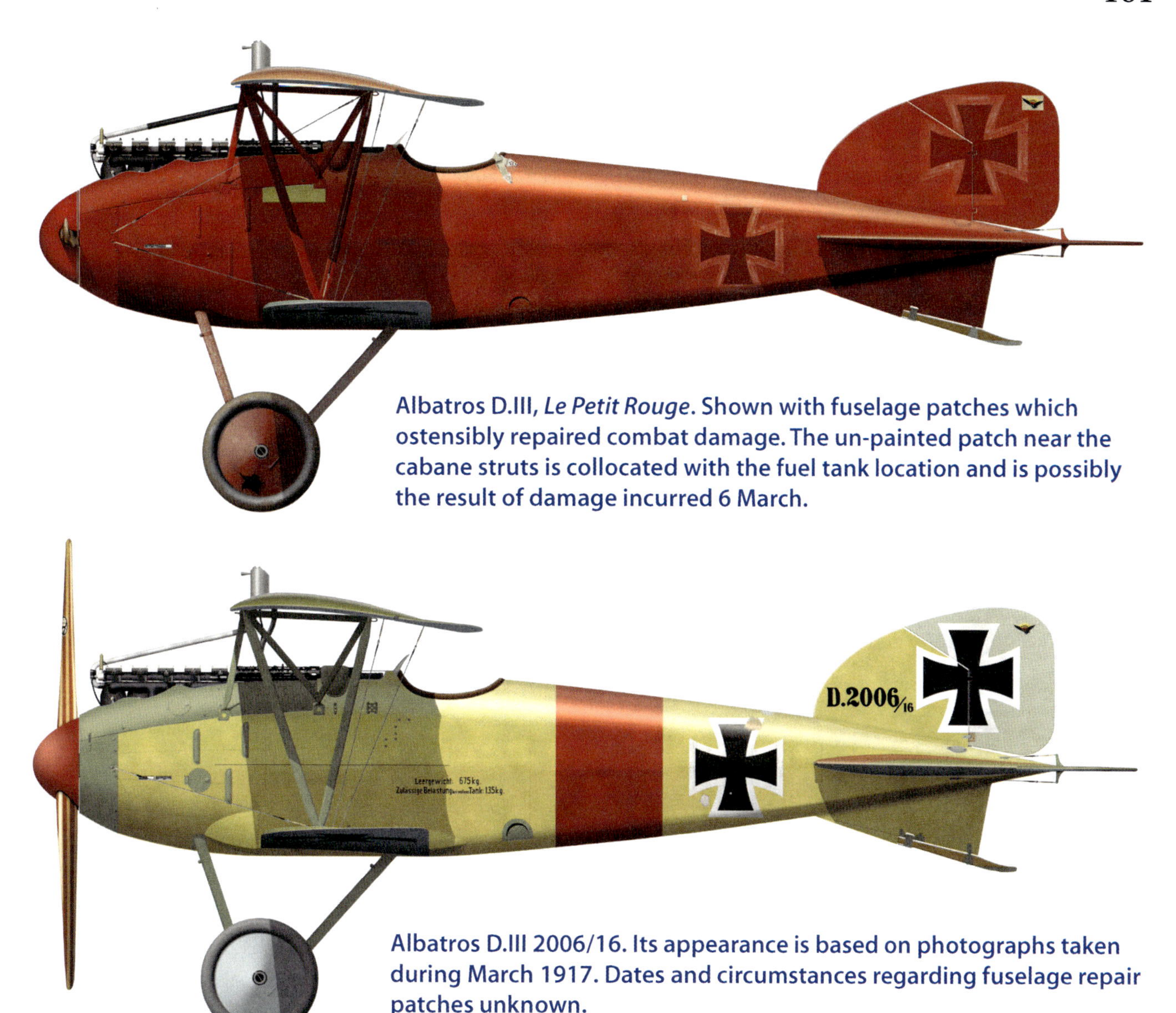

Albatros D.III, *Le Petit Rouge*. Shown with fuselage patches which ostensibly repaired combat damage. The un-painted patch near the cabane struts is collocated with the fuel tank location and is possibly the result of damage incurred 6 March.

Albatros D.III 2006/16. Its appearance is based on photographs taken during March 1917. Dates and circumstances regarding fuselage repair patches unknown.

Pushers:
2nd and 3rd of 5 April two-seater pusher crewmen KiA
4th and 5th of 7 1917 two-seater pusher crewmen KiA
14th and 15th of 17 total two-seater pusher crewmen KiA

Two-Seaters:
11th and 12th of 20 April two-seater crewmen KiA
26th and 27th of 44 1917 two-seater crewmen KiA
37th and 38th of 54 pre-wound two-seater crewmen KiA
37th and 38th of 63 total two-seater crewmen KiA

All:
12th and 13th of 23 total April crewmen KiA
30th and 31st of 53 total 1917 crewmen KiA
47th and 48th of 66 total pre-wound crewmen KiA
47th and 48th of 84 total crewmen KiA

Etc.
12th of 21 total April victories
28th of 48 total 1917 victories
43rd of 57 total pre-wound victories
26th victory in an Albatros D.III
22nd of 22 two-seater victories flying from La Brayelle
27th of 28 total victories flying from La Brayelle
22nd of 30 two-seater victories as *Staffelführer*
27th of 39 total victories flying as *Staffelführer*

Statistics Notes
* *All pusher victories are pre-wound*
Δ *All FE.2 victories are pre-wound*
± *All No.11 Squadron victories are pre-wound*
† *All No.11 Squadron deaths are pre-wound*

Victory No.44

Above: A6796's souvenired rudder fabric, as seen hanging in Richthofen's bedroom.

14 April
Nieuport 23 A6796
No.60 Squadron RFC

Day/Date: Saturday, 14 April
Time: 9.15AM
Weather: Fine morning; cloudy in the afternoon.
Attack Location: Above Harlex
Crash Location: 1 kilometer south of Bois Bernard
Side of Lines: Friendly

KOFL 6. Armee Weekly Activity Report: "9.15 vorm. 1 Nieuport Einsitzer bei Fresnoy (diess.) durch Rittm. Frhr. von Richthofen, J.St.11 (als 44.)

9.15AM. 1 Nieuport single-seater near Fresnoy (this side [of lines]) by *Rittm. Frhr.* von Richthofen, *Jagdstaffel* 11 (as 44th)."

MvR Combat Report: "Nieuport One-Seater. Occupant: Lieut.W.O.Russell, made prisoner. Plane No.6796. Motor No.8341/I.B. Rotation [rotary engine].

Above Harlex one of our observer planes was attacked by several Nieuports. I hurried to the place of action, attacked one of the planes and forced it to land 1 km. south of Bois Bernard.

Frhr. v. Richthofen.
Acknowledged."

RFC Combat Casualties Report: "Left aerodrome 8-30 am. Lt.W.O.Russell is unofficially reported by his brother to be unwounded and a prisoner of war. (3rd Bde ltr.3B/57/A d-13/6/17). German message states that Lt. W.O.Russell is unwounded."

RFC A/C:
Make/model/serial number: Nieuport 23 A6796 (SFA N3468)
Manufacturer: *Société Anonyme des Établissements Nieuport*
Unit: No.60 Squadron RFC
Aerodrome: Filescamp Farm
Sortie: Offensive Patrol East of DOUAI
Colors/markings: Generally, overall silver aluminum with b/w/r roundels on wings and fuselage; white-bordered black serial number on r/w/b rudder; likely an identifying number (u/k) aft of the fuselage roundel.
Engine: 120 hp Le Rhône 9J rotary
Engine Number: T8341
Gun: One Lewis. No. 15104
Manner of Victory: Gliding descent to dead-stick forced landing, during which machine overturned.
Items Souvenired: Serial number on swatch of port rudder fabric. Photographed on display in Richthofen's home bedroom in Schweidnitz.

RFC Crew:
Pilot: Lt. William Oswald Russell, 24 (DoB 11 June 1892), PoW
Repatriation: 2 January 1919

MvR A/C:
Make/model/serial number: Albatros D.III, possibly *Le Petit Rouge*, or 2006/16.
Unit: *Jasta* 11
Commander: *Rittm.* Manfred von Richthofen
Airfield: La Brayelle
Colors/markings: *Le Petit Rouge* – See Vic. #17. Albatros D.III 2006/16 – See Vic. #23.
Initial Attack Altitude: ca. 7,000 feet (Russell)
Gunnery Range: u/k
Rounds Fired: u/k
Known Staffel Participants: Ltn. Lothar von Richthofen (#6), *Ltn.* Kurt Wolff (#14), *Vzfw.* Sebastian Festner (#11).

Notes:
1. Russell recounted postwar he had been aloft for 1.5 hours when his flight made a diving attack on a pair of two-seaters. However, his engine failed, likely damaged from return fire; "petrol was flowing freely." While gliding he was attacked by "two enemy scouts," one of which "followed me to the ground...shooting all the way." This was Richthofen, prosecuting his normal no-quarter attack methodology.

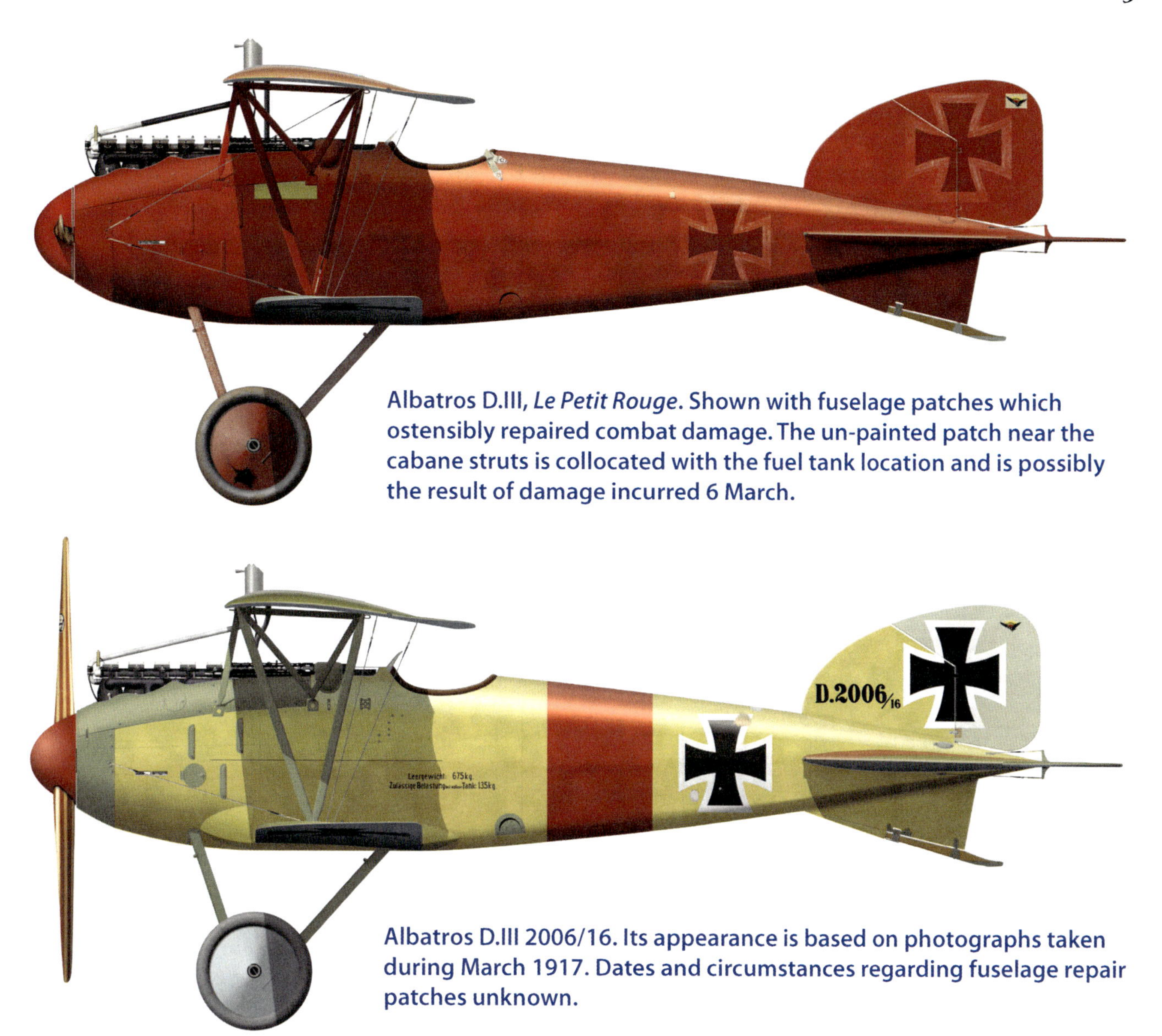

Albatros D.III, *Le Petit Rouge*. Shown with fuselage patches which ostensibly repaired combat damage. The un-painted patch near the cabane struts is collocated with the fuel tank location and is possibly the result of damage incurred 6 March.

Albatros D.III 2006/16. Its appearance is based on photographs taken during March 1917. Dates and circumstances regarding fuselage repair patches unknown.

2. Last MvR victory while flying from La Brayelle.

Victory No.44 Statistics:

Single-Seaters
2nd of 4 April single-seater victories
7th of 15 1917 single-seater victories
15th of 18 pre-wound single-seater victories
15th of 35 total single-seater victories

Nieuport *
2nd of 2 April Nieuport victories
3rd of 4 1917 Nieuport victories
3rd of 4 total Nieuport victories

No.60 Squadron ±
2nd of 2 No.60 Nieuport victories
2nd of 2 April No.60 Squadron victories
2nd of 2 1917 No.60 Squadron victories
2nd of 2 total No.60 Squadron victories

Etc.
13th of 21 total April victories
29th of 48 total 1917 victories
44th of 57 total pre-wound victories
27th victory in an Albatros D.III
6th of 6 single-seater victories flying from La Brayelle
28th of 28 total victories flying from La Brayelle
6th of 9 single-seater victories as *Staffelführer*
28th of 39 total victories flying as *Staffelführer*

Statistics Notes
* *Includes N.17 and N.23 models*
± *All No.60 Squadron victories are pre-wound*
† *All No.60 Squadron deaths are pre-wound*

Victory No.45

Albatros D.III, *Le Petit Rouge*. Shown with fuselage patches which ostensibly repaired combat damage. The un-painted patch near the cabane struts is collocated with the fuel tank location and is possibly the result of damage incurred 6 March.

16 April
BE.2e A3156
No.13 Squadron RFC

Day/Date: Monday, 16 April
Time: 5.30PM
Weather: Rain and low clouds all day.
Attack Location: Between Bailleul and Gavrelle
Crash Location: Between Bailleul and Gavrelle
Side of Lines: Enemy

RFC Communique No. 84: "Twenty targets were dealt with by artillery with aeroplane observation."

KOFL 6. Armee Weekly Activity Report:"5.30 nachm. 1 B.E. –D.D. bei Gavrelle (jens.) durch Rittm. Frhr. von Richthofen, J.St.11 (als 45.)

5.30PM. 1 B.E. biplane near Gavrelle (other side [of lines]) by *Rittmeister Frhr*. von Richthofen, *Jasta* 11 (as 45th)."

MvR Combat Report: "BE Two-Seater. No details, as plane fell on other side.

When pursuit-flying (height of clouds 1000 meters) I observed an artillery flyer at 880 meters altitude, approached him unnoticed, and attacked him, whereupon he fell down, smoking. The pilot caught the machine once more, but then lost control in 100 meters. The plane dashed down between Bailleul and Cavrelle [*sic*].

Frhr. v. Richthofen
Acknowledged."

RFC Combat Casualties Report: "Left aerodrome 2.50PM. Machine was attacked by H.A. Pilot and observer wounded."

RFC A/C:
Make/model/serial number: Royal Aircraft Factory BE.2e A3156
Unit: No.13 Squadron RFC
Aerodrome: Savy
Sortie: Artillery Patrol
Colors/markings: u/c. Generally, PC10 wings/ailerons/vertical and horizontal stabilizers/elevators; battleship gray engine cowl; b/w/r rudder; black serial number on vertical stabilizer. No.13 Sqn markings were a narrow horizontal bar running down the fuselage from observer's cockpit to the tail (black on CDL machines, white on PC10).
Engine: 90 H.P. R.A.F. 1a
Engine Number: 1742 WD 2679
Gun: Two Lewis. Nos. 13040, 18976
Manner of Victory: Gunfire precipitated forced landing
Items Souvenired: n/a

RFC Crew:
Pilot: Lt. Alphonso Pascoe, WiA
Manner of Injury: Either gunshot wounds or injury associated with forced landing.
Obs: 2nd Lt. Frederick Seymour Andrews, 27 or 28 (DoB 1889). DoW 29 April 1917
Cause of Death: u/c, likely gunshot wounds
Burial: Etaples Military Cemetery, France. Grave XVII. A. 11.

Albatros D.III 2006/16. Its appearance is based on photographs taken during March 1917. Dates and circumstances regarding fuselage repair patches unknown.

MvR A/C:
Make/model/serial number: Albatros D.III, possibly *Le Petit Rouge*, or 2006/16.
Unit: *Jasta* 11
Commander: *Rittm.* Manfred von Richthofen
Airfield: Roucourt
Colors/markings: *Le Petit Rouge* – See Vic. #17.
Albatros D.III 2006/16 – See Vic. #23.
Damage: n/a
Initial Attack Altitude: 800 metres
Gunnery Range: u/k
Rounds Fired: u/k
Known Staffel Participants: u/k

Note: First MvR victory while flying from Roucourt.

Victory No.45 Statistics:
Two-Seaters
12th of 17 April two-seater victories
23rd of 33 1917 two-seater victories
30th of 39 pre-wound two-seater victories
30th of 45 total two-seater victories

BE.2 Δ
4th of 7 April BE.2 victories
12th of 15 1917 BE.2 victories
14th of 17 total BE.2 victories

No.13 Squadron ±
3rd of 4 No.13 BE.2 victories
3rd of 4 April No.13 Squadron victories
3rd of 4 1917 No.13 Squadron victories
3rd of 4 total No.13 Squadron victories

Deaths †
No.13 Squadron:
3rd of 4 April No.13 Squadron crewmen KiA
3rd of 4 1917 No.13 Squadron crewmen KiA
3rd of 4 total No.13 Squadron crewmen KiA

Two-Seaters:
13th of 20 April two-seater crewmen KiA
28th of 44 1917 two-seater crewmen KiA
39th of 54 pre-wound two-seater crewmen KiA
39th of 63 total two-seater crewmen KiA

All:
14th 23 total April crewmen KiA
32nd of 53 total 1917 crewmen KiA
49th of 66 total pre-wound crewmen KiA
49th of 84 total crewmen KiA

Wounded
No.13 Squadron:
3rd of 4 April No.13 Squadron crewmen WiA
3rd of 4 1917 No.13 Squadron crewmen WiA
3rd of 4 total No.13 Squadron crewmen WiA

Two-Seaters:
8th of 11 April two-seater crewmen WiA
12th of 16 1917 two-seater crewmen WiA
13th of 16 pre-wound two-seater crewmen WiA
13th of 18 total two-seater crewmen WiA

All:
8th of 11 total April crewmen WiA
13th of 19 total 1917 crewmen WiA
15th of 18 total pre-wound crewmen WiA
15th of 26 total crewmen WiA

Etc.
8th of 14 victories behind enemy lines
14th of 21 total April victories
30th of 48 total 1917 victories

45th of 57 total pre-wound victories
28th victory in an Albatros D.III
1st of 6 two-seater victories flying from Roucourt
1st of 8 total victories flying from Roucourt
23rd of 30 two-seater victories as *Staffelführer*
29th of 39 total victories flying as *Staffelführer*

Statistics Notes
Δ *All BE.2 victories are pre-wound*
± *All No.13 Squadron victories are pre-wound*
† *All No.13 Squadron deaths are pre-wound; KiA includes DoW*

Above: Aerial view of Roucourt airfield. Landing grounds were to the right of the triangular-shaped woods, where three tent hangars are positioned along the rear wall but have not yet been erected. *Jasta* 11 billeted in the chateau at center, in front of which can be seen the stairs that served as the location of several famous *Jasta* 11 group photographs.

Victory No.46

Albatros D.III, *Le Petit Rouge*. Shown with fuselage patches which ostensibly repaired combat damage. The un-painted patch near the cabane struts is collocated with the fuel tank location and is possibly the result of damage incurred 6 March.

22 April
FE.2b 7020
No.11 Squadron RFC

Day/Date: Sunday, 22 April
Time: 5.30PM
Weather: Fine but cloudy.
Attack Location: "Above our side"
Crash Location: Near Lagnicourt
Side of Lines: Enemy

KOFL 6. Armee Weekly Activity Report: "5.10 N. 1 fdl. Flugzeug bei Cagnicourt (jenseits) dch. Rittm. Frhr. v. Richthofen, J.St.11 (als 46.)

5.10PM. 1 enemy airplane near Cagnicourt (other side [of lines]) by *Rittmeister Freiherr* von Richthofen, *Jasta* 11 (as 46th)."

MvR Combat Report: "Vikkers Two-Seater. No details, as plane fell on other side of line

When my *Staffel* was attacking an enemy squad, I personally attacked the last of the enemy planes. Immediately after I had discharged my first shots, the plane began to smoke. After 500 shots the plane dashed down and crashed to splinters on the ground. The fight had begun above our side, but the prevailing East wind had drifted the planes to the west.

Frhr. v. Richthofen
Was acknowledged."

RFC Combat Casualties Report: "Left aerodrome 2.45PM. Pilot badly wounded in head and arm and observer shot in leg, during aerial combat."

RFC A/C:
Make/model/serial number: FE.2b 7020
Manufacturer: Subcontracted to and built by Bolton & Paul, Ltd., Norwich.
Unit: No.11 Squadron RFC
Aerodrome: Izel-le-Hameau
Sortie: Photographic Reconnaissance
Colors/markings: Presentation a/c *Punjab No 34, Jhellum*. Generally, PC10 nacelle with white closed triangle squadron marking across nose; likely white "Punjab 34 Jhellum" on sides of nacelle; wing/ailerons/horizontal stabilizers/elevators uppersurfaces PC10 with CDL undersurfaces; likely upperwing roundels had white surrounds.
Engine: 160 hp Beardmore
Engine Number: 1049 WD 7659
Gun: Two Lewis. Nos. 22247, 21893
Manner of Victory: Gunfire precipitated crash landing
Items Souvenired: n/a

RFC Crew:
Pilot: Lt. William Fred Fletcher, 22 (DoB 6 December 1894), WiA
Manner of Injury: Gunshot wounds to head and arm; possible mild trauma associated with crash landing
Obs: Lt. Waldemar Franklin, 20 (29 December 1896), WiA
Manner of Injury: Gunshot wound to leg; possible

mild trauma associated with crash landing

MvR A/C:
Make/model/serial number: Albatros D.III, possibly *Le Petit Rouge*.
Unit: *Jasta* 11
Commander: *Rittm.* Manfred von Richthofen
Airfield: Roucourt
Colors/markings: *Le Petit Rouge* – See Vic. #17.
Initial Attack Altitude: u/k
Gunnery Range: u/k
Rounds Fired: "500 shots"
Known Staffel Participants: Ltn. Kurt Wolff (#19)

Note: From this point it is presumed Richthofen flew *Le Petit Rouge* during sorties, although the possibility exists he flew other *Jasta* 11 Albatrosses as demanded by the exigencies of machine availability.

Victory No.46 Statistics:

Pushers*
4th of 5 April pusher victories
4th of 5 April two-seater pusher victory
6th of 7 1917 two-seater pusher victories
11th of 12 total two-seater pusher victories
17th of 18 total pusher victories

Two-Seaters
13th of 17 April two-seater victories
24th of 33 1917 two-seater victories
31st of 39 pre-wound two-seater victories
31st of 45 total two-seater victories

FE.2 Δ
4th of 5 April FE.2 victories
6th of 7 1917 FE.2 victories
11th of 12 total FE.2 victories

No.11 Squadron ±
4th of 4 No.11 FE2 victories
2nd of 2 April No.11 Squadron victories
2nd of 2 1917 No.11 Squadron victories
4th of 4 total No.11 Squadron victories

Wounded
No.11 Squadron:
3rd and 4th of 4 April No.11 Squadron crewmen WiA
3rd and 4th of 4 1917 No.11 Squadron crewmen WiA
3rd and 4th of 4 total No.11 Squadron crewmen WiA

Pushers:
3rd and 4th of 4 April two-seater pusher crewmen WiA
4th and 5th of 5 1917 two-seater pusher crewmen WiA
4th and 5th of 5 total two-seater pusher crewmen WiA

Two-Seaters:
9th and 10th of 11 April two-seater crewmen WiA
13th and 14th of 16 1917 two-seater crewmen WiA
14th and 15th of 16 pre-wound two-seater crewmen WiA
14th and 15th of 18 total two-seater crewmen WiA

All:
9th and 10th of 11 total April crewmen WiA
14th and 15th of 19 total 1917 crewmen WiA
16th and 17th of 18 total pre-wound crewmen WiA
16th and 17th of 26 total crewmen WiA

Etc.
9th of 14 victories behind enemy lines
15th of 21 total April victories
31st of 48 total 1917 victories
46th of 57 total pre-wound victories
29th victory in an Albatros D.III
2nd of 6 two-seater victories flying from Roucourt
2nd of 8 total victories flying from Roucourt
24th of 30 two-seater victories as *Staffelführer*
30th of 39 total victories flying as *Staffelführer*

Statistics Notes
* *All pusher victories are pre-wound*
Δ *All FE.2 victories are pre-wound*
± *All No.11 Squadron victories are pre-wound*
† *All No.11 Squadron deaths are pre-wound*

Victory No.47

Above: 2nd Lt. Eric Arthur Welch's headstone, 2011.

Above: Sgt. Amos George Tollervey's headstone, 2011.

23 April
BE.2e A3168
No.16 Squadron RFC

Day/Date: Monday, 22 April
Time: 12.05PM
Weather: Fine.
Attack Location: Mericourt, "this side of the lines."
Crash Location: Near Mericout
Side of Lines: Enemy

KOFL 6. Armee Weekly Activity Report: "12.13 N. 1 B.E.D.D.bei Avion (jenseits) dch, Rtm. Frhr.v.Richthofen, J.St.11 (als 47.)

12.13PM. B.E. biplane near Avion (other side [of lines]) by *Rittmeister Freiherr* von Richthofen, *Jasta* 11 (as 47th)."

MvR Combat Report: "BE Two-Seater. No details, as plane broke in air and was scattered in falling.

I observed an artillery flyer, approached him unnoticed, and shot at him from closest range, until his left wing came off. The machine broke to pieces and fell down near Mericourt.

Frhr. v. Richthofen.
Acknowledged."

RFC Combat Casualties Report: n/a

RFC A/C:
Make/model/serial number: BE.2e A3168
Manufacturer: Subcontracted to and built by Wolseley Motors, Ltd., Adderley Park, Birmingham.
Unit: No.16 Squadron RFC
Aerodrome: Bruay

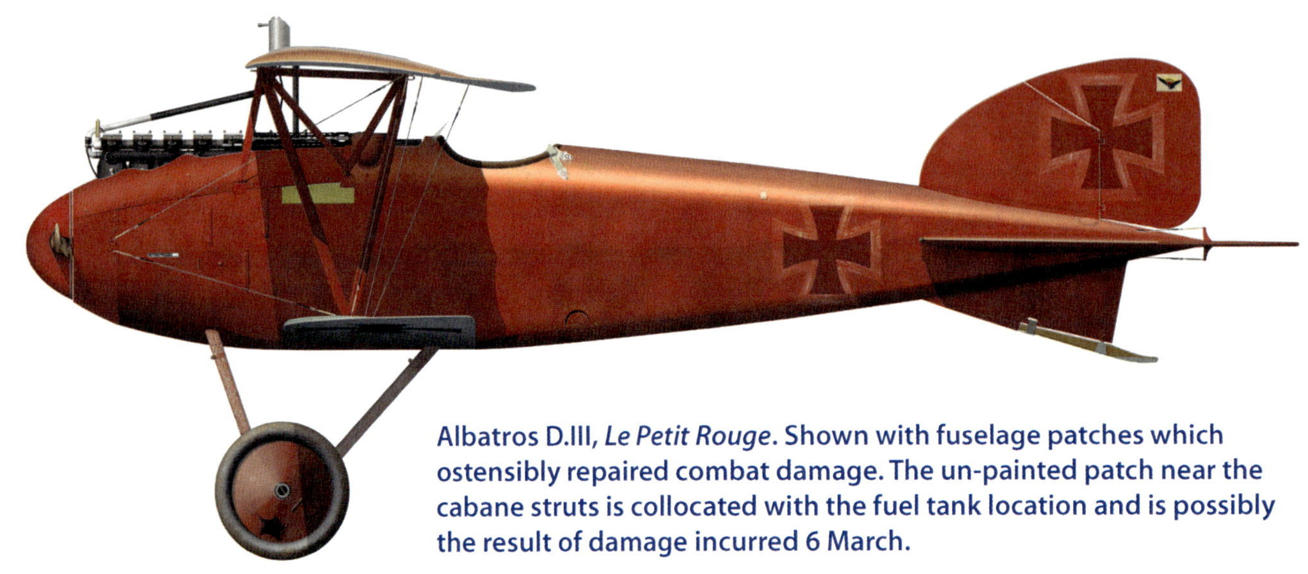

Albatros D.III, *Le Petit Rouge*. Shown with fuselage patches which ostensibly repaired combat damage. The un-painted patch near the cabane struts is collocated with the fuel tank location and is possibly the result of damage incurred 6 March.

Sortie: Photo Op
Colors/markings: Overall PC10, with Battleship Gray engine panels and CDL undersurfaces. White-bordered roundels on wings and fuselage. b/w/r (leading to trailing edge) tri-color rudder.
Engine: 90 hp RAF 1a
Engine Number: E2167 WD 10718
Gun: Two Lewis. Nos. 19782, 19783
Manner of Victory: Gunfire-precipitated structural compromise and in-air breakup.
Items Souvenired: n/a

RFC Crew:
Pilot: 2nd Lt. Eric Arthur Welch, 23 (DoB 1894, age indicated on headstone), KiA
Cause of Death: Either gunshot wounds, trauma associated with airplane crash, or post-free-fall terrain impact.
Burial: Petit Vimy British Cemetery, Vimy France, Grave A.5
Obs: Sgt. Amos George Tollervey, Service No. 3284, 21 (DoB 9 March 1896), KiA
Cause of Death: Either gunshot wounds, trauma associated with airplane crash, or post-free-fall terrain impact.
Burial: Petit Vimy British Cemetery, Vimy France, Grave A.4

MvR A/C:
Make/model/serial number: Albatros D.III, possibly *Le Petit Rouge*.
Unit: *Jasta* 11
Commander: *Rittm.* Manfred von Richthofen
Airfield: Roucourt
Colors/markings: *Le Petit Rouge* – See Vic. #17.
Initial Attack Altitude: u/k
Gunnery Range: "Closest range"
Rounds Fired: u/k
Known Staffel Participants: Ltn. Kurt Wolff (#19)

Note: Discrepancy exists regarding whether A3168 was a BE.2e or f, because RFC records alternately define it as each. Regarding design development, the BE.2e was a continuation of the BE.2 series and differed from previous variants via shorter spanned lower wings, reduced number of interplane wing struts, an added strut that connected the upper and lower ailerons, and a redesigned empennage. A photograph of A3168 clearly shows these features. To meet demands for BE.2e machines the RFC began modifying BE.2c and d variants to e standards. Resultant supply and maintenance issues led to an October 1916 directive that stated all modified BE.2c airplanes would be re-designated as BE.2f, and all BE.2d airplanes would be re-designated as BE.2gs.

RFC records show A3168 was manufactured by Wolseley under contract 87/A/625, issued 21 August 1916 for 120 BE.2e airplanes. A3168 is also listed as a BE.2e under contract A/3062, dated 21 August 1916 that included 20 machines from A3149–A3168. A search amongst casualty reports for individual machines in A/3062 reveals airplanes identified as BE.2e, f, and even g. Yet all manufacture records indicate the machines were built as e variants. Which dovetails with the question asked in the notes of victory 39: Since the BE.2f and g were BE.2c and d machines modified to e standards, why would airplanes built as a BE.2e in the first place need to be modified to its own standard?

Victory No.47 Statistics:

Two-Seaters
14th of 17 April two-seater victories
25th of 33 1917 two-seater victories
32nd of 39 pre-wound two-seater victories
32nd of 45 total two-seater victories

BE.2 Δ
5th of 7 April BE.2 victories
13th of 15 1917 BE.2 victories
15th of 17 total BE.2 victories

No.16 Squadron ±
6th of 6 No.16 BE.2 victories
2nd of 2 April No.16 Squadron victories
6th of 6 1917 No.16 Squadron victories
6th of 6 total No.16 Squadron victories

Deaths †
No. 16 Squadron:
3rd and 4th of 4 April No.16 Squadron crewmen KiA
11th and 12th of 12 1917 No.16 Squadron crewmen KiA
11th and 12th of 12 total No.16 Squadron crewmen KiA

Two-Seaters:
14th and 15th of 20 April two-seater crewmen KiA
29th and 30th of 44 1917 two-seater crewmen KiA
40th and 41st of 54 pre-wound two-seater crewmen KiA
40th and 41st of 63 total two-seater crewmen KiA

All:
15th and 16th of 23 total April crewmen KiA
33rd and 34th of 53 total 1917 crewmen KiA
50th and 51st of 66 total pre-wound crewmen KiA
50th and 51st of 84 total crewmen KiA

Etc.
10th of 14 victories behind enemy lines
16th of 21 total April victories
32nd of 48 total 1917 victories
47th of 57 total pre-wound victories
30th victory in an Albatros D.III
3rd of 6 two-seater victories flying from Roucourt
3rd of 8 total victories flying from Roucourt
25th of 30 two-seater victories as *Staffelführer*
31st of 39 total victories flying as *Staffelführer*

Statistics Notes
Δ All BE.2 victories are pre-wound
± All No.16 Squadron victories are pre-wound
† All No.16 Squadron deaths are pre-wound; KiA includes DoW

Below: A jocular Kurt Wolff looks on as Richthofen suits up for flight.

Victory No.48

Albatros D.III, *Le Petit Rouge*. Shown with fuselage patches which ostensibly repaired combat damage. The un-painted patch near the cabane struts is collocated with the fuel tank location and is possibly the result of damage incurred 6 March.

28 April
BE.2e 7221
No.13 Squadron RFC

Day/Date: Saturday, 28 April
Time: 9.30AM
Weather: Low clouds
Attack Location: "Wood east of Pelves, southeast corner of Square 6998, this side of line"
Crash Location: "Wood of Pelves"
Side of Lines: Friendly

KOFL 6. Armee Weekly Activity Report: "9.30 V. 1 fdl. Flugzeug bei Pelves (diess.) dch. Rittm. Frhr. v. Richthofen Jagdstaffel Richthofen (als 48.)

9:30AM. One enemy airplane near Pelves (this side [of lines]) by *Rittmeister* Freiherr von Richthofen, *Jagdstaffel* Richthofen (as 48th)."

MvR Combat Report: "BE 2. Lieut. Follitt [*sic*], killed; Observer, F.I. Kirckham [*sic*], wounded, harmlessly.

While on pursuit-flying, about 9.30 a.m. I attacked an enemy infantry or artillery flyer in 600 meters above the trenches. Above the wood of Pelves I caused the enemy plane to fall. The adversary from the beginning to the end of the fight was never able to get out of range of my gun.

Frhr. v. Richthofen.
Acknowledged."

RFC Combat Casualties Report: "Left aerodrome 7.20AM. The following is extract from letter received by O.C. No.13 Sqdn from 2/Lt. P.E.H. Van Baerle, a prisoner at Karlsruhe: 'A fellow who was with you called, I think, Merriman [*sic*], has just arrived here. Follet [*sic*] his pilot died of wounds.'"

RFC A/C:
Make/model/serial number: BE.2e 7221
Manufacturer: Subcontracted to and built by The British and Colonial Aeroplane Co., Ltd., Filton, Bristol.
Unit: No.13 Squadron RFC
Aerodrome: Savy
Sortie: Artillery Patrol. XVII Corps Front.
Colors/markings: Generally, overall PC10, with Battleship Gray engine panels and CDL undersurfaces. White-bordered roundels on wings and fuselage. b/w/r (leading to trailing edge) tri-color rudder.
Engine: 90 hp RAF 1a
Engine Number: E1017 WD 5168
Gun: Two Lewis. Nos. 27330, 6327
Manner of Victory: Gunshot wounds incapacitated pilot who fell forward on the control column, precipitating a diving descent until impacting trees.
Items Souvenired: n/a

RFC Crew:
Pilot: Lt. Reginald William Follit, 26 (DoB 3 September 1890), PoW/DoW less than an hour later.
Cause of Death: Gunshot wound to back, possibly exacerbated by trauma associated with airplane crash.

RFC Combat Casualties Report: "German message states that Lt. Follitt [sic] was killed."
Burial: In-field; location lost during turmoil of war.
Obs: 2nd Lt. Frederick James Kirkham, 22 or 23 (DoB 1894), WiA/ PoW
Manner of Injury: Bullet "splashes" to face; soreness, cuts, and bruises associated with airplane crash.
RFC Combat Casualties Report: "An unconfirmed report in a German publication states that...Lt. F.J. Kirkham slightly wounded."
Incarceration: Karlsruhe
Repatriation: 31 December 1918

MvR A/C:
Make/model/serial number: Albatros D.III, possibly *Le Petit Rouge*.
Unit: *Jasta* Richthofen
Commander: *Rittm*. Manfred von Richthofen
Airfield: Roucourt
Colors/markings: *Le Petit Rouge* – See Vic. #17.
Damage: n/a
Initial Attack Altitude: 600 metres
Gunnery Range: u/k
Rounds Fired: u/k
Known Staffel Participants:

Notes:

1. In a post-war interview Kirkham recalled the BE.2 had "two Lewis guns, one firing forward through the propeller, and the other fixed over the top plane [wing] firing backward." This is unlikely because Vickers rather than Lewis machine guns were used for synchronised forward firing through the rotating propeller disc, and on a BE.2 any guns mounted to the upper wing would be rendered unreachable from the observer's cockpit. Nor would there be any reason to mount a machine gun in that fashion.
2. Kirkham and Follit survived the airplane crash because the BE.2 "hit a clump of small trees," which cushioned the impact. Follit died an hour later but as a result of his gunshot wound, not the crash.
3. In his combat report Richthofen refers to his weapon as a "gun," singular. However, in late April *Jasta* 11's complement of one-gun Halberstadts was long gone. Thus, "gun" is likely a typo either in the original German document or more likely the translation.
4. Kirkham recalled that after Follit was hit and the BE.2 dived toward the ground, "the red plane just hung on my tail and kept firing all the time" and "the red scout stuck right there on the tail, and his two machine[guns] were pumping lead all the time." Another example of Richthofen's no-quarter *modus operandi*. Although "the sleeves and shoulders of my flying jacket had several dozen holes through them and then one bullet hit the barrel of the machine gun right under my nose," Kirkham only suffered bullet splashes (finely divided fragments produced by bullets impacting armor or hard objects [such as a machine gun]) to the face and hands.
5. As appears at the end of *Kofl* 6. *Armee* report Nr.50790:

"14.) Besonderes: Auf Befehl S.M hat die Jagdstaffel 11 den Namen Jagdstaffel "Richthofen" zu fuehren.

14.) Special: By command of *Seine Majestät* [His Majesty] the *Jagdstaffel* 11 has to bear the name *Jagdstaffel* 'Richthofen'."

Victory No.48 Statistics:
Two-Seaters
15th of 17 April two-seater victories
26th of 33 1917 two-seater victories
33rd of 39 pre-wound two-seater victories
33rd of 45 total two-seater victories

BE.2 Δ
6th of 7 April BE.2 victories
14th of 15 1917 BE.2 victories
14th of 17 total BE.2 victories

No.13 Squadron ±
4th of 4 No.13 BE.2 victories
4th of 4 April No.13 Squadron victories
4th of 4 1917 No.13 Squadron victories
4th of 4 total No.13 Squadron victories

Deaths †
No.13 Squadron:
4th of 4 April No.13 Squadron crewmen KiA
4th of 4 1917 No.13 Squadron crewmen KiA
4th of 4 total No.13 Squadron crewmen KiA

Two-Seaters:
16th of 20 April two-seater crewmen KiA
31st of 44 1917 two-seater crewmen KiA
42nd of 54 pre-wound two-seater crewmen KiA
42nd of 63 total two-seater crewmen KiA

All:
17th of 23 total April crewmen KiA
35th of 53 total 1917 crewmen KiA
52nd of 66 total pre-wound crewmen KiA
52nd of 84 total crewmen KiA

Wounded
No.13 Squadron:
4th of 4 April No.13 Squadron crewmen WiA
4th of 4 1917 No.13 Squadron crewmen WiA
4th of 4 total No.13 Squadron crewmen WiA

Two-Seaters:
11th of 11 April two-seater crewmen WiA
15th of 16 1917 two-seater crewmen WiA
15th of 16 pre-wound two-seater crewmen WiA
16th of 18 total two-seater crewmen WiA

All:
11th of 11 total April crewmen WiA
16th of 19 total 1917 crewmen WiA
18th of 18 total pre-wound crewmen WiA
18th of 26 total crewmen WiA

Etc.
17th of 21 total April victories
33rd of 48 total 1917 victories
48th of 57 total pre-wound victories
31st victory in an Albatros D.III
4th of 6 two-seater victories flying from Roucourt
4th of 8 total victories flying from Roucourt
26th of 30 two-seater victories as *Staffelführer*
32nd of 39 total victories flying as *Staffelführer*

Statistics Notes
Δ All BE.2 victories are pre-wound
± All No.13 Squadron victories are pre-wound
† All No.13 Squadron deaths are pre-wound; KiA includes DoW

Above: *Jasta* 11 takes it easy at Roucourt, 29 April 1917. L-R: *Oblt.* Wolfgang Plüschow, Richthofen (astride bicycle), *Ltn.* Lothar von Richthofen, *Ltn.* von Hartmann, u/k, *Ltn.* Karl Esser, *Ltn.* Otto Brauneck, *Maj.* Albrecht von Richthofen (Manfred and Lothar's father, present for a social visit), u/k. The offset radiator and nose footstep of the adjacent Albatros reveal it is not *Le Petit Rouge*. It does possess similar features of Kurt Wolff's new 632/17, but precise identification remains elusive. (Lance Bronnenkant)

Victory No.49

Right: Colorized photo of Richthofen climbing into *Le Petit Rouge*; using ladders eased cockpit ingress and egress. Note Richthofen's abundant clothing to stave the cold at altitude, including a thick scarf wrapped securely around his neck and face—not trailing down his back loosely to flap in the wind a la airshow and Hollywood cliché.

29 April
SPAD VII C.1 B.1573
No.19 Squadron RFC

Day/Date: Sunday, 29 April
Time: 12.05PM
Weather: Fine
Attack Location: "Swamps near Lecluse, this side of lines"
Crash Location: "The swamp near Lecluse"
Side of Lines: Friendly

KOFL 6. Armee Weekly Activity Report:"12.15 N. 1 fdl. Flugzeug bei Lecluse (diess.) dch. Rtm. Frhr. v. Richthofen Jagdstaffel Richthofen (als 49.)

12.15PM. 1 enemy airplane near Lecluse (this side [of lines]) by *Rittmeister Freiherr* von Richthofen *Jagdstaffel* Richthofen (as 49th)."

MvR Combat Report: "Spad one-Seater. No details concerning plane, as it vanished in a swamp.

With several of my gentlemen I attacked an English Spad group consisting of 3 machines. The plane I had singled out broke to pieces whilst curving and dashed, burning, into the swamp near Lecluse.

Frhr. v. Richthofen.
Was acknowledged."

RFC Combat Casualties Report: "From an unconfirmed report in a German publication 2/Lt. R.Applin is believed to be dead."

RFC A/C:
Make/model/serial number: SPAD (Société Pour L'Aviation et ses Dérivés) VII C.1 B.1573
Unit: No.19 Squadron RFC
Commander: Major H. D. Harvey-Kelly DSO (KiA this sortie by *Ltn.* Kurt Wolff)
Aerodrome: Vert Galant, France
Sortie: Offensive Patrol LENS-LE FOREST-FONTAINE-NOTRE DAME-NOREUIL.
Colors/markings: Generally, beige engine cowlings and forward fuselage up to cockpit, thence natural clear doped fabric surfaces aft with black serial number on vertical stabilizer. Wings and horizontal surfaces clear doped fabric. Tri-colored b/w/r rudder. White-bordered b/w/r roundels on wings and fuselage. No. 19 Squadron (unofficial) marking was a black dumbbell on the fuselage sides aft of the roundel.
Engine: Hispano-Suiza
Engine Number: 5768
Gun: One Vickers. No. L8479
Manner of Victory: In-flight break-up and burning descent until terrain impact.
Items Souvenired: none

RFC Crew:
Pilot: Lt. Richard Applin, 23 (DoB 3 June 1894), KiA.
Cause of Death: u/k.
RFC Combat Casualties Report: "German message states that Lt. R.APPLIN was killed."
Burial: Body never recovered from the swamp and presumably remains where fallen.

MvR A/C:
Make/model/serial number: Albatros D.III, possibly *Le Petit Rouge.*
Unit: *Jasta* Richthofen
Commander: *Rittm.* Manfred von Richthofen
Airfield: Roucourt
Colors/markings: *Le Petit Rouge* – See Vic. #17.
Damage: n/a
Initial Attack Altitude: u/k
Gunnery Range: u/k
Rounds Fired: u/k
Known Staffel Participants: Ltn. Kurt Wolff (#25), *Ltn.* Lothar von Richthofen (#13)

Notes:
1. The RFC Combat Casualties Report listed the SPAD serial number as "A.1573."
2. The RFC Combat Casualties Report listed B.1573's machine gun as a "Lewis."
3. Richthofen's autobiography *Der Rote Kampfflieger* described Applin's downing: "My opponent was the first to fall, after I had shot his engine to pieces. In any case, he decided to land near us. I no longer know any mercy, [and] for that reason I attacked him a second time, whereupon the aeroplane fell apart in my stream of bullets. The wings each fell away like pieces of paper and the fuselage went roaring down like a stone on fire." Again, this account is consistent with Richthofen's usual no-quarter *modus operandi.*
4. This was Richthofen's first of four victories for the day.

Victory No.49 Statistics:
Single-Seaters
3rd of 4 April single-seater victories
8th of 15 1917 single-seater victories
16th of 18 pre-wound single-seater victories
16th of 35 total single-seater victories

SPAD Δ
1st and only April SPAD victory

Albatros D.III, *Le Petit Rouge*. Shown with fuselage patches which ostensibly repaired combat damage. The un-painted patch near the cabane struts is collocated with the fuel tank location and is possibly the result of damage incurred 6 March.

2nd of 4 1917 SPAD victories
2nd of 4 total SPAD victories

No.19 Squadron
2nd of 3 No.19 SPAD victories
1st and only April No.19 Squadron victory
2nd of 3 1917 No.19 Squadron victories
3rd of 4 total No.19 Squadron victories

Deaths †
No.19 Squadron:
1st and only April No.19 Squadron crewman KiA
1st of 2 1917 No.19 Squadron crewmen KiA
2nd of 3 total No.19 Squadron crewmen KiA

Single-seaters:
2nd of 3 April single-seater crewmen KiA
5th of 9 1917 single-seater crewmen KiA
11th of 12 pre-wound single-seater crewmen KiA
11th of 21 total single-seater crewmen KiA

All:
18th of 23 total April crewmen KiA
36th of 53 total 1917 crewmen KiA
53rd of 66 total pre-wound crewmen KiA
53rd of 84 total crewmen KiA

Etc.
18th of 21 total April victories
34th of 48 total 1917 victories
49th of 57 total pre-wound victories
32nd victory in an Albatros D.III
1st of 2 single-seater victories flying from Roucourt
5th of 8 total victories flying from Roucourt
7th of 9 single-seater victories as *Staffelführer*
33rd of 39 total victories flying as *Staffelführer*

Below: One of the most famous and iconic photographs of *Jasta* 11, taken on the steps of Chateau Roucourt, 29 April 1917. L-R: (Top) *Ltn*. Carl Allmenröder, *Ltn*. Lothar von Richthofen, *Oblt*. Wolfgang Plüschow, *Ltn*. von Hartmann. (Bottom) *Ltn*. Georg Simon, *Ltn*. Kurt Wolff, Richthofen, *Maj*. Albrecht von Richthofen, *Ltn*. Konstantin Krefft, *Ltn*. Hans Hinsch.

Victory No.50

Albatros D.III, *Le Petit Rouge*. Shown with fuselage patches which ostensibly repaired combat damage. The un-painted patch near the cabane struts is collocated with the fuel tank location and is possibly the result of damage incurred 6 March.

29 April
FE.2b 4898
No.18 Squadron RFC

Day/Date: Sunday, 29 April
Time: 4.55PM
Weather: Fine
Attack Location: "South-west of Inchy, Hill 90 near Pariville [*sic*], this side of the line."
Crash Location: Southwest of Inchy, near Pronville
Side of Lines: Friendly

KOFL 6. Armee Weekly Activity Report: "4.55 N. 1 fdl. Flugzeug bei Inchy (diess.) dch. Rittm. Frhr. v. Richthofen. Jagdstaffel Richthofen (as 50.)

4.55PM. 1 enemy airplane near Inchy (this side [of lines]) by *Rittmeister Freiherr* von Richthofen. *Jagdstaffel* Richthofen (as 50th)."

MvR Combat Report: "Vikkers [*sic*] 2. Occupants: Capt. [*sic*] G.Stead, R.F.C. No details concerning plane, went down burning in first line.

I attacked together with 5 of my gentlemen an enemy group of 5 Vikkers [*sic*]. After along curve fight, during which my adversary defended himself admirably, I managed to put myself behind the enemy. After 300 shots the enemy plane caught fire. The plane burnt to ashes, and the occupants fell out."

RFC Combat Casualties Report: "Left aerodrome 2.20PM."

RFC A/C:
Make/model/serial number: FE.2b 4898
Manufacturer: Subcontracted to and built by G. & J. Weir, Ltd., Cathcart, Glasgow
Unit: No.18 Squadron RFC
Aerodrome: Bertangles
Sortie: Escort to photography
Colors/markings: Presentation a/c *Johore No. 14*. Generally, PC10 nacelle with white "JOHORE 14" on sides of nacelle; wing/ailerons/horizontal stabilizers/elevators uppersurfaces PC10 with CDL undersurfaces; likely upperwing roundels had white surrounds.
Engine: 160 hp Beardmore
Engine Number: 866/WD7476
Gun: Two Lewis. Nos. 17898, 20823
Manner of Victory: Uncontrolled flaming descent until terrain impact, prior to which the crew jumped overboard.
Items Souvenired: n/a

RFC Crew:
Pilot: Sgt. George Stead, 19, KiA.
Cause of Death: u/c, most likely blunt force trauma from post-free-fall terrain impact.
Burial: Either in-field or laid where fallen; location lost during turmoil of war.
Obs: A/Cpl. Alfred Beebee, 18 (DoB July 1898), KiA.
Cause of Death: u/c, most likely blunt force trauma from post-free-fall terrain impact.
Burial: Either in-field or laid where fallen; location lost during turmoil of war.

MvR A/C:
Make/model/serial number: Albatros D.III, possibly *Le Petit Rouge*.
Unit: *Jasta* Richthofen
Commander: *Rittm.* Manfred von Richthofen
Airfield: Roucourt
Colors/markings: *Le Petit Rouge* – See Vic. #17.
Damage: n/a
Initial Attack Altitude: 10,000 feet
Gunnery Range: u/k
Rounds Fired: 300
Known Staffel Participants: Ltn. Kurt Wolff

Notes:
1. This was Richthofen's second of four victories for the day.
2. RFC Communiques No. 85 notes a victory by Beebee on 24 April: "2nd Lieut. Traylen and Corpl. Beebee, No. 18 Squadron, 2nd Lieut. Hunt and Lieut. Partington, No. 18 Squadron, drove down hostile machines in spinning nose-dives, after having been engaged at close quarters."
3. This was Richthofen's last pusher victory.

Victory No.50 Statistics:

Pushers*
5th of 5 April pusher victories
5th of 5 April two-seater pusher victories
7th of 7 1917 two-seater pusher victories
12th of 12 total two-seater pusher victories
18th of 18 total pusher victories

Two-Seaters
16th of 17 April two-seater victories
27th of 33 1917 two-seater victories
34th of 39 pre-wound two-seater victories
34th of 45 total two-seater victories

FE.2 Δ
5th of 5 April FE.2 victories
7th of 7 1917 FE.2 victories
12th of 12 total FE.2 victories

No.18 Squadron
4th of 4 No.18 FE.2 victories
1st and only April No.18 Squadron victory
1st and only 1917 No.18 Squadron victory
4th of 4 total No.18 Squadron victories

Deaths †
No.18 Squadron:
1st and only April No.18 Squadron crewman KiA
1st and only 1917 No.18 Squadron crewman KiA
4th of 4 total No.18 Squadron crewmen KiA

Pushers:
4th and 5th of 5 April two-seater pusher crewmen KiA
6th and 7th of 7 1917 two-seater pusher crewmen KiA
16th and 17th of 17 total two-seater pusher crewmen KiA

Two-Seaters:
17th and 18th of 20 April two-seater crewmen KiA
32nd and 33rd of 44 1917 two-seater crewmen KiA
43rd and 44th of 54 pre-wound two-seater crewmen KiA
43rd and 44th of 63 total two-seater crewmen KiA

All:
19th and 20th of 23 total April crewmen KiA
37th and 38th of 53 total 1917 crewmen KiA
54th and 55th of 66 total pre-wound crewmen KiA
54th and 55th of 84 total crewmen KiA

Etc.
19th of 21 total April victories
35th of 48 total 1917 victories
50th of 57 total pre-wound victories
33rd victory in an Albatros D.III
5th of 6 two-seater victories flying from Roucourt
6th of 8 total victories flying from Roucourt
27th of 30 two-seater victories as *Staffelführer*
34th of 39 total victories flying as *Staffelführer*

Statistics Notes
* *All pusher victories are pre-wound*
Δ *All FE.2 victories are pre-wound*
± *All No.18 Squadron victories are pre-wound*
† *All No.18 Squadron deaths are pre-wound*

Victory No.51

Albatros D.III, *Le Petit Rouge*. Shown with fuselage patches which ostensibly repaired combat damage. The un-painted patch near the cabane struts is collocated with the fuel tank location and is possibly the result of damage incurred 6 March.

29 April
BE.2e 2738
No.12 Squadron RFC

Day/Date: Sunday, 29 April
Time: 7.25PM
Weather: Fine
Attack Location: "Near Roeux, this side of the line."
Crash Location: "Near the trenches near Roeux"
Side of Lines: Friendly

KOFL 6. Armee Weekly Activity Report: "7.25 N. 1 fdl. Flugzeug bei Roeux (zwischen den Linien) dch. Rittm. Frhr. v. Richthofen, Jagdstaffel Richthofen (als 51.)

7.25PM. 1 enemy airplane near Roeux (between the lines) by *Rittmeister Frhr.* von Richthofen, *Jagdstaffel* Richthofen (as 51st)."

MvR Combat Report: "BE DD 2. No details, as plane is under fire.

Together with my brother we each of us attacked an artillery flyer at a low altitude. After a short fight my adversary's plane lost its wings. When touching ground near the trenches near Roeux the plane caught fire.

Frhr. v. Richthofen.
Was acknowledged."

RFC Combat Casualties Report: "Left aerodrome 4.45PM."

RFC A/C:
Make/model/serial number: BE.2.e 2738
Manufacturer: Subcontract to and built by Ruston, Proctor & Co., Ltd., Lincoln
Unit: No.12 Squadron RFC
Aerodrome: Avesnes-le-Comte
Sortie: Artillery Observation
Colors/markings: Generally, overall PC10, with Battleship Gray engine panels and CDL undersurfaces. White-bordered roundels on wings and fuselage. b/w/r (leading to trailing edge) tri-color rudder. No.12 Sqn markings were two longitudinal black bands spanning the fuselage just below the upper longeron and above the lower longeron.
Engine: 90 hp RAF 1a
Engine Number: E674 WD 3023
Gun: Two Lewis. Nos. 15775, 14469
Manner of Victory: Machine gun bullets caused structural compromise and in-flight breakup, followed by an uncontrolled descent until terrain impact.
Items Souvenired: n/a

RFC Crew:
Pilot: Lt. David Evan Davies, 24 or 25 (DoB 1892), KiA.
Cause of Death: u/k. Either gunshot wound(s) or trauma associated with airplane crash.
RFC Combat Casualties Report: "Death of 2/ Lt.D.E.Davies accepted by the Army Council as having occurred on 29.4.17 on lapse of time (Part II orders G.H.Q. 3rd Echelon d/5.1.18)."
Burial: Either in-field or laid where fallen; location

lost during turmoil of war.
Obs: Lt. George Henry Rathbone, 22 (DoB 30 June 1895), KiA.
Cause of Death: u/k. Either gunshot wound(s) or trauma associated with airplane crash.
Burial: Either in-field or laid where fallen; location lost during turmoil of war.

MvR A/C:
Make/model/serial number: Albatros D.III, possibly *Le Petit Rouge*.
Unit: *Jasta* Richthofen
Commander: *Rittm.* Manfred von Richthofen
Airfield: Roucourt
Colors/markings: *Le Petit Rouge* – See Vic. #17.
Damage: n/a
Initial Attack Altitude: u/k
Gunnery Range: u/k
Rounds Fired: u/k
Known Staffel Participants: Ltn. Lothar von Richthofen (#14)

Note: This was Richthofen's third of four victories for the day.

Victory No.51 Statistics:
Two-Seaters
17th of 17 April two-seater victories
28th of 33 1917 two-seater victories
35th of 39 pre-wound two-seater victories
35th of 45 total two-seater victories

BE.2 Δ
7th of 7 April BE.2 victories
15th of 15 1917 BE.2 victories
17th of 17 total BE.2 victories

No.12 Squadron
2nd of 2 No.12 BE.2 victories
1st and only April No.12 Squadron victory
1st and only 1917 No.12 Squadron victory
2nd of 2 total No.12 Squadron victories

Deaths †
No.12 Squadron:
1st and only April No.12 Squadron crewman KiA
1st and only 1917 No.12 Squadron crewman KiA
2nd and 3rd of 3 total No.12 Squadron crewmen KiA

Two-Seaters:
19th and 20th of 20 April two-seater crewmen KiA
34th and 35th of 44 1917 two-seater crewmen KiA
45th and 46th of 54 pre-wound two-seater crewmen KiA
45th and 46th of 63 total two-seater crewmen KiA

All:
21st and 22nd of 23 total April crewmen KiA
39th and 40th of 53 total 1917 crewmen KiA
56th and 57th of 66 total pre-wound crewmen KiA
56th and 57th of 84 total crewmen KiA

Etc.
20th of 21 total April victories
36th of 48 total 1917 victories
51st of 57 total pre-wound victories
34th victory in an Albatros D.III
6th of 6 two-seater victories flying from Roucourt
7th of 8 total victories flying from Roucourt
28th of 30 two-seater victories as *Staffelführer*
35th of 39 total victories flying as *Staffelführer*

Statistics Notes
Δ All BE.2 victories are pre-wound
± All No.12 Squadron victories are pre-wound
† All No.12 Squadron deaths are pre-wound; KiA includes DoW

Victory No.52

Above: Sopwith Triplane.

29 April
Sopwith Triplane N5463
No.8 Squadron RNAS

Day/Date: Sunday, 29 April
Time: 7.40PM
Weather: Fine
Attack Location: "Between Billy-Montigny and Sallaumines, this side of the lines."
Crash Location: "North of Henin Liétard."
Side of Lines: Friendly

KOFL 6. Armee Weekly Activity Report: "7.45 N. 1 fdl. Dreidecker bei Lens (diess.) dch. Rittm. Frhr. v. Richthofen, Jagdstaffel Richthofen, (als 52.)

7.45PM. 1 enemy triplane near Lens (this side [of lines]) by *Rittmeister Freiherr* von Richthofen, *Jagdstaffel* Richthofen, (as 52nd)."

MvR Combat Report: "No details concerning enemy plane; it was burnt.

Soon after having shot down a BE near Roex [*sic*] we were attacked by a strong enemy One-seater squad of Nieuports, Spads and triplanes. The plane I had singled out caught fire after a short time, burned in the air and fell north of Henin Lietard."

RNAS A/C:
Make/model/serial number: Sopwith Triplane N5463
Manufacturer: Built by the Sopwith Aviation Company, Ltd., Canbury Park Road, Kingston-on-Thames.
Unit: 8 Squadron RNAS
Aerodrome: Auchel
Sortie: Offensive Patrol Henin-Liétard-Vitry
Colors/markings: Generally, overall PC-10 or PC-12 uppersurfaces, clear-doped fabric undersurfaces, with unpainted engine-turned metal engine cowl; clear-doped fabric vertical stabilizer with "The Sopwith Aviation Company Ltd., Kingston on Thames" stenciled in black; b/w/r tri-colored rudder, borderless roundels on wings and fuselage. Black serial number on rectangular white background located well aft on fuselage near horizontal stabilizer.

Albatros D.III, *Le Petit Rouge*. Shown with fuselage patches which ostensibly repaired combat damage. The un-painted patch near the cabane struts is collocated with the fuel tank location and is possibly the result of damage incurred 6 March.

Engine: 130 hp Clerget
Engine Number: 1491
Gun: One Vickers.
Manner of Victory: In-flight fire precipitated by machine gun bullets, eventuating uncontrolled descent and terrain impact.
Items Souvenired: n/a

RNAS Crew:
Pilot: Flight Sub-Lt. Albert Edward Cuzner, 26 (DoB 30 August 1890), KiA.
Cause of Death: u/k. Either gunshot wounds, immolation, trauma associated with airplane crash, or post-free-fall terrain impact.
Burial: Either in-field or laid where fallen; location lost during turmoil of war.

MvR A/C:
Make/model/serial number: Albatros D.III, possibly *Le Petit Rouge*.
Unit: *Jasta* Richthofen
Commander: *Rittm*. Manfred von Richthofen
Airfield: Roucourt
Colors/markings: *Le Petit Rouge* – See Vic. #17.
Damage: n/a
Initial Attack Altitude: 9,000 feet
Gunnery Range: u/k
Rounds Fired: u/k
Known Staffel Participants: Ltn. Lothar von Richthofen

Notes:
1. This was Richthofen's fourth of four victories for the day.
2. This was Richthofen's last victory flying from Roucourt.
3. Presuming Richthofen flew *Le Petit Rouge*, this was the last victory attained flying that airplane.

Victory No.52 Statistics:
Single-Seaters
4th of 4 April single-seater victories
9th of 15 1917 single-seater victories
17th of 18 pre-wound single-seater victories
17th of 35 total single-seater victories

Sop. Triplane Δ
1st and only April Sop. Triplane victory
1st and only 1917 Sop. Triplane victory
1st and only total Sop. Triplane victory

No.8 Squadron
1st and only No.8 Sop. Triplane victory
1st and only April No.8 Squadron victory
2nd of 2 1917 No.8 Squadron victories
2nd of 2 total No.8 Squadron victories

Deaths †
No.12 Squadron:
1st and only April No.8 Squadron crewman KiA
2nd of 2 1917 No.8 Squadron crewmen KiA
2nd of 2 total No.8 Squadron crewmen KiA

Single-seaters:
3rd of 3 April single-seater crewmen KiA
6th of 9 1917 single-seater crewmen KiA
12th of 12 pre-wound single-seater crewmen KiA
12th of 21 total single-seater crewmen KiA

All:
23rd of 23 total April crewmen KiA
41st of 53 total 1917 crewmen KiA
58th of 66 total pre-wound crewmen KiA
58th of 84 total crewmen KiA

Etc.
21st of 21 total April victories
37th of 48 total 1917 victories
52nd of 57 total pre-wound victories
35th victory in an Albatros D.III
2nd of 2 single-seater victories flying from Roucourt
8th of 8 total victories flying from Roucourt
8th of 9 single-seater victories as *Staffelführer*
36th of 39 total victories flying as *Staffelführer*

Statistics Notes

Δ *All Sop. Triplane victories are pre-wound*
± *All No.8 Squadron victories are pre-wound*
† *All No.8 Squadron deaths are pre-wound*

Left: Richthofen suits up pre-flight at Roucourt, presumably next to *Le Petit Rouge*. Photographs reveal Albatroses occasionally had their serial numbers chalked onto the ailerons, and this appears to have been done with this machine. Tantalizingly, the serial number is illegible.

Victory No.53

Albatros D.III 789/17 in standard factory finish.

18 June
RE.8 A4290
No.9 Squadron RFC

Day/Date: Monday, 18 June
Time: 1.15PM
Weather: Fine in the morning, but heavy storm in the afternoon.
Crash Location: "Struywe's farm"
Side of Lines: Friendly

KOFL 4. Armee Weekly Activity Report: "1,15 Nachm. 1 R.E. DD zwischen den Linien östl. Ypern durch Rittmeister Frhr. v. Richthofen (Jasta 11).

1.15PM. 1 R.E. biplane between the lines east of Ypres by *Rittmeister Freiherr* von Richthofen (*Jasta* 11)."

MvR Combat Report: "Alb. D III 789, R.E.2 [*sic*]. (Burnt).

Accompanied by my *Staffel* I attacked in 2500 meters north of Ypres on this side of the line an English artillery R.E. I fired from shortest distance some 200 shots, whereafter I jumped over the enemy plane. In this moment I noticed that both pilot and observer were lying dead in their machine. The plane went without falling in uncontrolled curves to the ground. Driven by the wind it fell into Struywe's farm where it began to burn after touching the ground.

Frhr. v. Richthofen.
Was acknowledged."

RFC Combat Casualties Report: "Left aerodrome 11AM. An unconfirmed report in a German publication states that Lt.R.W.Ellis was killed."

RFC A/C:
Make/model/serial number: Royal Aircraft Factory RE.8 A4290
Manufacturer: Subcontracted to/built by Austin Motor Co. (1914), Ltd., Northfield, Birmingham.
Unit: No.9 Squadron RFC
Aerodrome: Proven
Sortie: Photographic Reconnaissance. Sh.28.c.8.
Colors/markings: u/c. Generally, PC10 wings/ailerons/vertical and horizontal stabilizers/elevators; battleship gray engine cowl; b/w/r rudder; white serial number on vertical stabilizer. No.9 Sqn marking was a 12–15 inch wide vertical black band around the fuselage, aft of the roundel.
Engine: 150 hp RAF 4a
Engine Number: 887 WD 4754
Gun: One Vickers, one Lewis. Nos. 17045, A3903
Manner of Victory: Crew presumably killed by machine gun bullets, precipitating uncontrolled curved decent until terrain impact.
Items Souvenired: n/a

RFC Crew:
Pilot: Lt. Ralph Walter Elly Ellis, KiA.
Cause of Death: u/c. Either machine gun wounds or trauma associated with airplane crash.
RFC Combat Casualties Report: "Death of Lt.R.W.Ellis officially accepted as having occurred on 18.6.17. (Part II orders G.H.Q. 3rd Echelon d/5.1.18)."

Burial: Either in-field or laid where fallen; location lost during turmoil of war.
Obs: Lt. Harold Carver Barlow, 24 or 25 (DoB 1891), KiA.
Cause of Death: u/c. Either machine gun wounds or trauma associated with airplane crash.
RFC Combat Casualties Report: "Death of Lieut.H.C.Barlow accepted. (Part II orders d/16.3.18)."
Burial: Either in-field or laid where fallen; location lost during turmoil of war.

MvR A/C:
Make/model/serial number: Albatros D.III 789/17
Unit: *Jasta* 11
Commander: *Rittm*. Manfred von Richthofen
Airfield: Harlebeke, Belgium
Colors/markings: u/k. 789/17 was manufactured toward the end of the third and final Albatros D.III production batch and arrived brand new to *Jasta* 11 on 2 May, just after Richthofen departed on leave, when the *Staffel* was based at Roucourt. Although speculative it is unlikely that a brand new war machine would have remained idle for six weeks until Richthofen returned. Likely its ownership had been assumed by another pilot and upon Richthofen's return—as he often did—he flew whatever machine was available, which on this day was 789/17.
However, its appearance and any photograph of this machine are unknown to the author. Photographs taken in May of Albatros D.III and D.V lineups at Roucourt show the majority of their machines were painted overall red, with spinners, tails, and/or colored fuselage bands denoting pilot identification. Based on these photographs it is safe to presume 789/17 also was overpainted red, although any secondary colors, if any, are unknown. In any event, the profile is speculative possibility only and should not be considered an accurate representation of 789/17's appearance on 18 June 1917. Research is ongoing.
Damage: n/a

Initial Attack Altitude: "2,500 meters"
Gunnery Range: u/k
Rounds Fired: "Some 200 shots"
Known Staffel Participants: u/k

Notes:
1. In the book *Under the Guns of the Red Baron*, Straywe's Farm is identified as "probably Stray Farm, map reference C 3c., ½ Km east of Pilckem, five Km north of Ypres."
2. This was Richthofen's last victory flying an Albatros D.III and his first after returning from a six-week leave.
3. *Jasta* 11 was now in the *KOFL* 4. *Armee* sector.
4. *KOFL Armee* report *Staffel* nomenclature had switched from *Jagdstaffel* Richthofen back to *Jasta* 11.

Victory No.53 Statistics:

Two-Seaters
1st of 3 June two-seater victories
29th of 33 1917 two-seater victories
36th of 39 pre-wound two-seater victories
36th of 45 total two-seater victories

RE.8
1st of 2 June RE.8 victories
2nd of 5 1917 RE.8 victories
2nd of 7 total RE.8 victories

No.9 Squadron
1st and only No.9 RE.8 victory
1st and only June No.9 Squadron victory
1st and only 1917 No.9 Squadron victory
1st and only total No.9 Squadron victory

Deaths †
No.9 Squadron:
1st and 2nd of 2 June No.9 Squadron crewmen KiA
1st and 2nd of 2 1917 No.9 Squadron crewmen KiA
1st and 2nd of 2 total No.9 Squadron crewmen KiA

Two-Seaters:
1st and 2nd of 6 June two-seater crewmen KiA
36th and 37th of 44 1917 two-seater crewmen KiA
47th and 48th of 54 pre-wound two-seater crewmen KiA
47th and 48th of 63 total two-seater crewmen KiA

All:
1st and 2nd of 6 total June crewmen KiA
42nd and 43rd of 53 total 1917 crewmen KiA
59th and 60th of 66 total pre-wound crewmen KiA
59th and 60th of 84 total crewmen KiA

Etc.
1st of 4 total June victories
38th of 48 total 1917 victories
53rd of 57 total pre-wound victories
36th victory in an Albatros D.III
1st of 3 two-seater victories flying from Harlebeke
1st of 4 total victories flying from Harlebeke
29th of 30 two-seater victories as *Staffelführer*
37th of 39 total victories flying as *Staffelführer*

Statistics Notes
± *All No.9 Squadron victories are pre-wound*
† *All No.9 Squadron deaths are pre-wound*

Victory No.54

Above: Two Albatros D.Vs and Gotha bombers parked at the Zeppelin hangar at Gontrode, Belgium, late June or early July 1917. Richthofen is seen in other photos at this location which suggests these machines belong to *Jasta* 11. The one at left is suspected to be 1177/17.

23 June
SPAD VII C.1 B1530
No.23 Squadron RFC

Day/Date: Saturday, 23 June
Time: 9.30PM
Weather: Cloud, with bright intervals; visibility very good at times.
Attack Location: North of Ypres
Crash Location: n/a (MvR claimed "2 kilometers north of Ypres")
Side of Lines: Enemy

KOFL 4. Armee Weekly Activity Report: "9,15 Nachm. 1 Spad jenseits nordl. Jeperen.

9.15PM. 1 Spad other side [of lines] north of Ypres."

MvR Combat Report: "North of Ypres. Spad One-Seater. Alb. D V 1177 red body.

I attacked together with several of my gentlemen an enemy One-Seater squad on the enemy's side. During the fight I fired at a Spad some 300 shots from shortest distance. My adversary did not start to curve and did nothing to evade my fire. At first the plane began to smoke, then it fell, turning and turning, 2 kilometers north of Ypres to the ground without having been caught.

Frhr. v. Richthofen.
Was acknowledged."

RFC Combat Casualties Report: n/a

RFC A/C:
Make/model/serial number: SPAD (Société Pour

"Red body." An Albatros D.V that is believed to be 1177/17.

L'Aviation et ses Dérivés) VII C.1 B1530
Unit: No.23 Squadron RFC
Aerodrome: La Lovie
Sortie: Offensive Patrol
Colors/markings: Generally, beige engine cowlings and forward fuselage up to cockpit, thence natural clear doped fabric surfaces aft with black serial number on vertical stabilizer. Wings and horizontal surfaces clear doped fabric. Tri-colored b/w/r rudder. White-bordered b/w/r roundels on wings and fuselage. Flight markings consisted of red/white or red/blue, etc., striped nose cowls, with a mixture of bands and numbers on the fuselage. The particular markings of B1530 presently are undetermined.
Engine: 140 hp
Engine: Hispano-Suiza
Engine Number: n/a
Gun: One Vickers.
Manner of Victory: Smoking spiral descent.
Items Souvenired: n/a

RFC Crew:
Pilot: 2nd Lt. Robert Wallace Farquhar, 19 (DoB 4 February 1898), uninjured.
Victories at time of encounter: 4 (4 out of control‡). One flying an FE.2b, three flying a SPAD VII C.1.

Victory#	Date	Notes
1	4 Feb. 1917	Albatros D.II‡ (flying FE.2b A5460, with obs. 2/Lt C.N. Blennerhasset)
2	4 May 1917	Albatros D.III‡
3	13 May 1917	Albatros D.III‡ (shared with Capt. C.K.C Patrick)
4	26 May 1917	Aviatik C‡

On 7 July Farquhar was credited with his fifth victory, an Albatros shot down in flames on 23 June, shortly before his encounter with Richthofen.

MvR A/C:
Make/model/serial number: Albatros D.V 1177/17
Unit: *Jasta* 11
Commander: *Rittm.* Manfred von Richthofen
Airfield: Harlebeke, Belgium
Colors/markings: "Red body." A photograph of what is believed to be 1177/17 shows the entire airplane was painted red, although it is unknown if this overpainting included the undersurfaces of the wings. The upper wing mauve/dark green factory camouflage pattern can be discerned through the red overpainting, but the lower wing appears as one solid color. This matches a confirmed photograph of 1177/17 that reveals its factory camouflage on the lower wings was dark green only, across the entire span.
Damage: n/a
Initial Attack Altitude: ca. 8,000 feet
Gunnery Range: "Shortest distance"
Rounds Fired: "Some 300 shots"
Known Staffel Participants: u/k

Notes:
1. The exact manner of Farquhar's downing is unclear. His SPAD suffered engine and radiator damage, yet not enough for a damage report, and it is unknown if he force landed or was able to reach his base at La Lovie.
2. This was Richthofen's first victory flying an Albatros D.V.

Victory No.54 Statistics:
Single-Seaters
1st and only June single-seater victory

10th of 15 1917 single-seater victories
18th of 18 pre-wound single-seater victories
18th of 35 total single-seater victories

SPAD
1st and only June SPAD victory
3rd of 4 1917 SPAD victories
3rd of 4 total SPAD victories

No.23 Squadron ±
1st and only No.23 SPAD victory
1st and only June No.23 Squadron victory
1st and only 1917 No.23 Squadron victory
1st and only total No.23 Squadron victories

Etc.
11th of 14 victories behind enemy lines
2nd of 4 total June victories
39th of 48 total 1917 victories
54th of 57 total pre-wound victories
1st of 8 victories in an Albatros D.V
1st of 3 victories in 1177/17
1st and only single-seater victory flying from Harlebeke
2nd of 4 total victories flying from Harlebeke
9th of 9 single-seater victories as *Staffelführer*
38th of 39 total victories flying as *Staffelführer*

Statistics Notes
± *All No.23 Squadron victories are pre-wound*

Above: A row of brand new first-production batch Albatros D.Vs. The second machine from bottom is D.1177/17, ultimately flown by Richthofen. Note it had been built with a factory headrest, and the lower wings of all these machines appear to be one solid color, ostensibly dark green.

Victory No.55

Above: Aircraft Manufacturing Company DH.4.

24 June
DH.4 A7473
No.57 Squadron RFC

Day/Date: Sunday, 24 June
Time: 9.10AM
Weather: Fine but cloudy, visibility good in early morning and again in the evening.
Attack Location: Near Koelenbergmolen, south of Becelaere
Crash Location: "A hangar between Keibergmelen and Lichtensteinlager"
Side of Lines: Friendly

KOFL 4. Armee Weekly Activity Report: "9,30 Vorm. 1 Bristol DD. Diesseits bei Becelaere.

9.30AM. 1 Bristol biplane this side [of lines] near Becelaere."

MvR Combat Report: "Between Keibergmelen and Lichtensteinlager, this side of the line. De Havilland DD. Alb. D.V. 1177.

With six machines of my *Staffel* I attacked enemy squad consisting of 2 planes of reconnissance and 10 pursuit planes. Unimpeded by the enemy pursuit planes I managed to break one of the reconnissance plane by my fire. The body fell with the inmates into a hangar between Keibergmelen and Lichtensteinlager, this side of our lines. The plane exploded when crashing on the ground and destroyed the hangar.

Frhr. v. Richthofen.
Was acknowledged."

RFC Combat Casualties Report: "Left aerodrome 7.40. Last seen 9000 ft. over BECELAERE about 8.30 am in combat with E.A."

RFC A/C:
Make/model/serial number: AMC DH.4 A7473
Manufacturer: Aircraft Manufacturing Co., Ltd., Hendon, London, N.W.
Unit: No.57 Squadron RFC
Aerodrome: Droglandt
Sortie: Photographic Reconnaissance. BECELAERE.
Colors/markings: u/c. Generally, PC10 upper surfaces, CDL undersurfaces; b/w/r rudder with white-bordered black serial number. No.57 Squadron markings were a flight letter and individual number painted on both sides of the nacelle.
Engine: 375 hp Rolls-Royce Eagle VII
Engine Number: 2275173 WD 16052
Gun: One Vickers, one Lewis, Nos. 20473, A3353.

"Red body." An Albatros D.V that is believed to be 1177/17.

Manner of Victory: Uncontrolled descent until terrain impact.
Items Souvenired: n/a

RFC Crew:
Pilot: Capt. Norman George McNaughton, 27 (DoB May 1890), KiA.
Victories at time of encounter: 5 (1 destroyed*, 3 out of control‡)

Victory#	Date	Notes
1	4 March 1917	u/k‡ (flying FE.2b A1955, with obs. 2/Lt H.G. Downing)
2	4 March 1917	u/k‡ (flying FE.2b A1955, with obs. 2/Lt H.G. Downing)
3	29 April 1917	u/k* (flying FE.2d A6365, with obs. 2/Lt H.G. Downing)
4	21 June 1917	Albatros Scout‡ (flying DH.4)
5	u/k	

Cause of Death: u/k Either gunshot wounds, trauma associated with airplane crash, or post-terrain-impact explosion.
RFC Combat Casualties Report: "Death of Capt. N.G.McNaughton accepted by Army Council (Part II Orders d/23.3.18)."
Burial: Either in-field or laid where fallen; location lost during turmoil of war.
Obs: Lt. Angus Hughes Mearns, 21 (DoB 13 December 1895), KiA.
Cause of Death: u//k Either gunshot wounds, trauma associated with airplane crash, or post-terrain-impact explosion.
RFC Combat Casualties Report: "Death of Lt. A.H. Mearns officially accepted (Part II Orders 13.4.18)."
Burial: Either in-field or laid where fallen; location lost during turmoil of war.

MvR A/C:
Make/model/serial number: Albatros D.V 1177/17
Unit: *Jasta* 11
Commander: *Rittm.* Manfred von Richthofen
Airfield: Harlebeke, Belgium
Colors/markings: See Vic. #54.
Damage: n/a
Initial Attack Altitude: ca. 9,000 feet
Gunnery Range: u/k
Rounds Fired: u/k
Known Staffel Participants: u/k

Notes:
1. Richthofen's translated combat report is erroneously dated as "June 26th, 1917."
2. Richthofen's translated combat report records an erroneous time of "9.10PM."
3. From RFC Communique No. 86: "April 29th. Hostile Aircraft—A patrol of No. 57 Squadron observed several hostile aircraft engaging two S.E.5's. The F.E.'s joined in the combat, and a pilot of one of the hostile machines, which was engaged by Capt. McNaughton and 2nd Lieut. Downing, was seen to fall from his machine, which crashed."
4. From RFC Communique No. 86: "April 30th. Hostile Aircraft—Major L.A. Pattinson and Lieut. A. H. Mearns, No. 57 Squadron, observed a formation of their F.E.'s fighting with German machines, and immediately joined in the combat. They drove down one of the hostile machines in a damaged condition…"
5. From RFC Communique No. 93: "June 21st. Enemy Aircraft—Capt. McNaughton, No.57 Squadron, drove down an Albatross [*sic*] Scout out of control."

Victory No.55 Statistics:

Two-Seaters

2nd of 3 June two-seater victories
30th of 33 1917 two-seater victories
37th of 39 pre-wound two-seater victories
37th of 45 total two-seater victories

DH.4

1st and only June DH.4 victory
1st and only 1917 DH.4 victory
1st and only total DH.4 victory

No.57 Squadron ±

1st and only No.57 DH.4 victory
1st and only June No.57 Squadron victory
1st and only 1917 No.57 Squadron victory
1st and only total No.57 Squadron victory

Deaths †

No.57 Squadron:
1st and 2nd of 2 June No.57 Squadron crewmen KiA
1st and 2nd of 2 1917 No.57 Squadron crewmen KiA
1st and 2nd of 2 total No.57 Squadron crewmen KiA

Two-Seaters:

3rd and 4th of 6 June two-seater crewmen KiA
38th and 39th of 44 1917 two-seater crewmen KiA
49th and 50th of 54 pre-wound two-seater crewmen KiA
49th and 50th of 63 total two-seater crewmen KiA

All:

3rd and 4th of 6 total June crewmen KiA
44th and 45th of 53 total 1917 crewmen KiA
61st and 62nd of 66 total pre-wound crewmen KiA
61st and 62nd of 84 total crewmen KiA

Etc.

3rd of 4 total June victories
40th of 48 total 1917 victories
55th of 57 total pre-wound victories
2nd of 8 victories in an Albatros D.V
2nd of 3 victories in D.1177/17
2nd of 3 two-seater victories flying from Harlebeke
3rd of 4 total victories flying from Harlebeke
30th of 30 two-seater victories as *Staffelführer*
39th of 39 total victories flying as *Staffelführer*

Statistics Notes

± *All No.57 Squadron victories are pre-wound*
† *All No.57 Squadron deaths are pre-wound*

Above: Richthofen engages in conversation with *Kampfgeschwader* 3 commander *Hptm*. Rudolf Kleine (left) and his adjutant *Oblt*. Gerlich before the Zeppelin hangar at Gontrode, Belgium, late June or early July 1917. The Albatros at left is believed to be D.1177/17, now sans headrest and completely overpainted red.

Victory No.56

"Red body." An Albatros D.V that is believed to be 1177/17.

25 June
RE.8 A3847
No.53 Squadron RFC

Day/Date: Monday, 25 June
Time: 6.40PM
Weather: Fine, clouding over towards evening.
Attack Location: Near Le Bizet
Crash Location: Between the trenches near Le Bizet
Side of Lines: No Man's Land

KOFL 4. Armee Weekly Activity Report: "7,20 Nachm. 1 R.E.-jennseits östl. Ploegsteert Wald.

7.20PM. 1 R.E. – other side [of lines] east Ploegsteert wood."

MvR Combat Report: "Above trenches near Le Bizet other side of line. RE plane. Alb. D V 1177, red body. I was flying together with Lieut. Allmenroeder. We spotted an enemy artillery flyer whose wings broke off in my machine-gun fire. The body dashed to the ground between the trenches, burning.

Frhr. v. Richthofen
Was acknowledged."

RFC Combat Casualties Report: "Left aerodrome 4.35PM. Machine was brought down by E.A. It nosedived and crashed near front-line trenches at U.16.A.Sh.28. Pilot and observer killed."

Right: The cockpit of a captured Albatros D.V, sporting the usual additional British instrumentation. The D.V was the only Albatros model to route the aileron cables vertically from the cockpit to the upper wings, through which they fed to the ailerons, and these cables can be seen to the back-left of the control column. Note how the D.V cockpit sides curved deeper toward the upper longerons than did the cockpits of the D.I–D.III. Ostensibly, this eased pilot entry and exit.

RFC A/C:
Make/model/serial number: Royal Aircraft Factory RE.8 A3847
Manufacturer: Subcontracted to/built by D. Napier & Son, Ltd., Acton, London, W.
Unit: No.53 Squadron RFC
Aerodrome: Bailleul (Town Ground)
Sortie: Artillery Co-operation
Colors/markings: u/c. Generally, PC10 wings/ailerons/vertical and horizontal stabilizers/elevators; battleship gray engine cowl; b/w/r rudder; white serial number on vertical stabilizer. No.53 Sqn marking was a white crescent on the fuselage, aft of the roundel.
Engine: 150 hp RAF 4a
Engine Number: 569 WD 3685
Gun: One Vickers, one Lewis, Nos. 23595, A3544.
Manner of Victory: Uncontrolled nosedive until terrain impact.
Items Souvenired: n/a

RFC Crew:
Pilot: Lt. Leslie Spencer Bowman, 20 (DoB 21 June 1897), KiA.
Cause of Death: u/k. Either gunshot wounds or trauma associated with airplane crash.
Burial: Either in-field or laid where fallen; location lost during turmoil of war.
Obs: 2nd Lt. James Edward Power-Clutterbuck, 22 or 23 (DoB 1894), KiA.
Cause of Death: u/k. Either gunshot wounds or trauma associated with airplane crash.
Burial: Strand Military Cemetery, Grave IX. I. 7. Located on the N365, one-half mile north of Ploegsteert, Belgium.

MvR A/C:
Make/model/serial number: Albatros D.V 1177/17
Unit: *Jagdgeschwader Nr. 1*
Commander: *Rittm.* Manfred von Richthofen
Airfield: Harlebeke, Belgium
Colors/markings: See Vic. #54.
Damage: n/a
Initial Attack Altitude: u/k
Gunnery Range: u/k
Rounds Fired: u/k
Known Staffel Participants: Ltn. Karl Allmenröder (#29. KiA two days later, 27 June 1917)

Note: MvR's first victory as JG1 commander.

Victory No.56 Statistics:
Two-Seaters
3rd of 3 June two-seater victories
31st of 33 1917 two-seater victories
38th of 39 pre-wound two-seater victories
38th of 45 total two-seater victories

RE8
2nd of 2 June RE.8 victories
3rd of 5 1917 RE.8 victories
3rd of 7 total RE.8 victories

No.53 Squadron ±
1st of 2 No.53 RE.8 victories
1st and only June No.53 Squadron victory
1st of 2 1917 No.53 Squadron victories
1st of 2 total No.53 Squadron victories
Deaths †
No.53 Squadron:
1st and 2nd of 2 June No.53 Squadron crewman KiA
1st and 2nd of 4 1917 No.53 Squadron crewmen KiA
1st and 2nd of 4 total No.53 Squadron crewmen KiA

Two-Seaters:
5th and 6th of 6 June two-seater crewmen KiA
40th and 41st of 44 1917 two-seater crewmen KiA
51st and 52nd of 54 pre-wound two-seater crewmen KiA
51st and 52nd of 63 total two-seater crewmen KiA

All:
5th and 6th of 6 total June crewmen KiA
46th and 47th of 53 total 1917 crewmen KiA
63rd and 64th of 66 total pre-wound crewmen KiA
63rd and 64th of 84 total crewmen KiA

Etc.
4th of 4 total June victories
41st of 48 total 1917 victories
56th of 57 total pre-wound victories
3rd victory in an Albatros D.V
3rd of 3 victories in D.1177/17
3rd of 3 two-seater victories flying from Harlebeke
4th of 4 total victories flying from Harlebeke
1st of 8 two-seater victories as *Geschwaderkommandeur*
1st of 25 total victories flying as *Geschwaderkommandeur*
Statistics Notes
± All No.53 Squadron victories are pre-wound
† All No.53 Squadron deaths are pre-wound

Victory No.57

Above: Richthofen next to the Albatros D.V believed to be the one described the combat report for victory 57, and the one in which he was wounded 6 July.

2 July
RE.8 A3538
No.53 Squadron

Day/Date: Monday, 2 July
Time: 10.20AM
Weather: Cloudy but clear at times
Attack Location: Deulemont
Crash Location: Deulemont, between the lines (MvR combat report); "About 1,000 yards northeast of Messines" (letter to pilot Sgt. Whatley's mother).
Side of Lines: No Man's Land

KOFL 4. Armee Weekly Activity Report: "10,20 Vorm. Ein R.E. jenseits in Gegend Waasten.

10.20AM. An R.E. other side [of lines] in the area of Warneton."

MvR Combat Report: "R.E. No details, as fell burning. Alb. D V, hood, tail, decks [wings] red.

I attacked foremost plane of enemy squad. After the first shots the observer collapsed. Shortly thereafter the pilot was wounded mortally. The RE fell and I fired at it at a distance of 50 meters. The plane caught fire and dashed to the ground.

Frhr. v. Richthofen
Was acknowledged."

RFC Combat Casualties Report: "Machine brought down in flames by E.A. behind COMINES."

RFC A/C:
Make/model/serial number: Royal Aircraft Factory RE.8 A3538
Manufacturer: Subcontracted to/built by The Daimler Co., Ltd., Coventry.

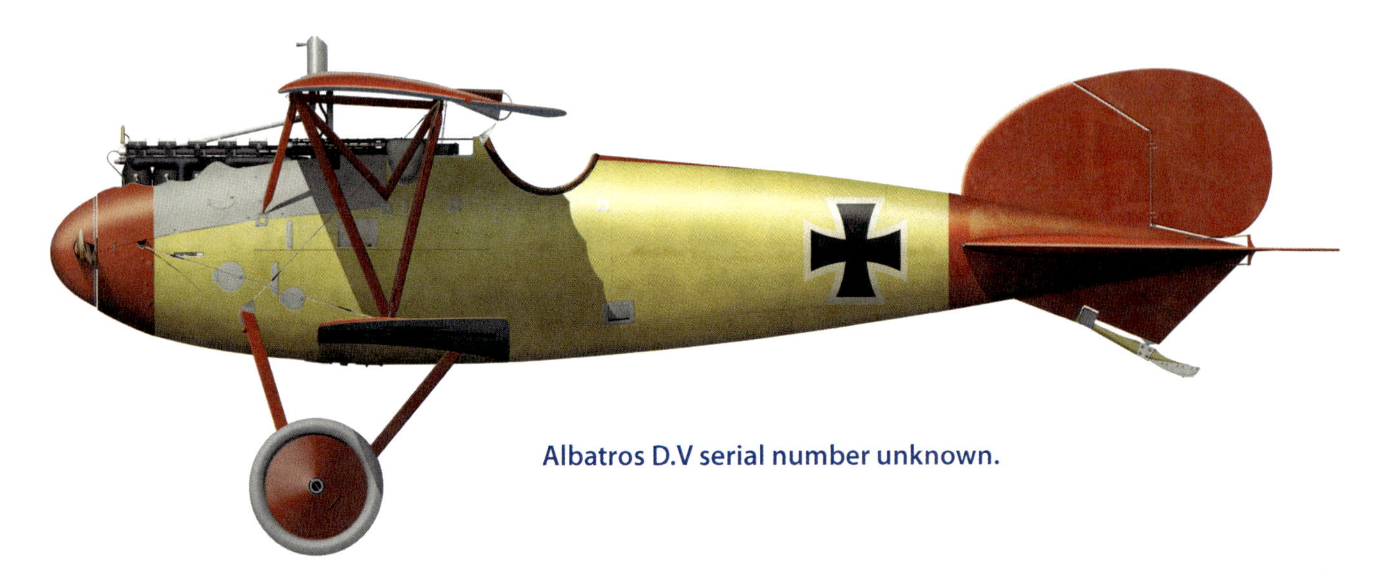

Albatros D.V serial number unknown.

Unit: No.53 Squadron RFC
Aerodrome: Bailleul (Town Ground)
Sortie: Escort to Photographic Reconnaissance
Colors/markings: u/c. Generally, PC10 wings/ailerons/vertical and horizontal stabilizers/elevators; battleship gray engine cowl; b/w/r rudder; white serial number on vertical stabilizer. No.53 Sqn marking was a white crescent on the fuselage, aft of the roundel.
Engine: 150 hp R.A.F. 4a
Engine Number: 29460 WD 12293
Gun: One Vickers, one Lewis, Nos. A1695, 24051.
Manner of Victory: Uncontrolled descent in flames until terrain impact.
Items Souvenired: n/a

RFC Crew:

Pilot: Sgt. Hubert Arthur Whatley, 19, KiA.
Cause of Death: u/c. Gunshot wounds, immolation, or trauma associated with airplane crash.
Burial: In-field; location lost during turmoil of war.
Obs: 2nd Lt. Frank Guy Buckingham Pascoe, 19 or 20 (DoB 1897), KiA.
Cause of Death: u/c. Gunshot wounds, immolation, or trauma associated with airplane crash.
Burial: In-field; location lost during turmoil of war.

RFC Combat Casualties Report: "Now reported killed in action (2nd Bde tel: A.785 dated 4/7/17.)

MvR A/C:

Make/model/serial number: Albatros D.V, serial no. u/k
Unit: *Jagdgeschwader Nr. 1*
Commander: *Rittm.* Manfred von Richthofen
Airfield: Marckebeke, Belgium
Colors/markings: "Hood, tail, decks [wings] red." Red spinner, nose cowl, and empennage, between which the unpainted fuselage retained its shellacked and varnished "warm straw yellow" wood. Uppersurfaces of both wings red, as well as the wheel covers and all interplane/cabane/landing gear struts. Wing undersurfaces light blue. Triangular "insert" on fuselage spine aft of cockpit (where the factory fitted headrest had been removed) was possibly red. Fittings, upper cowls, vents and hatches greenish-gray. White-bordered crosses on wings and rudder overpainted yet still visible through the red "wash," the black portion of the crosses especially so. Fuselage crosses were not overpainted yet appeared "subdued" and not as resplendent as normally seen on Albatros D.Vs.

Markings details include: Garuda propeller; rigging label on port fuselage near aft cabane strut; square cut-out in port cowl, adjacent to and midway along port Maxim; rigging between landing gear strut attachment point and upper-aft interplane strut; fuselage, especially the starboard side, possessed an unusually rough and weathered appearance with a "subdued" white outline of the fuselage cross. The beginning of the red empennage extends forward the leading edges of the horizontal and vertical stabilizers, where a "stripe" of darker red paint reveals multiple inadvertent coats likely applied when attaining a straight demarcation line.
Damage: n/a

Initial Attack Altitude: u/k
Gunnery Range: 50 meters
Rounds Fired: u/k
Known Staffel Participants: Ltn. Gisbert-Wilhelm Groos (#3)

Notes:

1. The author suggests a plausible yet unproven explanation of this subdued white fuselage cross outline could be the result of fuselage sanding during a keying process, whereby the high-gloss factory finish was scuffed to promote paint adhesion for a planned-to-be-applied "wash" of red paint over the entire fuselage that—because of his wounding 6 July—never came. This speculation contends the sanding could have reduced the paint depth of the white outline—which also may have received more sanding than the overall fuselage—and allowed the warm-straw-colored wood underneath to become subtly visible. This is just one of many unproven speculations.
2. This victory is another example of Richthofen's no-quarter *modus operandi*. His combat report clearly indicates he believed he had shot the crew to death, yet he continued to attack an obviously doomed and falling airplane until it also burst into flames.
3. This was Richthofen's first victory from Marckebeke.
4. This was Richthofen's final victory before being wounded in the head four days later on 6 July.
5. *KOFL 4. Armee Weekly Activity Report 29.6 – 5.7.1917: Durch Verfügung des Chefs des Generalstabes des Feldheeres vom 23.6.17 ist aus den Jagdstaffel 4, 6, 10 und 11 das Jagdgeschwader I gebildet worden.*

   By decree from the Chief of the General Staff of the Field Army from 23.6.17, the [new] *Jagdgeschwader* I was formed from the *Jagdstaffel*[*n*] 4, 6, 10 and 11."

Victory No.57 Statistics:

Two-Seaters
1st and only July two-seater victory
32nd of 33 1917 two-seater victories
39th of 39 pre-wound two-seater victories
39th of 45 total two-seater victories

RE8
1st and only July RE.8 victory
4th of 5 1917 RE.8 victories
4th of 7 total RE.8 victories

No.53 Squadron ±
2nd of 2 No.53 RE.8 victories
1st and only July No.53 Squadron victory
2nd of 2 1917 No.53 Squadron victories
2nd of 2 total No.53 Squadron victories

Deaths †
No.53 Squadron:
1st and 2nd of 2 July No.53 Squadron crewmen KiA
3rd and 4th of 4 1917 No.53 Squadron crewmen KiA
3rd and 4th of 4 total No.53 Squadron crewmen KiA

Two-Seaters:
1st and 2nd of 2 July two-seater crewmen KiA
42nd and 43rd of 44 1917 two-seater crewmen KiA
53rd and 54th of 54 pre-wound two-seater crewmen KiA
53rd and 54th of 63 total two-seater crewmen KiA

All:
1st and 2nd of 2 total July crewmen KiA
48th and 49th of 53 total 1917 crewmen KiA
65th and 66th of 66 total pre-wound crewmen KiA
65th and 66th of 84 total crewmen KiA

Etc.
1st and only July victory
42nd of 48 total 1917 victories
57th of 57 total pre-wound victories
4th victory in an Albatros D.V
1st of 2 two-seater victories flying from Marckebeke
1st of 5 total victories flying from Marckebeke
2nd of 8 two-seater victories as *Geschwaderkommandeur*
2nd of 25 total victories flying as *Geschwaderkommandeur*

Statistics Notes
± *All No.53 Squadron victories are pre-wound*
† *All No.53 Squadron deaths are pre-wound*

Richthofen's Wounding, 6 July 1917

Above: No.20 Squadron FE.2d A6516. Similar in appearance to sister 6512, No.20 Squadron FE.2ds were armed with one fixed and two flexible Lewis machine guns and carried racks of 25lb bombs mounted on rails underneath the lower wings. As occurred on 6 July, No.20 Squadron sorties often began with a high-altitude bomb raid that morphed into an offensive patrol to hunt enemy airplanes.

I. How

On Friday 6 July 1917, Manfred von Richthofen was wounded attacking a flight of six RFC No.20 Squadron FE.2ds near Comines, France. Although this event is well known generally, and despite the author having written about this subject extensively elsewhere, the details of Richthofen's wounding and its repercussions on his future are still widely unknown or misunderstood. These collective and persistent misunderstandings are so vast and so deep as to warrant examination anew, via detail far beyond that normally given 6 July 1917. A comprehensive work of this magnitude would be bereft completeness were it to exclude this subject.

Events began that day at about 1030[1] (German time, one hour ahead of British time) when *Jagdgeschwader* 1 received an alert of incoming infantry support planes, which precipitated *Jagdstaffel* 11's immediate takeoff. Led by Richthofen, *Jasta* 11 flew the better part of an hour between Ypres and Armentieres without enemy contact until they happened upon the No.20 FE.2ds approaching the lines. These six machines were commanded by four-victory Captain Douglas Charles Cunnell and had departed St. Marie Cappel France between 0950–0955 for an Offensive Patrol above Comines, Warneton, and Frelinghien, along the French/Belgian border. Under orders to attack any enemy aircraft they encountered—a task about which none of the twelve men held any illusions, since dozens of previous sorties had demonstrated how German fighters could out-maneuver their two-seater pushers and "shoot hell out of us from that blind spot under our tails,"[2] Cunnell's observer/gunner Second Lieutenant Albert Edward Woodbridge opined the FE.2s were like "butterflies sent out to insult eagles... We were 'cold meat' and most of us knew it."[3]

Regardless, they sallied across the lines to bomb an oft-targeted ammunition dump in Houthem before reaching their assigned patrol area. Richthofen shadowed "the Big Vickers" as they went, content to bide his time and let them fly deeper into German territory, but soon Cunnell's maneuvering prior to No.20's bomb run fooled Richthofen into believing the English had detected *Jasta* 11 and were turning away to avoid combat. To counter this, Richthofen led his machines south toward the pushers to position west of the English formation and "cut off their retreat," ensuring the presumably timid FE.2s would have no choice but engage the Germans blocking their way back to St. Marie Cappel. Moments after bombing Houthem

No.20 saw the Albatrosses behind them, approaching from the north and "making for lines West of F.E. formation."[4] Cunnell immediately banked right and led the pushers "behind E.A. so as to engage them"[5]—to be absolutely clear, since this detail is one that is often overlooked, at this point the FE.2s were in pursuit of *Jasta* 11's Albatros D.Vs—yet this chase hardly had begun when "before you could say Jack Robinson"[6] an estimated 30 additional Albatrosses swarmed in "from all sides, also from above and below."[7] Within seconds, No.20 Squadron had gone from quarry to hunter to becoming so tactically disadvantaged that they had little recourse but form a defensive circle.[8]

Far below the developing battle, Air Defense Officer *Leutnant der Reserve* Hans Schröder was assigned to an observation post on la Montagne ("The Mountain"), an area of high ground just south of Werviq-sud, France, from which he observed aerial activity and alerted nearby *Jagdstaffeln* of incoming enemy airplanes. Due to this relationship Schröder had visited various *Staffeln* and was familiar with the top German airman and their gaily-colored airplanes. He could identify them in flight, even when using field glasses to observe them at high altitudes.

It was during such observation that Schröder witnessed the struggle overhead:

"There was a mighty battle taking place in the air between Werwick [*sic*] and Comines, somewhere close to us. Richthofen had pitted himself against the famous English 'merry go round' squadron.

"Eight [*sic*] F.E.s...were revolving round one another in couples...The technique and tactics of the English were amazing, their main principle being that each machine should not look after itself but its partner. Each one therefore protected the other against any attack by their German opponents...

"The Englishmen refused to be rushed, and their steadiness gave them an absolute superiority. Meanwhile our machines tried to break their formation by a series of advances and retreats, like dogs attacking a hedgehog. They pirouetted and spiraled, but their movements exposed them to more risks than their opponents, who appeared to be invulnerable and unassailable."[9]

Far from invulnerable, the FE.2s were in the middle of all they could handle. In A6512, Woodbridge effected a nearly continuous return fire, switching repeatedly between the fore and aft machine guns as Cunnell "ducked dives from above and missed head-on collisions by bare margins of feet."[10] He had never seen "so many Huns in the air at one time"[11] and claimed a flamer after firing "a whole drum into him."[12] Cunnell claimed another two Albatrosses after firing "large bursts...from back gun" that "entered [each] fuselage under pilot's seat,"[13] and A6498 observer Second Lieutenant A. E. Wear's spirited fire led to a claim of "one E.A. out of control" after "a large burst at a range of about 20 yards...entered E.A. from underneath, entering between engine and pilot."[14] Yet the Germans "went to it hammer and tongs"[15] and inevitably their fire found its mark. A6376 had its oil tank and epicyclic gear shot through, A1963 suffered a damaged magneto and severed tail boom, and observer Second Lieutenant S. F. Trotter was mortally wounded defending A6419.

Aboard 6512, as Cunnell banked through "the damnedest scrimmage imaginable,"[16] Woodbridge spotted two approaching Albatrosses—the first of which was an "all-red scout."[17] This was Richthofen, who at some point after passing behind No.20 had reversed course and then led *Jasta* 11 back east toward the melee. Singling out A6512—to which Richthofen later referred as "the last plane," suggesting the FE.2s' defensive circle had widened considerably, had become ragged, or had even disintegrated altogether—Richthofen flew in from far enough astern to provide himself ample time to "consider a means of attacking."[18] However, he was unable to gain firing position before the FE.2 turned back toward him and opened fire in a head-on run—a tactical situation he disliked because "one almost never makes [the two-seater] incapable of fighting"[19] when attacking it head-on. Yet he did not disengage and instead checked his fire and bore-sighted the FE.2, planning to pass beneath it before hauling his Albatros around to attack from its six o'clock low.[20] He ignored Cunnell and Woodbridge's continuous gunfire as he came in, confident that "at a distance of 300 meters [984 feet] and more, the best marksmanship is helpless. One does not hit one's target at such a distance."[21]

This is another crucial detail that is persistently misunderstood: "300 meters" marks the *beginning* of A6512's gunfire—not the end—and therefore defines not the distance at which Richthofen was hit but the length of the head-on run, during which the two airplanes converged nearly 79 meters (260 feet) per second at a combined speed of approximately 281 km/h (175 mph).[22] Thus, two seconds after Richthofen saw A6512 open fire, the converging combatants had already covered more than half the distance between them. One second later the initial 300 meter range had dwindled to 63 meters (207 feet)—72% less than it had been two seconds previously—and approximately a half-second after that only 19 meters (60 feet) separated the airplanes. Woodbridge recalled that as the FE.2 and Albatros

Above: Closeup of 6516. FE.2d pilots sat before a large exposed radiator and manned the fixed, forward-firing Lewis machine gun. A bomb sight is mounted on the nacelle to his right. The observer manned the two flexible Lewis machine guns but this view reveals a large portion of rear airspace was obstructed by wings, struts, propeller, rigging, fuel tank, radiator, engine, booms, empennage, etc., in large part preventing effective rearward defensive fire. The pole-mounted Lewis was at least useful to counter high attacks from behind but German pilots endeavored to attack from six o'clock low, as Richthofen did during his first victory, where the observer's defensive fire could not reach. Note the nacelle-mounted camera on the observer's right.

converged, he and Cunnell "kept a steady stream of lead pouring into the nose of that machine"[23] and saw his own fire "splashing along the barrels of his Spandaus."[24] After the war Woodbridge stated return fire struck the cockpit around him, yet Richthofen recalled neither firing on the FE.2 (he later wrote that his guns were still in safety) nor his Albatros taking any hits.

In any event, at some point during the 3.5 to 4 second head-on run—Richthofen's recollection suggests early on, while Woodbridge's suggests toward the end—a single bullet struck the left rear side of Richthofen's head and caromed off his skull. He was immediately rendered blind and paralyzed.[25] Dazed, his limbs fell from the controls and Woodbridge watched his Albatros hurtle underneath the FE.2 before rolling into a spiral dive. Cunnell immediately banked the pusher to thwart an expected stern attack but instead he and Woodbridge watched as Richthofen's plane "turned over and over and round and round. It was no maneuver. He was completely out of control."[26]

Inside the Albatros, the still-conscious Richthofen felt his machine falling but could do nothing. His "arms [hung] down limply beside me"[27] and his "legs [flopped] loosely beyond my control."[28] The engine noise seemed very distant, and it occurred to him that "this is how it feels when one is shot down to his death."[29] Realizing the increasing airspeed would eventually tear off the wings, he resigned himself to the inevitable.

Within moments, however, he regained use of his extremities and seized the flight controls. Shutting down the engine, he tore away his goggles and forced his eyes open, willing to himself, "I must see—I must—I must see."[30] It was useless. Without vision—and likely experiencing some degree of spatial disorientation—he could not control the falling Albatros. Apparently it began a phugoid motion, whereby the airplane's diving airspeed increased lift and caused it to climb, which then decayed airspeed and lift until it nosed over into another dive to repeat the motion: "From time to time," Richthofen recalled, "my machine had caught itself, but only to slip off again."[31]

After falling an estimated two to three thousand meters Richthofen's vision returned—first as black and white spots, and then with increased normality. Initially it was similar to "looking through thick black goggles" but soon he saw well enough to regain spatial orientation and recover the Albatros from its unusual attitude. After recognizing he was over friendly territory, he established a normal glide east and as he descended was relieved to see two of his *Jasta* 11 comrades providing protective escort. Yet at 50 meters he could not find a suitable landing field amongst the cratered earth below, forcing him to restart his engine and continue east along the southern side of the Lys River until waning consciousness forced the issue, cratered earth or no: he had to get down immediately.

Fortunately, he had flown far enough east to spot a field free of shell holes and so he brought the Albatros in, flying through some telephone lines before landing in a field of tall floodplain grasses and thistles in far northeast Comines, France. This location is confirmed via a post-landing photograph in which the 14th century church *Sint Medarduskerk* is visible through the Albatros' starboard wing gap. Located on the northern Lys River bank in Wervik Belgium, *Sint Medarduskerk's* orientation with respect to the photographed Albatros verifies the landing sight was indeed in Comines. (See *Sidebar: Richthofen's Emergency Landing.*)

Where he landed made little difference to Richthofen—afterwards, he could not even remember the location. Rolling to a stop, he released his seatbelt and shoulder harnesses and attempted egress. Standing proved to be too much; he staggered and then fell to the ground. Landing on a thistle, he lay there without the strength to roll off. Less than a

Above: Richthofen's Albatros D.V after his forced landing in Comines. The airplane's starboard lean and closer-than-normal ground proximity of the empennage and lower starboard wing are apparent. Each is below the knees of nearby personnel—and almost as low as those of the dog. This view affords a good view of the fuselage cross and its "subdued" white border, the reason for which eludes certainty. This photograph was taken at Point "c" as depicted in the aerial view of Comines and the surrounding area, seen later in this chapter.

half-mile away, Hans Schröder and his corporal were "puffing and panting" down the side of la Montagne as they ran to administer first aid to the wounded airman, whose descent and subsequent landing they just witnessed. They found Richthofen lying on the ground "with his head resting on his leather helmet, while a stream of blood trickled from the back of his head. His eyes were closed and his face was as white as a sheet."[32] The pair managed to bandage his head and then Schröder sent his corporal to call an ambulance. While waiting, Richthofen drank some cognac procured from an onlooking soldier and then asked for water—a ubiquitous request of those with gunshot wounds.

Upon ambulance arrival Richthofen was placed on a stretcher and then driven toward Courtrai, his requested destination. Schröder rode with him, opening and closing the ambulance window as Richthofen complained alternately of being too hot and then too cold, but otherwise the pair rode in silence. Initially they stopped in Menin, whose medical facility was closer than Courtrai, but this was unacceptable to Richthofen, who commanded, "I want to go to Courtrai—at once. Please don't stop here any longer!"[33] Dutifully the ambulance drove on until arriving at 16 Infantry Division Feldlazarett 76 in St. Nicholas's Hospital, Courtrai.

Richthofen's diagnosis upon admittance was "ricochet to the head from machine gun",[34] the location of which was on the left side of his head, "on the border between the occiput and the parietal bone."[35] Although the bullet was a non-penetrating ricochet it created what doctors noted was a "Mark-sized" scalp wound with slightly gray, irregular margins.[36] His temperature was 37.2°C (99°F), his pulse 74 and "strong", and although there was "no sign of internal bleeding or of an injury to the inner surface of the bone"[37] Richthofen—not surprisingly—complained of headache. After medical personnel shaved his head and administered clorethyl anesthesia, Obergeneralarzt Prof. Dr. Kraske operated to determine the nature and severity of the wound:

"On the base of the wound is still some musculature with periosteum [dense fibrous membrane covering bone surfaces except at the joints and serving as a muscle and tendon attachment] and galea [sheetlike fibrous membrane that connects the occipitofrontal muscle to form the epicranium {membrane covering the skull}]. Incision [is] to the bone. The bone shows only superficial roughness, no other injury. The cranium

Above: Richthofen and Wolff (with arm in sling) take a daytrip from the hospital to visit their comrades at Marckebeke, 20 July 1917. From left to right: (Top) Eberhard Monicke, Wilhelm Reinhardt, Nurse Weinstroth, Richthofen, Albrecht von Richthofen (Manfred's father), Guido Scheffer. (Middle) Franz Müller, Karl Bodenschatz, Joachim Wolff, Alfred Niederhoff. (Bottom) Konstantin Krefft, Otto Brauneck (who would be KiA six days later), Wilhelm Bockelmann, Wolff, Prof. Arnold Busch, Karl Meyer, Carl August von Schönebeck.

is not opened as there is no sign of injury to its contents. Then the entire wound is excised within health tissue.[38] Fairly strong bleeding. Several catgut sutures through the galea, skin sutures with silk."[39]

Dr. Kraske sutured Richthofen's wound as completely as possible but a portion 3 cm long and 2 cm wide remained open, exposing Richthofen's bare skull. The wound was dressed with an iodoform[40] gauze tamponade and a pressure bandage, and then his entire head above the ears was swaddled in bandages. He also received a tetanus shot.

Afterwards, Richthofen wrote about his injury:

"I had quite a respectable hole in my head, a wound of about ten centimeters [four inches] across which could be drawn together later; but in one place clear white bone as big as a *Taler* [coin similar to U.S. silver dollar] remained exposed. My thick Richthofen head had once again proved itself. The skull had not been penetrated. With some imagination, in the X-ray photos one could notice a slight swelling. It was a skull fracture that I was not rid of for days..."[41]

Richthofen was bed-ridden during his initial recovery—to Bodenschatz he appeared "pale and uncharacteristically weak"[42]—and at times complained of headaches. He read reports and wrote letters to combat "the boredom that torments me amply here in bed,"[43] and soon shared a room with Kurt Wolff after the latter was shot through the left wrist 11 July. On 13 July Richthofen's sutures were removed and although his wound looked "fine" he felt poorly that evening—doctors recorded, "temperature rises to 38.2°C (almost 101°F). Slight constipation. Tongue is coated."[44] Morphine was administered, after which Richthofen had "good sleep" and felt well again the next morning. His

Above: Colorized photo of Richthofen with Nurse Otersdörf. The date is unknown but likely between 20–25 July.

diet improved from the initial "milk, tea, eggs and soup"[45] to "roast, potato, vegetable, butter, bread, sausage, wine,"[46] and by 17 July he felt well, with lessened headaches and "no other problems, especially no unsteadiness when standing up with closed eyes".[47] Further x-rays revealed nothing negative.

On 20 July Richthofen's wound looked clean, although "in the center the bone is visible, the size of an almond".[48] Regardless, he had regained sufficient strength—and no doubt inspired by restlessness and boredom—to visit his comrades at Marckebeke. This he did, although to his mild annoyance he was forced to endure a nurse chaperone. Richthofen paid for this excursion, as the following day doctors noted: "He does not look so great today. Therefore, he is advised to keep more rest."[49]

On 25 July, after having felt well since 21 July, doctors deemed further hospitalization unnecessary. Richthofen's wound had changed little, although they noted a slight increase in granulation tissue.[50] The still-exposed bone was covered with boric acid ointment[51] and the entire wound was covered with black ointment.[52] Consulting surgeon *Oberstabsarzt Prof. Dr.* Läven advised Richthofen not to fly until the wound healed completely, because "there is no doubt that there was a strong concussion of the brain (commotion cerebri) associated with the wounding, even more likely, associated with an internal bleed. Therefore, it could happen during a flight, that the sudden changes in air pressure could cause disturbances of consciousness".[53] This contradicted the earlier diagnosis upon admittance that Richthofen showed "no sign of internal bleeding." Regardless, having been informed of this possibility, Richthofen promised not to fly until he received medical permission—a promise that turned out to be so much chin music—and shortly thereafter he was discharged. (See *The Supposed PTSD* for comprehensive analysis regarding the long-term effects of this wound.)

II. Where

Although history has long-credited Cunnell and Woodbridge with firing the wounding shot, many believe researcher Ed Ferko's postulation that Richthofen was actually hit by German "friendly fire" emanating behind him. This theory is supported by the beliefs that 1) the 300 meter distance at which Cunnell and Woodbridge opened fire was too great for accurate gunnery, and/or 2) the wound's rearward location excludes a frontal shot. I.e., how could an airplane in *front* of Richthofen shoot him in the *back* of the head?

Prior to any conclusions regarding who shot Richthofen, *where* he was shot must be determined with as much anatomical precision as possible. Unfortunately, direct evidence is lacking. There are no known photographs of the wound and the head x-rays were destroyed in the 1970s to create storage room for modern records.[54] Thus, the closest direct evidence comes via Richthofen's medical history, whereupon hospital admittance surgeons described the wound as being located "left on the border between the occiput and the parietal bone."

"Border" refers to a *suture*, which is a line of junction or an immovable joint between the bones of the skull, where the bones are held together tightly by fibrous tissue. Specifically regarding Richthofen, this "border" description refers to the *Lamboid suture* between the left *parietal bone* (one of two large bones which form the sides and top of the skull) and the *occipital bone* (the curved, trapezoidal bone that forms the lower rear skull; i.e., the *occipital*). This suture runs at a 120 degree angle off

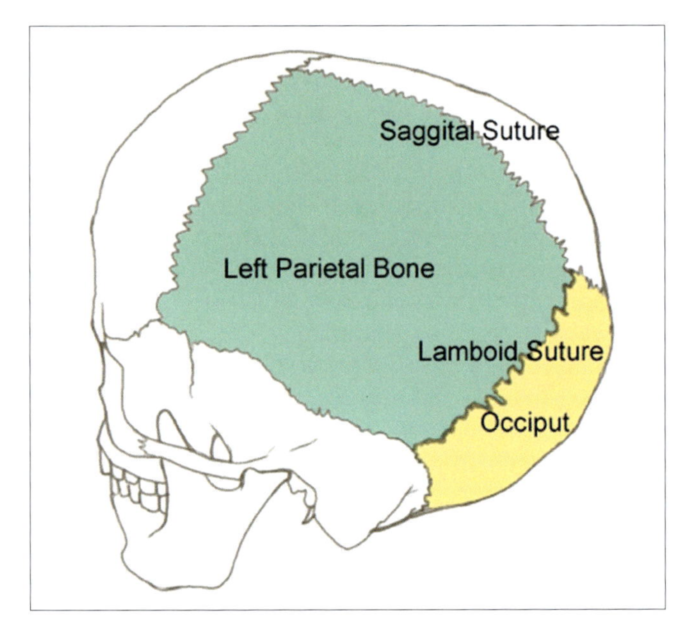

Above: This sketch reveals the general location of skull sutures. Richthofen's medical records indicate his wound was situated "left on the border between the occiput and the parietal bone," known as the Lamboid suture.

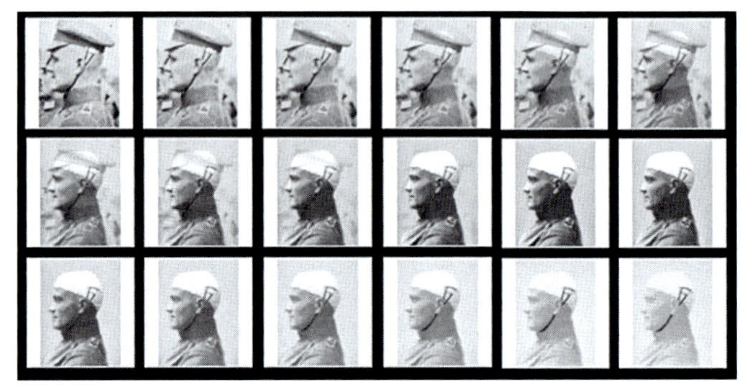

Above: A montage of disappearing photographs that shows the location of the localized bandage on Richthofen's head.

the *Sagittal suture*, which runs front-to-back directly up the center of the skull between the parietal bones. For a person sitting upright, the Lamboid suture runs downward from back-to-front at a 30 degree angle to the horizontal.[55]

Despite that specificity, each skull is different. Some skulls have squat occipital bones while others are quite high, depending upon general skull shape, and so the suture line between the occipital and parietal bones does not necessarily identify the same location on every person.[56] But it does support the general assertion that when laterally viewing the left side of Richthofen's head, the wound was right of an imaginary line drawn vertically through the left ear.

This location is corroborated circumstantially via photographs taken of Richthofen after his initial head "swaddling" was removed sometime between 20–31 August (possibly the 27th, after bone splinters were removed from the wound) and replaced by a smaller, more localized dressing. Unfortunately, in most photographs it is all but obscured by Richthofen's flight helmet or other head cover, yet in at least two photographs and one cine film these obstructions are absent which provides a clear view of the dressing and its restraining chin strap. Beginning above and slightly behind the left earlobe, it ran vertically up and then across the top of the head to approximately as far right of the sagittal suture as the right eye—in a photograph in which Richthofen faces the camera, the edge is at approximately 11 o'clock. It was secured via a strap that wrapped under Richthofen's chin and then up behind the left earlobe, where it branched into two near-vertical and parallel straps that continued across the top of the dressing, on the far side of which they rejoined into a single strap that descended vertically in front of the right earlobe before passing back under the chin, thereby encircling Richthofen's entire head.[57]

Above: Richthofen sitting at the top of the Castle de Bethune stairs, Marckebeke, his head swaddling in full view.

Having established a general location, the next determination is whether Richthofen's wound was parallel, perpendicular, or oblique to "the border between the occiput and the parietal bone." Determining this orientation is paramount because bullets that produce wounds such as

Above: Richthofen during his forced convalescent leave, standing with brother Lothar on the front stairs of the family home in Schweidnitz, mid-September 1917. This view enables a clear view of the entire strapped bandage.

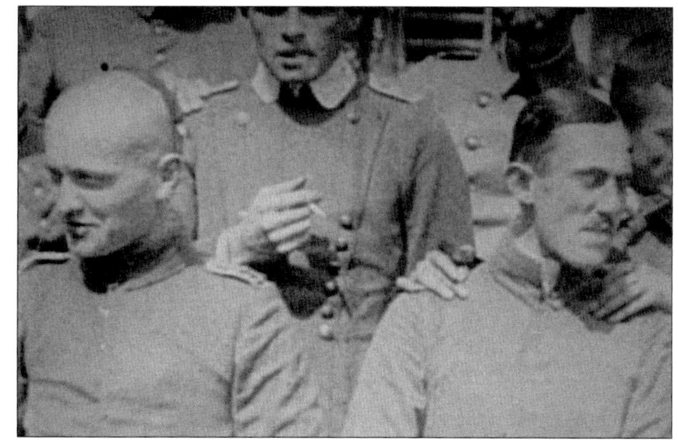

Above: This view reveals the straps the encircled Richthofen's head were either on top of or more likely attached to a piece of unknown material (e.g., leather) that held the wound bandage in place. Clearly, the wound orientation was vertical and not horizontal.

Above: Taken within seconds of the previous photograph, this view shows how the single strap branched into two straps to better secure the bandage.

Richthofen's traverse these wounds lengthwise—i.e., in Richthofen's case, along its 10cm axis. Thus, *determining wound orientation determines direction of fire.*

The first step requires examining the injury itself, which medical descriptions reveal was a *non-penetrating tangential gunshot wound.* Although not life-threatening, Richthofen's injury was much worse than the cavalier "graze" or "crease" descriptions normally ascribed. The difference is

Above: Profile that clearly shows the strapped bandage's orientation. Note the single strap passes in front of the right ear, while other photos show it passes behind the left ear.

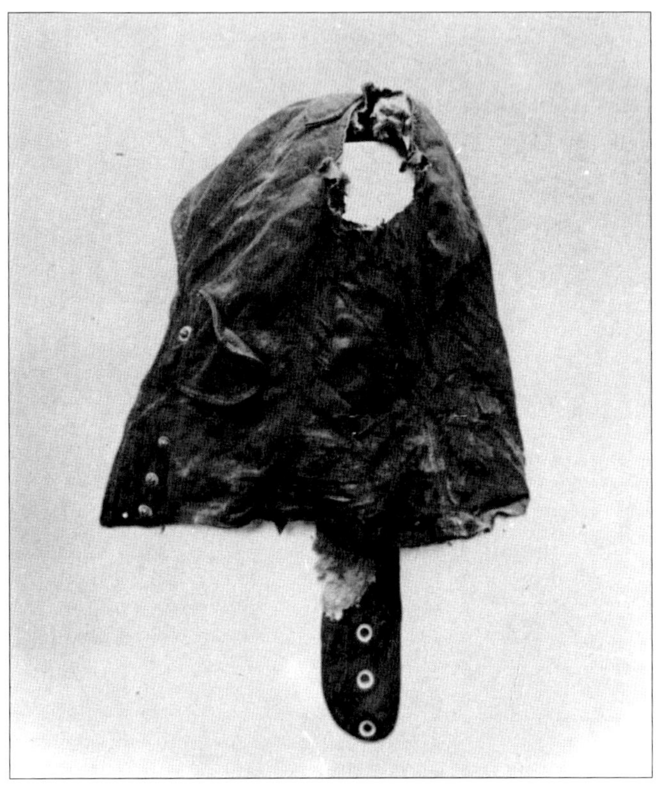

Above: Richthofen's rabbit fur-lined leather flight helmet worn 6 July. The bullet tear is located along two seams and without question is oriented vertically.

noteworthy. With *graze* gunshot wounds, a bullet strikes the skin at a shallow angle and creates an elongated abrasion without actual skin penetration. But with *tangential* gunshot wounds, although the bullet still strikes the skin at a shallow angle it creates a lacerating injury that extends down through the subcutaneous tissue.(58) In Richthofen's case, all the way down to the cranium, from which the bullet ricocheted (hence *non-penetrating*) to create a somewhat gaping oval "Mark-sized" scalp wound approximately 10 x 6 cm(59) in area and 3.5 to 4.0 mm deep.(60)

Additionally, this injury may have been accompanied by a *first-degree gutter fracture* of the skull, caused when a bullet grooves the outer table of the skull(61) and carries away small bone fragments, driving them with great violence into the surrounding tissue.(62) Although X-rays revealed no skull fracture, surgeons observed "superficial roughness" on the cranium (a bullet groove?) and it is known that for at least seven weeks afterward Richthofen endured the removal of numerous bone splinters. Modern wound ballistics expert Dr. Gary J. Ordog(63) supports the possibility of fracture, writing "[if] bone fragments were removed days later then there was obviously a skull fracture, even though it may have only been the outer table. If [a] bullet grooves the outer table of the skull...it is considered a skull fracture. Nowadays, that is well seen on CT scanning..."(64)

In any event, *if* Richthofen was either shot frontally by A6512 or from the rear by another Albatros; and presuming he focused on the onrushing FE.2 to avoid a head-on collision and judge his planned course reversal (i.e. sitting normally and looking forward—there would be little-to-no reason for him to look elsewhere during those scant 3.5 to 4 seconds); and knowing that bullets which create tangential wounds have a shallow impact angle with an almost parallel convergence between the bullet and the surface it strikes; then Richthofen's bullet wound should have been oriented more or less horizontally along the left side of his head, with at least some part of this wound crossing the Lamboid suture.

However, at least two if not three reasons render a horizontal wound orientation unlikely. The first is Richthofen's strapped-on localized dressing, which via the photographs and cine film noted previously was unquestionably aligned vertically rather than horizontally. Every doctor this writer consulted agreed that using a vertical bandage the size of Richthofen's would have been inconsistent with dressing a 10 cm horizontally-oriented wound because the ends of the laceration would have remained exposed. Rather, dressing a vertical wound entirely with a vertical dressing would have protected the still healing wound from dirt, sweat, the rabbit fur-lined flight helmet and the cold temperatures at altitude.(65) It would have covered any pustules and incisions associated with bone splinters and their removal, and it would have kept

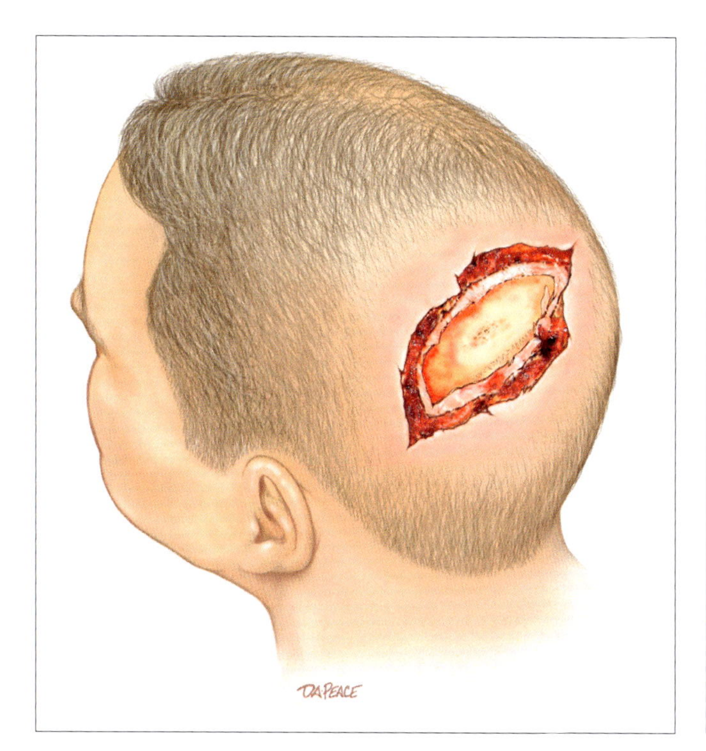

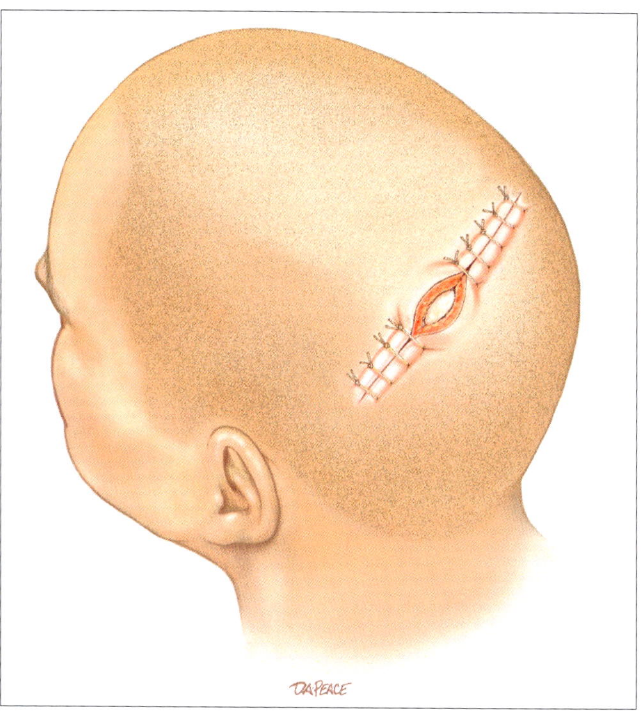

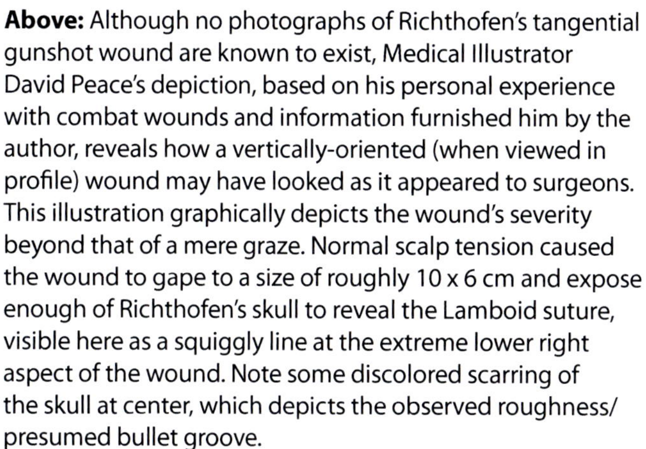

Above: Although no photographs of Richthofen's tangential gunshot wound are known to exist, Medical Illustrator David Peace's depiction, based on his personal experience with combat wounds and information furnished him by the author, reveals how a vertically-oriented (when viewed in profile) wound may have looked as it appeared to surgeons. This illustration graphically depicts the wound's severity beyond that of a mere graze. Normal scalp tension caused the wound to gape to a size of roughly 10 x 6 cm and expose enough of Richthofen's skull to reveal the Lamboid suture, visible here as a squiggly line at the extreme lower right aspect of the wound. Note some discolored scarring of the skull at center, which depicts the observed roughness/ presumed bullet groove.

Above: Richthofen's wound after surgery, fully sutured with an "almond sized" portion of the skull exposed at center.

any topical ointments free of dirt and other septic impurities. Partially dressing a horizontal wound with a vertical dressing provides either no such protection or partial protection at best.

Secondly, if the wound were located horizontally and partly above some portion of the Lamboid suture, "on the border between the occiput and the parietal bone" could mean anywhere along the suture's full length, from near the top of Richthofen's head to below/behind his left ear and anywhere in between. As such, "on the border" is an anatomically imprecise locator of a horizontally oriented wound and although speculative, it seems unlikely doctors would document Richthofen's wound so imprecisely.

Less speculative is a photograph of Richthofen's flight helmet worn 6 July that shows clearly a wide jagged tear beginning (or ending) above and behind the left ear flap that parallels a vertical seam extending up toward the top of the helmet. On either side of this tear the helmet is undamaged—strong documentary evidence supporting vertical bullet travel.

This and all presented forensic evidence reveals Richthofen's wound was oriented vertically rather than horizontally, more or less parallel and slightly forward the Lamboid suture, over which the "mark-sized" wound initially gaped to allow surgeons its visual eyewitness.

As noted previously, since bullets which cause tangential gunshot wounds traverse these wounds lengthwise along their long axes, then the bullet which inflicted Richthofen's vertically oriented wound must have been traveling vertically as well. Conclusion: *Richthofen was shot from neither the front nor the rear.*

Then from where? Unfortunately, determining the bullet's exact origin and impact angle is impossible, as is determining the precise angle at which any bullet strike ceases to be a ricochet and instead becomes penetrating. There are far too many variables (such as speed, direction, trajectory, range, air pressure, air temperature, head movement, biological composition, projectile speed at impact, tumbling, and intermediate barriers) to identify an absolute angular demarcation between ricochet and penetration. Until the availability of wound

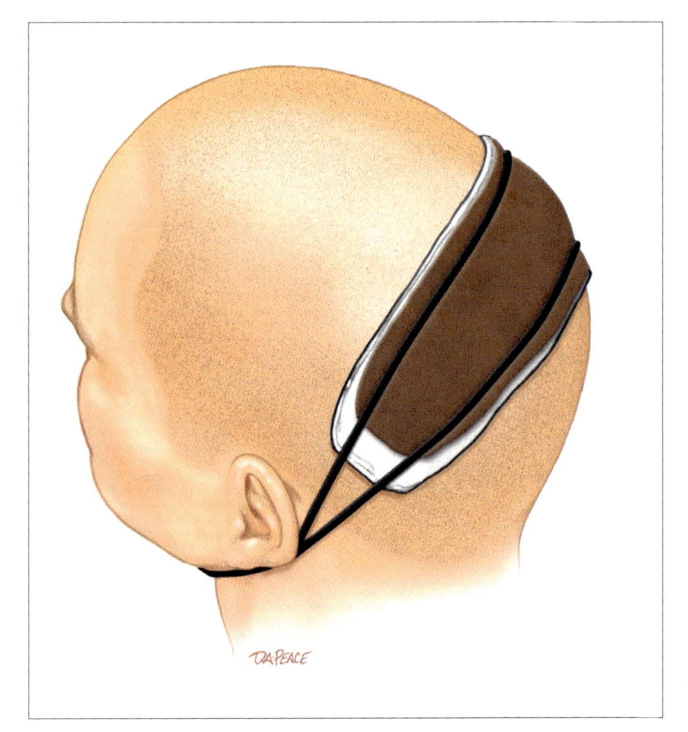

Above: The wound as covered by the localized dressing. In this view the straps lay across the wide portion atop the dressing but they may have been integrated within it; known photographs do not depict this area in detail. In any event, though the wound is covered, the position of the dressing reveals the vertical orientation.

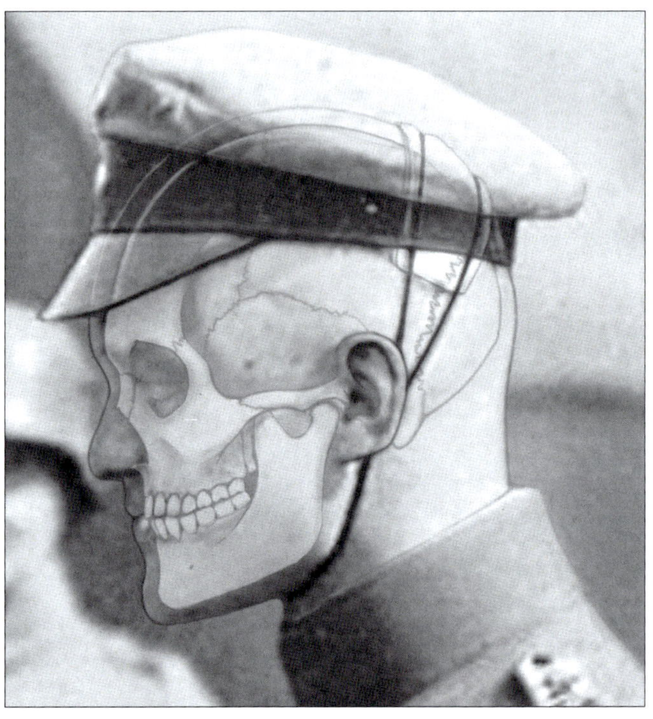

Above: An average skull of a mid-twenties European male is superimposed atop Richthofen's profile, with average tissue depths represented. The bandage and thus wound orientation atop the Lamboid suture is evident.

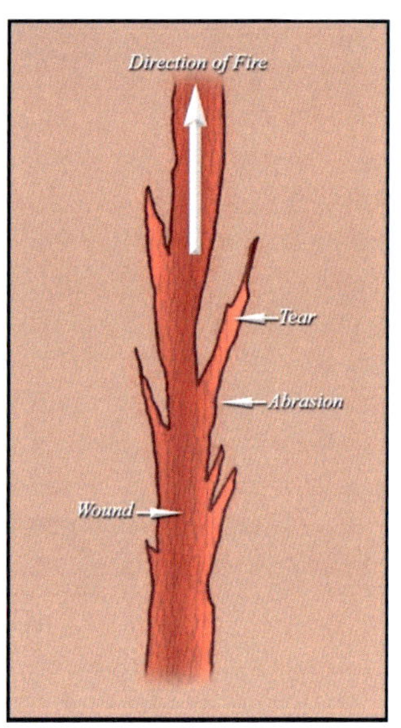

Above: Skin tags and indicated direction of bullet travel.

ballistics studies which concern headshot ricochet angles, absolutes do not apply beyond the general principle that the flatter the impact angle the greater the likelihood of a non-penetrating ricochet.[66] Additionally, although we know bullets which produce tangential gunshot wounds traverse these wounds lengthwise, it is difficult to establish direction—i.e., from left-to-right or right-to-left—without direct wound examination for *skin tags*. Skin tags are created when an impacting bullet stretches the skin until its elasticity is overcome and the margins of the resultant wound trough are multiply lacerated with the formation of these "tags", or tears. The lacerated borders of these tags are located on the side of the skin projection nearer the weapon—i.e., they point in the direction the bullet traveled.[67]

Without such precise directional evidence we are left with two possibilities. Since the Lamboid suture angles downward approximately 30-degrees from horizontal and forward approximately 30-degrees from vertical, then to inflict a tangential gunshot wound along this suture after a nearly parallel convergence and subsequently shallow impact angle, the bullet that struck Richthofen must have arrived from either his 1) ten o'clock and approximately 30-degrees below the Albatros's lateral axis—directly in the blind spot created by the lower port wing—or 2) from four o'clock and approximately 30-degrees above the Albatros's lateral axis—outside Richthofen's peripheral field of vision. Allowing for possible head rotation 45-degrees left and right of center does not affect the 30-degree impact angles but would expand the azimuth slightly from ten and four o'clock to ranges of nine to eleven o'clock low and three to five o'clock high. However, the author believes Richthofen was endeavoring to avoid a head-on collision and was most likely sitting upright and facing forward when struck by the bullet.

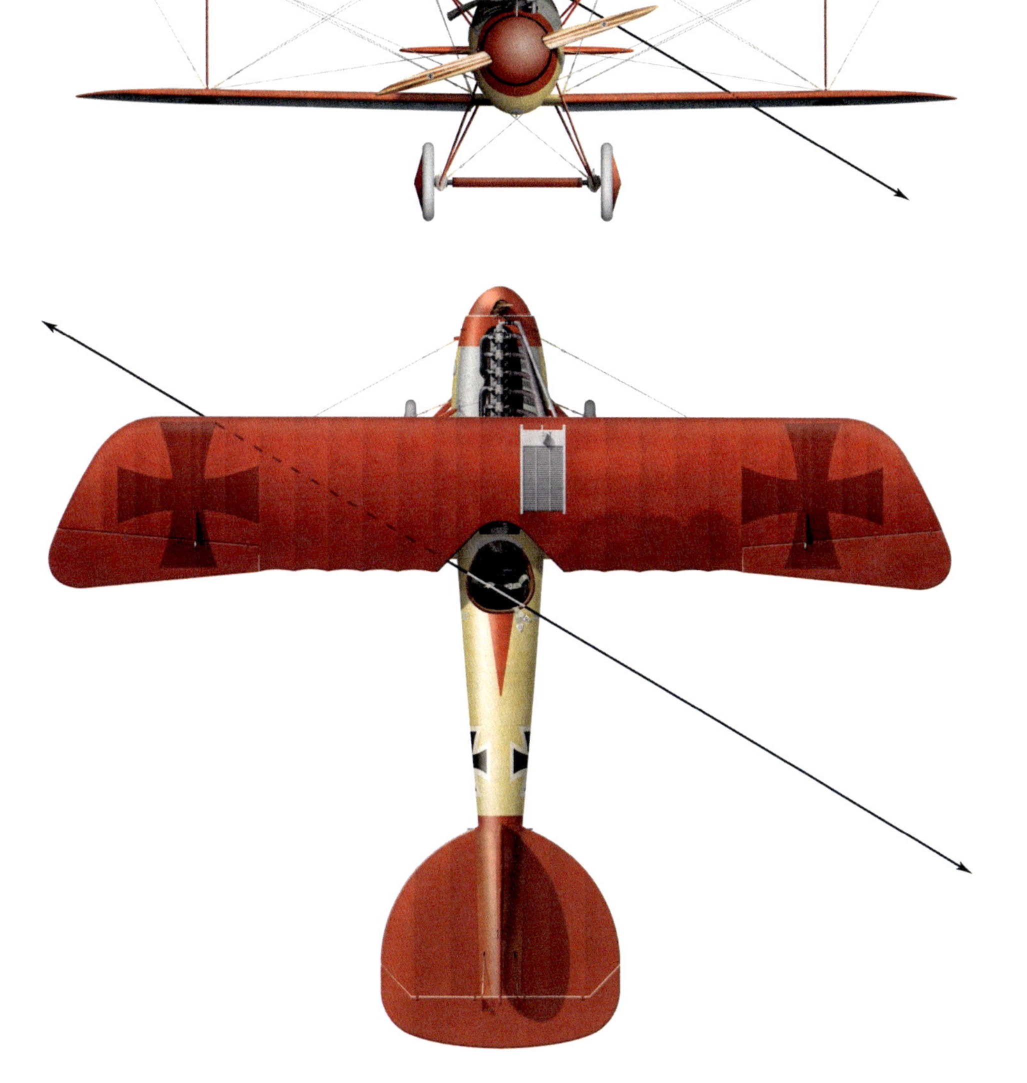

Left: Elevation of the bullet that struck Richthofen, travelling at 30 degrees above/below the horizontal. Note that if fired from below, the bullet must have passed through the lower port wing before striking the head.

Left: Azimuth of the bullet that struck Richthofen, fired from either 10 or 4 o'clock. Either would have arrived from a blind spot. The dashed line indicates the bullet flight path is below—not through—the upper port wing.

III. Who

If neither A6512 nor an Albatros behind Richthofen fired the wounding shot, then who did? The short answer—we will never know. The long answer—there are three possibilities:

1. *Richthofen was shot by another Albatros.* Friendly fire still cannot be discounted, considering the type of whirling battle as described by Cunnell, Woodbridge, and Schröder. It is not unreasonable to postulate, for instance, that an unseen Albatros tracked A6512 from the latter's four o'clock low and opened fire from this position as the FE.2 began its head-on firing run at Richthofen. Recall Woodbridge stated he and Cunnell came under fire at this time ("lead came whistling past my head and rip[ped] holes in the bathtub"[68] [euphemism for the FE.2's fuselage]) but presumed it was from Richthofen. Such a deflection shot would require the unseen Albatros continuously adjust aim ahead of the FE.2; perhaps one of its bullets struck Richthofen when he suddenly appeared from the right and flew into this line of fire.

Of course, this illustrative speculation is but one of many possibilities. It is just as likely Richthofen flew into bullets fired by Albatros above him and aimed at another FE.2 that missed the English plane and struck Richthofen instead. The possibilities are as many as one can conjure.

2. *Richthofen was shot by an FE.2d other than A6512.* It is possible Richthofen came under fire from several FE.2s at once, especially if they were

still in a defensive circle. No.20 combat reports note "several...E.A. were engaged from favourable positions and at close ranges and driven down,"[69] and recall that A6498 "brought down one E.A. out of control, firing a large burst at a range of about 20 yards, and tracers entered E.A. from underneath, entering between engine and pilot."[70] None of these claims can be linked to Richthofen, but they do illustrate the frequency of multiple close-range firing encounters.

3. *Richthofen was shot by No.10 Squadron Royal Naval Air Service Sopwith Triplanes*. Heretofore unaddressed in this work, four No.10 RNAS Sopwith Triplanes happened upon the battle when above Deûlémont and entered the fray at 1100.[1*]

Having departed Droglandt France at 0940, this offensive patrol consisted of four Triplanes from B Flight[71] (Flt. Lieut. Raymond Collishaw; Flt. Lieut. William Melville Alexander; FSL Ellis Vair Reed; FSL Desmond Fitzgerald Fitzgibbon).[72] After flying for over an hour Collishaw spotted "an encounter between some F.E's and a number of enemy scouts"[73] below; Reid counted "15 E.A. at 8,000 ft."[74] Regardless of their numerical inferiority B Flight "dived and went into the fight,"[75] after which a "general engagement ensued" as the four Tripes tangled with a horde of aggressive Albatrosses. When it was all said and done the four B Flight pilots returned to Droglandt claiming nine Albatrosses. Eventually, they were credited with four OOC.[76]

But were any of them Richthofen? Despite B Flight's claims, Richthofen's was the only Albatros that never returned from that battle (as far as is noted in the surviving records for that area and time of day)—any of the "OOC" claims could refer to him. Yet Richthofen did not mention Triplanes in his account, nor did he portray the kind of intense dogfighting as is described in B Flight's combat reports. Therefore, it seems that if a B Flight pilot fired the wounding shot it would have most likely occurred during their initial dive at 1100, before Richthofen was aware of their presence.

Examining the timeline, B Flight's 1100 attack was approximately fifteen to twenty minutes after No.20 Squadron was first attacked between 1040-1045,[77] but because it is unknown when Richthofen reversed course back east after "cutting off" No.20 Squadron the specific time of his attack on A6512 is also unknown. However, recall that as Schröder watched from la Montagne "the aerial battle lasted for a good quarter of an hour" before "Richthofen's red machine went suddenly on to its nose and shot down out of the throng of combatants."[78] Based on when the battle began, this estimation marks the time of Richthofen's fall at approximately either 1055 or 1100. The latter time matches B Flight's engagement time exactly.

Crosschecking these timelines requires comparing the combatants' reported altitudes. Since No.20 Squadron was first attacked at 12,000 feet and then fifteen to twenty minutes later B Flight had to dive to 8,000 feet to attack, obviously the combatants lost altitude as the battle progressed. Given that No.20 Squadron's combat reports state they fought from 12,000 feet to 3,000 feet between 1040–45 and 1120, there was an average altitude loss of either 225 feet per minute (fpm) or 257 fpm;[79] again, depending when the battle started. Based on these rates, when No.10 Squadron made their initial diving attack at 1100, the aerial battle had descended to either 7,500 or 8,145 feet altitude—the latter a close match with No.10 Squadron's reported 8,000 foot attack. (See Fig.1)

However, none of this matches Richthofen's account. In it, he states his altitude "at the beginning" was 4,000 meters (13,123 feet). The beginning of what? Stalking No.20 Squadron? His head-on run with A6512? His uncontrolled fall? The first seems most likely, as only No.20 Squadron recorded being close to this altitude (12,000 feet), and it coincides with Richthofen's comment that *Jasta* 11 had a "greater altitude" than No.20 Squadron. After turning back east Richthofen traded this altitude for airspeed to close on the FE.2s but he did not state to what altitude he descended before entering the head-on run with A6512. He only estimated that after being shot he fell "two or three thousand meters" before recovering at 800 meters, which he read off the altimeter.

Presuming this 800 meter recovery altitude is accurate, then Richthofen's "two to three thousand meter" fall reveals his attack altitude was either 2,800 or 3,800 meters (9,186 to 12,467 feet). Neither matches No.10 Squadron's 8,000 foot attack altitude. The former comes closest but the latter is way off—as noted previously, it is higher than the highest altitude flown by No.20 Squadron at the start of the attack ten to fifteen minutes earlier. That Richthofen's estimations varied so widely is understandable, considering he had been shot, concussed, dazed, paralyzed, blinded, and then spatially disoriented as his airplane spiraled, swooped and dived. He did not know or did not remember his attack altitude and simply had no useful frame of reference with which to measure his altitude loss.

Regardless, presuming the deduced 2,800 meter estimate of Richthofen's attack altitude is accurate, then based on the battle's 225–257 fpm average

[1*] From this point forward only British time is used. All German time is converted accordingly.

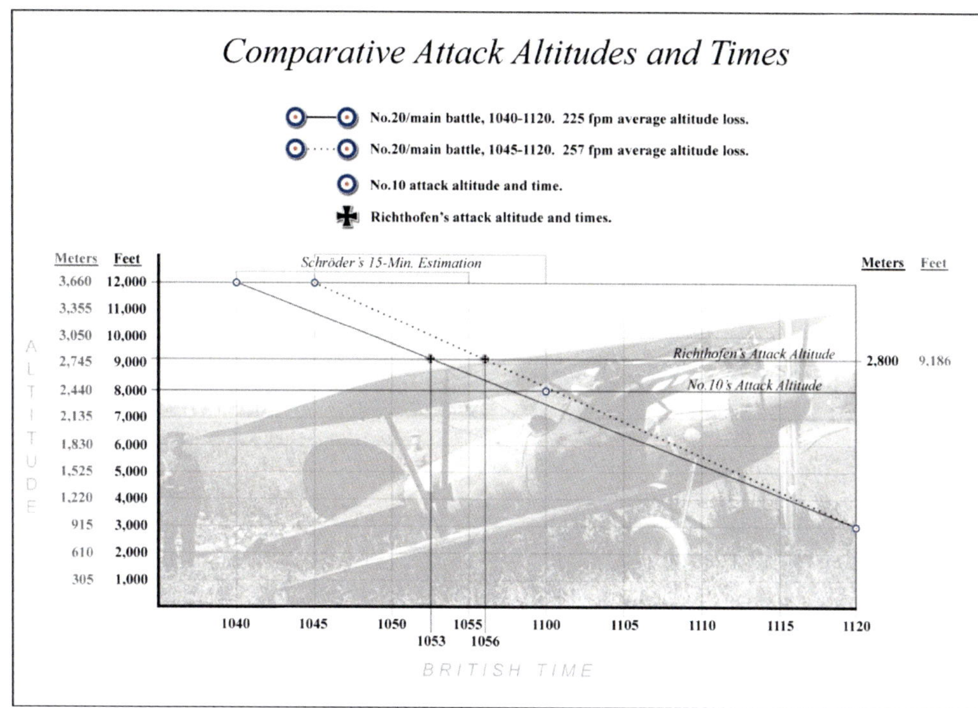

Left: Graphic depiction of the 6 July aerial battle. Rate of average altitude loss is dependent upon the time of the German attack and varies between accounts; thus, two are shown. Crosses mark Richthofen's estimated attack altitude and are plotted against each No.20 Squadron altitude loss time frame, as well as Air Defense Officer *Ltn.* Hans Schröder's estimation of Richthofen's fall 15 minutes after the battle began. The roundel reveals that in both time frames No.10 Squadron Triplanes arrived too late and too low to attack Richthofen before his wounded descent.

altitude loss between 12,000 and 3,000 feet, the FE.2s would have reached 2,800 meters at either 1053 or 1056. This reveals rough estimates of Richthofen's attack time at that altitude: 1053 if the battle began at 1040—within two minutes of Schröder's estimation that Richthofen fell[2*] 15 minutes after the battle started, but seven minutes prior to the Tripes' 1100 attack—or 1056 if it began at 1045, which would be a little further (four minutes) from Schröder's 15 minutes estimation but three minutes closer to No. 10 Squadron's 1100 attack time—which in this timeline matches Schröder's 15 minute estimation of 1100.

Yet Richthofen's possible 2,800 meter attack altitude is 362 meters (1,186 feet) too high to match the triplanes' initial attack at 8,000 feet. Thus, the circumstances of Richthofen's wounding must be compared with B Flight's individual pilot accounts to detect any matches or similarities. Specifically, with those portions referencing the Tripes' 1100 dives:

Flt. Lieut. Collishaw – "At the beginning of the fight, I attacked and drove down one scout entirely out of control, the pilot appearing to be hit."

Flt. Lieut. Alexander – "I dived on one E.A. and closed to within about 75 feet behind him, firing about 25 rounds. I could see all my tracers going into the pilot's back and he fell against the side of the fuselage and the machine nose-dived completely out of control."

Flt. Sub-Lieut. Reid – "I attacked one and after firing a good burst, the E.A. nose-dived and then turned over on its back and went down to about 4,000 ft. when it again nose-dived and then side-slipped after which I lost sight of him, he was completely out of control."

Flt. Sub-Lt. Fitzgibbon – "We dived down on several scouts. I fired a long burst at one broadside on at close range. I saw tracers going into him but he appeared to carry on."[80]

Of these four examples, Fitzgibbon's account is the furthest from matching Richthofen's experience. His attack was ineffective and the "broadside" firing angle—i.e., at or near a 90-degree deflection shot—was too lateral to have caused Richthofen's wound. Collishaw and Alexander claimed OOCs after each believed their fire struck and incapacitated the pilot. Alexander's account is most interesting, inasmuch as he fired from a range close enough to believe his tracers struck the pilot, albeit in the back, not the head. This target then immediately nose-dived out of control, as did Reid's claim—both accounts match Woodbridge's that "the Albatross [*sic*] pointed her nose down suddenly" before it "turned over and over and round and round...completely out of control."[81] Yet Collishaw stated Alexander shot this Albatros off his tail, in which case the event could not have occurred during the Tripes' initial diving attack, and although Reid's account matches Richthofen's description of "from time to time my machine had caught itself, but only to slip off again," Reid's eyewitness could have beheld *any* of the maneuvering Albatrosses they claimed as falling out of control.

The most tantalizing part of No.10 Squadron's

2* Richthofen's attack and fall occurred so close together that they can be considered simultaneous as regards time.

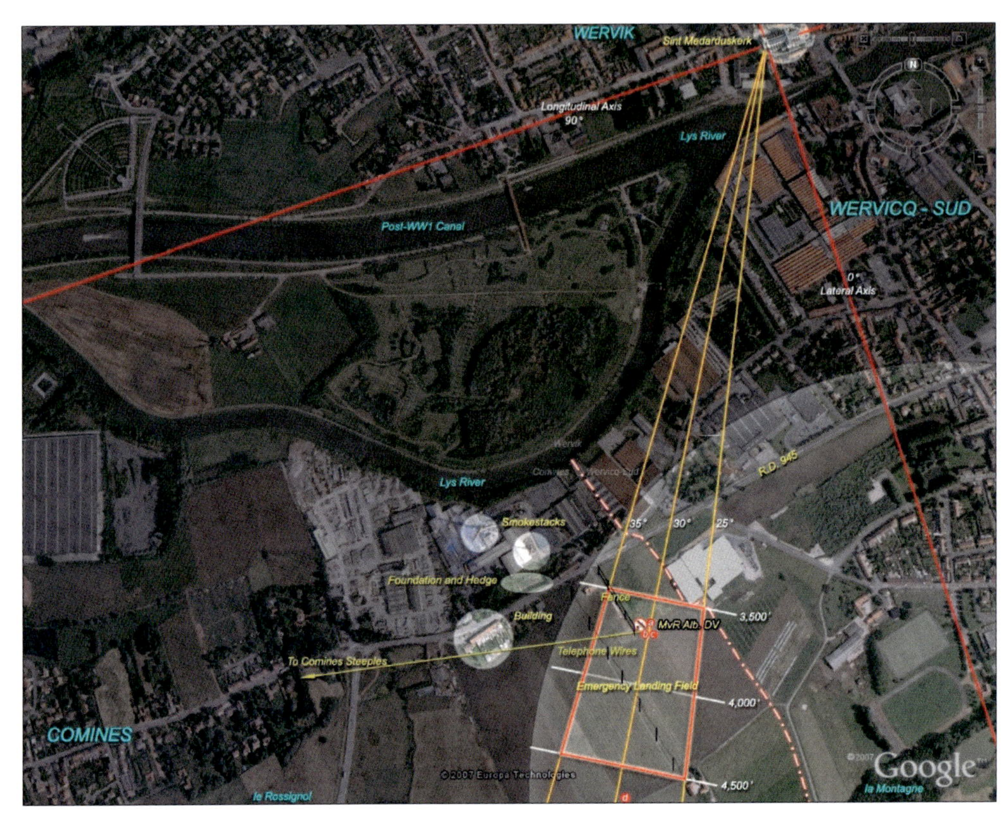

Right: A modern aerial view of Richthofen's landing site in Comines, France. The 14th-century church Sint Medarduskerk is visible at upper right. La Montagne is lower right at the center of a highlighted circular area that depicts the 1 km radius within which Hans Schröder estimated the emergency landing field was located from his observation post. Points "a," "b," and "c" depict locations from which photos of Richthofen's downed Albatros were taken in 1917. "d" depicts the locations from which a view of the landing ground was taken in 2004. (Underlying geographic image courtesy of Google Earth©)

combat report is Collishaw's statement that after diving into the fight he "saw one of my flight get an E.A. and observed it crash on the ground." If one accepts his eyewitness as accurate then it could only be Richthofen he observed on the ground since presumably Richthofen was the only Albatros brought down. Yet "crash" is a too strong description of Richthofen's emergency landing, and undoubtedly several minutes elapsed between Richthofen's wounding and his landing—several minutes in which Collishaw had his hands too full battling the swarming Albatrosses to allow his continuous observation of this particular stricken and falling airplane. "In a situation of this sort things happened quickly," Collishaw wrote. "You might get in a good shot and see the hostile fighter fall off one wing and go down but you would not be able to follow up your attack for a pair of his fellows would be on your own tail."(82)

In any event, it appears none of No.10 Squadron's combat reports offers conclusive evidence that one of their Tripes fired the wounding shot. The reports certainly can not be considered on their own merit lest the tail wag the dog, and binding them with B Flight's attack time and altitude still does not provide conclusive evidence. More than likely, Richthofen was hit prior to their arrival.

Conclusion

Despite the possibilities suggested by this work's presented evidence, there is no definitive answer as to who shot Richthofen 6 July 1917. Although gunshot wound ballistics exclude Woodbridge and Cunnell (regardless of their point-blank gunnery) as well as any German pilot flying with or directly behind Richthofen, none of the various combatants' timelines and altitudes match well enough to state conclusively who fired the telling shot. That is, not beyond the generality that Richthofen was struck by either an errant shot fired by another Albatros or a deliberate shot fired by an FE.2 in his blindspot. Either is just as likely but across the decades any definitive answer has vanished into historical vapor—if it could have ever been determined at all.

Richthofen's Emergency Landing

Approach, Touchdown, and Rollout

Once Richthofen recovered from the initial impact trauma of being shot and had regained control of his plunging Albatros, he understood the immediate need to land and receive medical attention. With waning consciousness he flew east along the southern side of the Lys River until spotting a suitable landing field. About his approach he later wrote:

"I had no idea where I was… Nothing but shell holes was below me. A big block of forest came before my vision, and I recognized that I was within our lines.

"First, I wanted to land immediately, for I didn't

Above: The classic view of Richthofen's Albatros D.V after his forced landing in Comines, taken at Point "a" in the aerial view. Several signs of damage can be noted. Sagging leading edge tape is visible on the lower port wing, and the rear horizontal flying surfaces appear to be sitting directly on the ground. The starboard wheel appears pushed up towards the wing, and close examination of the original photo reveals a wire—perhaps one of the telephone wires Richthofen took out while landing—protruding from behind the spinner and extending back beyond the gear struts. An object, possibly a fence post, can be seen lying on the ground behind the soldier on the far left. This could have caused some of the damage the Albatros sustained upon landing. Note hedged building to the right of the spinner. At this approximate location today a building foundation and hedge can still be seen. The Comines steeples (apparent here as a distant smudge) are visible beyond the trees above the starboard aileron.

know how long I could keep up consciousness and my strength; therefore, I went down to fifty [meters] but could not find amongst the many shell holes a spot for a possible landing. Therefore, I again speeded up the motor and flew to the east at a low height. At this, the beginning, I got on splendidly, but, after a few seconds, I noticed that my strength was leaving me and that everything was turning black before my eyes. Now it was high time."

Fortunately, Richthofen was already flying into the wind, which increased his descent angle and reduced his groundspeed and eventual landing distance. Had there been the more common westerly wind that day he most likely would have landed with a tailwind because the urgency to land before losing consciousness would have eclipsed the normal flying procedure of maneuvering *into* the wind, which for Richthofen would have involved an extremely low-altitude (150 feet or less) 180° course reversal while struggling to retain consciousness. Since tailwinds increase groundspeed, landing distance, and can precipitate porposing and ground loop for the unwary (or in this case, semi-conscious), the east wind was one of the few breaks afforded Richthofen that day.

Richthofen recalled he landed "without any particular difficulties" but by his own admission "tore down some telephone wires." Post-landing photographs reveal airplane damage consistent with a hard landing. Contrastingly, his post-victory landing 17 September 1916 is serially described as "poor," based entirely on a self-deprecating statement regarding what had just been one of his first single-seater landings after almost a year of flying two-seaters exclusively.[83] There is no evidence of any aircraft damage and a universal disregard of his ability to takeoff again minutes later and fly away without incident (*if* he even really landed at all), yet his landing 6 July is considered "good" despite ample photographic evidence to the contrary. Perhaps a better description of that landing is it was good *under the circumstances.*

The Albatros D.V[84] rolled to a stop facing east-northeast in a field of tall floodplain grasses

Above: This undated photograph shows Richthofen's Albatros sometime after 6 July. The elevator, horizontal stabilizers, and wings have been removed, as was common with and likely in preparation for towed transport. The airplane attitude reveals the tailskid has been repaired, and a temporary braces has been fitted to the starboard landing gear struts and wing stub to which the struts appear lashed via straps and perhaps rope. Poor photo quality prevents discerning such details absolutely, and it is unknown if there was a similar arrangement to port. Final disposition of this airplane is unknown to the author. (Greg VanWyngarden)

and native thistles.[85] All known post-landing photographs of this machine feature its starboard side and show it sitting tail-low in the weeds, leaning to starboard. At first glance the airplane appears normal. The leading edges of the wings show the usual flaking paint and insect accumulation common in summer months, although the lower wings are more affected than the upper, probably due to their closer proximity to any dirt, mud, pebbles and stones kicked up by the prop. The fuselage is intact and without evidence of battle damage, nor is there any visible damage to what can be seen of the engine, spinner, propeller, exhaust manifold, radiator and associated plumbing. The machine guns are obscured mostly by shadow, but all struts, control surfaces and rigging appear normal.

However, upon close inspection it can be seen that the tailskid and housing had collapsed, allowing the empennage to rest directly on the ground, and after noticing that the starboard wing's outboard trailing edge was just twelve inches above the ground it can be seen that the landing gear's axle appears to have shorn away from the starboard strut, suggesting sprung rubber bungee shock cords. This would leave the axle restrained by only the strut's steel safety limit cable to create the noticeable right lean. Additionally, the right wheel is angled slightly inward ("pigeon-toed") rather than ninety-degrees to the axle, and the right tire is flat. A slack wire or cable protrudes from the gap between the engine and spinner back-plate and dangles across the starboard forward gear leg and back under the lower wings—likely a prop-severed phone line that became entangled—and an estimated six feet of leading edge tape had detached and sagged several inches below the port wing, although full view of this damage is partially obstructed by one of the Garuda propeller blades.

The collapsed tailskid, partially collapsed landing gear and the flat tire are hallmarks of a too-hard landing—no doubt precipitated by Richthofen's fading faculties and urgency to land before becoming unconscious and perhaps precipitated by his impact with the telephone lines. Combat damage cannot be ruled out either, although another possible cause is the various fence posts located throughout the area—one photograph shows an apparently sheared fence post lying near the Albatros' empennage. The photographs reveal no evidence of ground loop, supporting Schröder's testimony that the airplane landed and then taxied to a stop, and the airplane was aligned more or less in the same easterly direction as Richthofen reported flying prior to landing. Cause of the lower port wing's leading edge damage is unknown, although possible candidates are battle damage incurred during the head-on run; excessive airspeed during the out-of-control spiral dive; impact with the telephone wires (although it is undocumented as to what part of the Albatros actually hit the wires, beyond the photographic

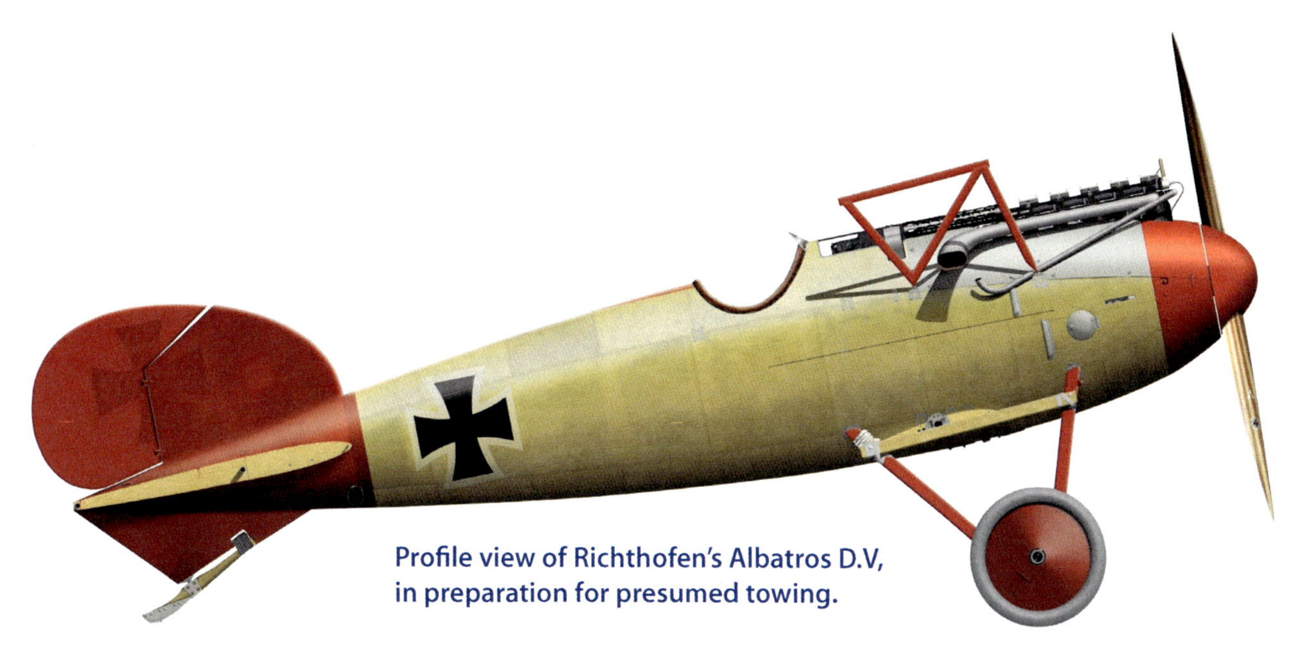

Profile view of Richthofen's Albatros D.V, in preparation for presumed towing.

evidence that suggests the spinner and/or propeller were involved); or high-speed weed impacts incurred during the landing rollout.

In any event, the Albatros was damaged to such an extent that apparently it was not flown out of the field. A subsequent photograph taken at an undetermined later date (although obviously still within the summer months, judging by field's taller foliage) reveal the Albatros still in the field with its landing gear and tail skid repaired, although both sets of wings had been completely removed from the airplane, leaving the naked cabane struts jutting from the fuselage. The propeller and spinner were still present, as were the Maxims, exhaust manifold and radiator plumbing.

The Landing Field

Determining the precise location of this landing required methodical investigative research. Road maps supplied by the cities of Wervik and Wervicq-Sud, as well as liberal use of Google Earth's high-resolution aerial photographs, revealed *Sint Medarduskerk's* exact location and orientation with respect to north and confirmed Richthofen landed to its south-southwest. These certainties became reference data used to find the emergency landing field.

First, longitudinal and lateral axis lines were drawn across a map of Wervik/Wervicq-Sud, with the axes intersecting at *Sint Medarduskerk's* steeple. A 3-D paper mock-up of the church was then oriented along these axes until the paper church's perspective angularly matched the real *Sint Medarduskerk's* perspective as it appears in the 1917 post-landing photograph. Once visually identical, the mock-up's apparent angular divergence was measured against the real church's lateral axis and then this process was repeated several times to ensure accuracy and consistency. Each measurement derived the same angle: 30 degrees. A line representing this angle was drawn south-southwestward from *Sint Medarduskerk's* steeple, as were lines at 25 degrees and 35 degrees to afford a margin of error—after all, the measurements were based on visual observation and not precisely surveyed. The result was a slender wedge emanating from *Sint Medarduskerk* that cut a long swath through extreme northwest Wervicq-Sud and down through west Comines.

To confirm the azimuth and determine range, the writer recruited 25-year architect Christopher D. Cordry from Rees Associates, Inc., in Oklahoma City. After being furnished photographs of the Albatros, church, and the dimensions of each, Chris estimated *Sint Medarduskerk's* apparent rotation with respect to the Albatros was 30 degrees—dovetailing the earlier calculations—and he estimated the range to be 4,000 feet, "plus or minus 500 feet." Plotting this range information on the map's azimuth wedge created an approximately 1,000 by 700 foot (305 by 213 meters) trapezoid—not in Wervicq-Sud, but just across its border near le Rossignol in the far northeast corner of west-neighboring Comines, between Rue Aristide Briand (R.D. 945) and what is essentially a paved, one-lane farm path off Chemin de Bois. Somewhere inside this area Richthofen must have made his emergency landing.

Next an Albatros D.V mock-up was constructed and used to measure the angular relationship between the real Albatros and landmarks visible in the post-landing photographs. The airplane mock-

Above: Post landing photograph taken at Point "b," looking north/northeast. Orientation of nearby smokestacks, the steeple of Sint Medarduskerk (visible through the starboard wing gap and shown in the insert), and cross-referencing maps pinpoint the location as Comines, France, just across the border from Wervicq-sud. La Montagne is out-of-view to the right.

up was rotated until the paper *Sint Medarduskerk* appeared directly on a line that bisected the aft end of the starboard aileron control shroud and the lower wing's fifth rib, as is seen in the photographs, and then this line was measured against the Albatros' longitudinal axis. The angular relationship of photographed smokestacks, buildings and distant steeples were also measured and then all of this related information was plotted onto a modern aerial photograph of Wervik/Comines/Wervicq-Sud. When the D.V mock-up was then placed just west of the 30-degree radial from *Sint Medarduskerk,* with an angular orientation as shown in the 1917 photograph and within the ranges specified by Chris Cordry, the angular relationship between the Albatros and nearby landmarks in 1917 matched those in the modern aerial photograph nearly perfectly.

This placement revealed that at a range of approximately 3,700 feet (1,128 meters) *Sint Medarduskerk* would be visible off the spinner; the phone lines and fence would be behind the Albatros, where one would expect if the Albatros had encountered them during the landing; and although the hedged building visible off the nose and port wings in the 1917 forward starboard-quarter view is not in the modern aerial photograph, there is still a hedge and foundation visible at a location that angularly matches that in the 1917 photograph. Nearby smokestacks are of newer construction and do not appear identically located as those photographed near the Albatros, but their similar proximities to the presumed landing field are undeniable since the south-meandering Lys River shepherds the only industrial sites into the area just across R.D. 945. Additionally, a line drawn between and connecting the two prominent steeples in Comines leads straight to the landing site, from which the steeples would appear one behind the other as seen in the forward port-quarter view of Richthofen's Albatros, just above the starboard aileron.

These findings are corroborated by modern aerial photographs and First World War trench maps which illustrate there was nowhere else Richthofen could have landed and still have *Sint Medarduskerk* appear as it did in the 1917 photographs. The area immediately east of the 30-degree radial was developed in World War 1, and a building complex—also noted on a 1917 trench map—would have

Above: Richthofen's emergency landing field as seen looking north/northeast in September 2004, taken by the author at Point "d." The exact location of the Albatros was amongst the distant corn stalks (the height of which prevented constructive photography from that location) and between the two telephone poles bracketing the Sint Medarduskerk steeple, visible in the distance at right. Smokestacks and industry, although changed slightly from 1917, appear in the same relative location at left along the River Lys, which at this location meanders close to the landing field. La Montagne is out-of-view to the right.

partially or entirely obstructed the view of *Sint Medarduskerk* off the D.V's nose. Further west of the 30-degree radial and *Sint Medarduskerk's* appearance would not match that of the photo, and there are no sufficient landing fields along this radial north of R.D. 945, only industry. Further south on the radial the land becomes rolling and is bisected by a small stream—the 1917 photographs clearly shows the landing field as being very flat, as is the field adjacent R.D. 945 in the Lys River flood plain. Additionally, the further south one travels on the 30-degree radial the more side-by-side the Comines steeples appear off to the west, rather than in a straight line as photographed in 1917, and the location is easily within running distance from la Montagne and falls within Schröder's estimated distance of one kilometer from his observation post.

The author's personal visit to the area confirmed these findings were accurate. Even though much of the area was covered by 8-foot tall corn stalks, the angular appearance of *Sint Medarduskerk* on the 30-degree radial matched the 1917 photograph. R.D. 945 was within a stone's throw ("By a lucky chance, I had landed my machine beside a road"[86]) and the nearby phone lines ("I...tore down a few telephone wires"[87]) were in the same location and oriented identically as were the only phone lines depicted on the 1917 trench map. Old-posted barbed wire fences traversed the area and the nearest one (which surrounded the nearby building complex) matched the location and orientation of the fence visible in the post landing photographs. If not the exact spot, the above calculations certainly pinpointed it to within a few airplane lengths or wingspans.

Endnotes

1. Bodenschatz (translated by Hayzlett), op. cit., p.17
2. Quoted in Gibbons, *The Red Knight of Germany*, (1927), p.291.
3. Ibid., pp.290–291.
4. *No. 20 Squadron Combat Report*, Capt. D.C. Cunnell, 2nd Lt. A.E. Woodbridge, 6 July 1917
5. Ibid.
6. Quoted in Gibbons, op. cit., p.294.
7. *No. 20 Squadron Combat Report,* Capt. D.C. Cunnell, 2nd Lt. A.E. Woodbridge, 6 July 1917.
8. Maneuver designed to enable each airplane to cover defensively the airplane directly ahead, virtually eliminating the FE.2's notorious 5-to-7 o'clock-low blindspot.
9. Schröder, *An Airman Remembers*, circa 1936, p.229
10. Quoted in Gibbons, op. cit., p.291.
11. Ibid.
12. Ibid., p.294.
13. *No. 20 Squadron Combat Report,* Capt. D.C. Cunnell, 2nd Lt. A.E. Woodbridge, 6 July 1917. "Back gun" refers to FE.2 pilot's fixed machine gun.
14. Ibid.
15. Quoted in Gibbons, op. cit., p.292.
16. Ibid., p.291.
17. Ibid., p.292. "All-red scout" is erroneous. Although the D.V Richthofen flew 6 July had red gear, struts, wheels, spinner, front cowl, upper wings, and empennage, the fuselage remained in factory clear-varnish finish that appeared a "warm straw-yellow." However, it very well could have appeared as "all-red" in the few seconds Woodbridge saw it approaching head-on.
18. Ibid., p.297.

Above: Closeup of Sint Medarduskerk as it appeared in 2004. This beautiful 14th Century church serves as an excellent landmark for Richthofen's 6 July landing site.

19. Kilduff, op. cit., p.238. Portion of Richthofen's Air Combat Operations Manual.
20. Kilduff, op. cit., p.238. Portion of Richthofen's Air Combat Operations Manual: "If one is attacked from the front by a two-seater, that is no reason for one to have to pull away; rather, in the moment when the opponent flies over and away, one can try to make a sudden sharp turn below the enemy aeroplane."
21. Quoted in Gibbons, op. cit., p.297.
22. This speed is based on performance specifications. The exact airspeed will never be known and cannot be determined precisely.
23. Quoted in Gibbons, op. cit., p.292.
24. Quoted in Gibbons, *The Red Knight of Germany*, (1927), p.292.
25. Although his blindness is often attributed to a "shock" to the optic nerve, it was likely cortical blindness resulting from *transient vasospasm*, a sudden constriction of the blood vessels caused by (in Richthofen's case) bullet impact trauma. This temporarily restricted blood flow to the brain's occipital lobes, which control vision and color recognition. "The precise mechanism of dysfunction is not well understood," although "the natural history of the condition is to resolve." (http://www.parkhurstexchange.com/qa/A.php?q=/qa/Emergency/2001-09-14.qa)
26. Quoted in Gibbons, op. cit., p.292.
27. Ibid.
28. Ibid.
29. Ibid.
30. Ibid., p.298.
31. Ibid., pp.297, 298.
32. Schröder, op. cit., p.230.
33. Ibid.
34. Field Hospital Nr.76 Medical Records, Ferko Collection, UTD, Box 1, Folder 4.
35. Ibid.
36. Ibid.
37. Ibid.
38. Likely refers to *debridement*, which is the removal of dead, contaminated, or adherent tissue or foreign material.
39. Field Hospital Nr. 76 Medical Records, Ferko Collection, UTD, Box 1, Folder 4.
40. Iodine compound, used as antiseptic.
41. Kilduff, op. cit., pp.133–134.
42. Ibid., p.134.
43. Ibid., p.138.
44. Field Hospital Nr. 76 Medical Records, Ferko Collection, UTD, Box 1, Folder 4.
45. Ibid.
46. Ibid.
47. Ibid.
48. Ibid.

49. Ibid.
50. The perfused, fibrous connective tissue that grows from the base of a wound and is able to fill wounds of almost any size.
51. Used for its antiseptic and antipyic properties.
52. A salve combined from beeswax, herbs and oils, used to draw foreign debris from wounds and promote quicker healing.
53. Field Hospital Nr.76 Medical Records, UTD, Box 1, Folder 4.
54. Via telephone conversation with Dr. Henning Allmers.
55. Via email correspondence with Dr. Gary J. Ordog, 28 May 2005.
56. Via email communication with Forensic Imaging Specialist N. Eileen Barrow, of the LSU FACES (Forensic Anthropology and Computer Enhancement Services) Laboratory.
57. Since this dressing was changed frequently there are subtle differences in appearance between photos but generally appeared as described.
58. Di Maio, *Gunshot Wounds*, (1999), p.96.
59. Size of the 1914 paper denomination Mark.
60. Based on the average tissue depth, at presumed impact location, of a 25 year-old Caucasian European male. Via email communication with Eileen Barrow of the LSU FACES Laboratory, she revealed that the presumed wound location—as related to her by me—likely missed the nearby temporalis muscle and only involved fibrous membrane, fat and skin; the tissue depth of which is approximately 3.5 to 4.0mm.
61. The skull is separated into hard outer and inner layers (*tables*) between which there is a spongy middle layer (*diploë*).
62. LaGarde, *Gunshot Injuries*, (1916), p.170.
63. MD, FACEP, DABMT, FACFM, FACFE.
64. Email correspondence with Dr. Gary J. Ordog, 16 January 2006.
65. Average temperature lapse rate is 3.5 degrees Fahrenheit per 1,000 feet (2 degrees Celsius per 305 meters). Thus, if upon takeoff the ambient air temperature were 70 degrees Fahrenheit (21 degrees Celsius), at 18,000 feet the temperature would be 7 degrees Fahrenheit (-14 degrees Celsius at 5,486 meters.)
66. Email correspondence with the American Sniper Association, 15 March 2004.
67. Di Maio, op. cit., p.97.
68. Quoted in Gibbons, op. cit., p.292.
69. *No. 20 Squadron Combat Report*, Capt. D.C. Cunnell, 2nd Lt. A.E. Woodbridge, 6 July 1917.
70. Ibid.
71. Initially three Triplanes from C Flight accompanied B Flight but turned back for Droglandt after Flt. Lieut. Sharman developed engine trouble and his pilots misinterpreted his hand signals to continue without him.
72. Claims as of 6 July: Collishaw (25); Alexander (3); Reed (9); Fitzgibbon (4). Source: Shores, Franks, Guest, *Above the Trenches*, (1996).
73. *No. 10 Naval Combat Report*, 6 July 1917.
74. Ibid. (Both Reid quotes.)
75. Ibid.
76. "Out of control." Collishaw (1); Reid (1); Alexander (2). Footnoted in Collishaw's entry in *Above the Trenches* (page 116): "The RNAS credited Collishaw with 6 out of control, but the RFC Communiqué noted 1 out of control, and 5 apparently out of control. Collishaw's score, like all others, includes apparently out of control claims."
77. "Around" is used because bombing times in the combat records vary from 1040 to 1045.
78. Schröder, op. cit., p.229. (Both quotes.)
79. 9,000 foot altitude loss in 40 minutes, from 1040 to 1120. 9000 ÷ 40 = 225 fpm. 9,000 foot altitude loss in 35 minutes, from 1045 to 1120. 9000 ÷ 35 = 257 fpm.
80. *No. 10 Naval Combat Report*, 6 July 1917. (All four listed reports.)
81. Quoted in Gibbons, op. cit., p.292.
82. Collishaw, *Air Command*, (1973), p.111.
83. Richthofen began flight training in a two-seater October 1915 and had flown them exclusively ever since, including the fast yet cumbersome LFG Roland C.II. Save for a very brief stint in spring 1916 when he occasionally flew the less-than-nimble Fokker E.III (which after only a few sorties was destroyed via a post-takeoff engine-failure and subsequent crash landing), the Albatros D.I was Richthofen's first single-seater machine and by far the most control-responsive. Although speculative in Richthofen's case, it would not be uncommon for a career two-seater pilot to over-control a more nimble single-seater on what very well could have been his first landing ever in that make and model (*Jasta 2* had just received these planes the day before).
84. Serial number unknown.
85. *Weed Science Society of America*, http://www.wssa.net/ According to David Pike, Weed Science PHD, the grass in front of Richthofen's Albatros is most likely a sod-forming pasture type grass, such as a fescue, ryegrass or timothy, with the weeds behind the empennage being thistles, most likely *Cirsium arvense*, or Canadian thistle. Despite the name "Canadian" thistle ("that's just the way names of plants happen..."), the USDA reports it is native to France and Belgium.
86. Gibbons, op. cit., p.299.
87. Gibbons, op. cit., p.298.

Bibliography

BOOKS

Above the Lines, N Franks, F Bailey, R Guest, Grub Street, 1998
Above the Trenches, Shores, Franks, Guest, Grub Street, 1990
Ace of the Iron Cross, E Udet, Doubleday & Co., 1970
Airco, The Aircraft Manufacturing Company, M Davis, Crowood Press, 2001
Albatros D.I – D.II, J F Miller, Osprey, 2012
Albatros D.III, P M Grosz, Albatros Productions Ltd., 2003
Albatros D.III, J F Miller, Osprey, 2014
British Aeroplanes 1914-18, J M Bruce, Funk & Wagnalls, 1957
British Aviation Squadron Markings of World War I, L Rogers, Schiffer Publishing Ltd, 2001
De Havilland Aircraft of World War I, Vol. 2: DH5-DH15, C Owers, Flying Machine Press, 2001
Der rote Kampfflieger, M v Richthofen, Ullstein, 1917
Der rote Kampfflieger, M V Richthofen, Ullstein, 1933
DH 2 vs Albatros D I/D II, J F Miller, Osrpey, 2012
Ein Heldenleben, M V Richthofen, Ullstein, 1920
FE2b/d and Variants in RFC, RAF, RNAS & AFC Service, Cross and Cockade International, 2009
FE 2b/d vs Albatros Scouts, J F Miller, Osprey, 2014
Flying Fury, J T B Mc Cudden, Greenhill Books, 2000
Gunshot Injuries, 2nd Edition, L La Garde, 1991
Gunshot Wounds, 2nd Edition, V Di Maio, CRC Press, 1999
Hunting with Richthofen, J Hayzlett, Grub Street, 1996.
In the Footsteps of the Red Baron, O'Connor, Franks, Pen & Sword, 2004
Jagdstaffel 2 'Boelcke', G VanWyngarden, Osprey, 2004
Manfred von Richthofen, The Aircraft, Myths and Accomplishments of 'The Red Baron,' J F Miller, Chevron, 2009
Mother of Eagles, S H Fischer, Schiffer Publishing Ltd, 2001
Nieuports in RNAS, RFC and RAF Service, M. O'Connor and M Davis, Cross and Cockade International, 2007
RAF Squadrons 2nd Edition, CG Jefford, Airlife, 2001
Richthofen, A E Ferko, Albatros Productions Ltd, 1995
Richthofen, A True History of the Red Baron, Burrows, Harcourt, Brace & World, Inc., 1969
Richthofen, Beyond the Legend of the Red Baron, P Kilduff, Arms and Armour Press, 1993
Richthofen's Circus, Jagdgeschwader Nr 1, G VanWyngarden, Osprey, 2004
Sopwith Aircraft, M Davis, Crowood Press, 1999
The Blue Max Airmen, Vol. 5, Manfred von Richthofen, L Bronnenkant, Aeronaut Books, 2014
The Day the Red Baron Died, D M Titler, Ian Allan, 1973
The Jasta Pilots, N Franks, F Bailey, R Duiven, Grub Street, 1996
The Jasta War Chronology, N Franks, F Bailey, R Duiven, Grub Street, 1998
The Many Deaths of the Red Baron, F McGuire, Bunker to Bunker Publishing, 2001
The Red Baron, M v Richthofen, Doubleday & Co., 1969
The Red Baron Combat Wing, P Kilduff, Arms and Armour Press, 1997
The Red Baron's Last Flight, N Franks and A Bennet, Grub Street, 1997
The Red Knight of Germany, F Gibbons, Garden City Publishing, 1927
The Sky Their Battlefield, T Henshaw, Grub Street, 1995
Under the Guns of the Red Baron, Franks, Giblin, McCrery, Grub Street, 2000.
Von Richthofen and the Flying Circus, H J Nowarra and K S Brown, Harleyford Publications Ltd, 1959
Who Killed the Red Baron? P Carisella, J Ryan, Avon Books, 1979

JOURNALS, MAGAZINES

Over the Front, Vol 15, No. 3, Fall 2000
Over the Front, Vol 21, No. 2, Summer 2006
Over the Front, Vol 23, No. 3, Autumn 2008
Popular Flying, January 1934
Popular Flying, February 1934
Popular Flying, May 1935

WEBSITES (as of publication)

The Commonwealth War Cemeteries, Communal Cemeteries & Churchyards in Belgium & France
http://www.inmemories.com/index.htm